SURFACE WATER TREATMENT
FOR COMMUNITIES IN
DEVELOPING COUNTRIES

SURFACE WATER TREATMENT FOR COMMUNITIES IN DEVELOPING COUNTRIES

Christopher R. Schulz
and
Daniel A. Okun

Department of Environmental Sciences and Engineering
School of Public Health
University of North Carolina at Chapel Hill

PREPARED FOR THE
WATER AND SANITATION FOR HEALTH (WASH) PROJECT
OF THE
UNITED STATES AGENCY FOR INTERNATIONAL DEVELOPMENT

A Wiley-Interscience Publication
JOHN WILEY & SONS
New York · Chichester · Brisbane · Toronto · Singapore

Library of Congress Cataloging in Publication Data:

Schulz, Christopher R.
    Surface water treatment for communities in developing
countries.

    "A Wiley-Interscience publication."
    Bibliography: p.
    Includes index.
    1. Water-Purification.   2. Water treatment plants—
Developing countries.   I. Okun, Daniel Alexander.
II. Title.
TD430.S37   1984   628.1′62   84-3570
ISBN 0-471-80261-1

Printed in the United States of America

10 9 8 7 6 5 4 3 2 1

To
the memory
of
Robert F. Schulz

PREFACE

This text is intended for planners and engineers responsible for the design of water treatment plants to be built in Africa, Asia, and Latin America. In particular, the contents are addressed to treatment of surface waters for communities that, by virtue of being small or being located where supporting technical services are not readily available, should employ technologies that avoid the mechanization, instrumentation, and automation now common in the industrialized world.

Engineers educated in the industrialized world are taught to use technologies that are characterized as "capital-intensive." Their texts and references focus on the latest "high" technology, which is marketed locally and supported with maintenance services and stocks of spare parts. In developing countries, these maintenance services and spare parts are often difficult or impossible to obtain. When personnel do become trained to provide maintenance services, their knowledge and skills are increased to a point where the private sector, or other higher-paying jobs, entice them away.

Engineers educated in the industrialized countries are, furthermore, not familiar with technologies that minimize the need for support facilities and highly skilled technicians. Some of these technologies are identified in this text. Information concerning their performance is provided where available. Otherwise they must be considered experimental. Readers who have experience with such technologies are urged to communicate their data, including especially cost data, to the authors (Department of Environmental Sciences and Engineering, University of North Carolina, Chapel Hill, North Carolina 27514 USA) for inclusion in later editions of this work. One purpose of this volume is to stimulate investigations into appropriate methods for surface water treatment.

Additional, related information is available from many national and international agencies, and references to them appear in the text. Appendix F is a glossary of individuals and institutions referred to in the text to which inquiries can be addressed.

This volume was supported by two projects sponsored by the U.S. Agency for International Development: Christopher R. Schulz served as an engineering research assistant for a workshop conducted by the University of North Carolina under the Environmental Training and Management Project in Nairobi, Kenya, much of which was devoted to appropriate design for water treatment. He used this material as the basis for his master's report, conducted under the direction of Daniel A. Okun. Camp, Dresser, and McKee, Inc. of Boston, Massachusetts, which conducts the Water and Sanitation for Health (WASH) project, provided the funds that permitted the preparation of this volume and the formal reviews conducted by Messrs. Roland Burlingame, senior consultant of CDM, Inc., and Octavio Cordon, consulting engineer from Guatemala.

The authors are indebted to the following individuals for supporting this project: David Donaldson, Associate Director of the Water and Sanitation for Health (WASH) project; John Austin, senior training officer of the Office of Health of the U.S. Agency for International Development; Roland Burlingame and John Willis, senior consultant and vice president, respectively, of Camp, Dresser, and McKee, Inc. of Boston, Massachusetts; Herbert E. Hudson* and E. G. Wagner, chairman and director, respectively, of Water and Air Research, Inc.; Susumu Kawamura, senior project engineer of James M. Montgomery Consulting Engineers of Pasadena, California; John Briscoe and Francis A. DiGiano of the faculty of the University of North Carolina; Kazayoshi Kawata of the faculty of The Johns Hopkins University; and Mrs. Phyllis Centry of the University of North Carolina for the painstaking effort involved in preparing the text for publication. Felix Filho helped in the translation of the Spanish and Portuguese documents. Ann Jennings drew many of the figures in the manual.

Special thanks are owed to engineers from around the world who have been innovative in developing practices appropriate to the needs in their countries, practices which should be useful in other parts of the world. Jorge Arboleda Valencia from Colombia; Jose Azevedo-Netto, Renhato Pinheiro and Carlos Richter from Brazil; Octavio Cordon from Guatemala; Jose Perez from Peru; A.G. Bhole and J.N. Kardile from India; and Samia Al Azharia Jahn from the Sudan were particularly helpful. Attributions in the text without dates refer to personal communications received during the preparation of this volume. Responsibility for the material in this text rests solely with authors.

CHRISTOPHER R. SCHULZ**
DANIEL A. OKUN

Washington, D. C.
Chapel Hill, North Carolina
May 1984

---

*Mr. Hudson died while this book was going to press. He had reviewed several drafts and his substantial contributions are greatly appreciated.

**At the time of publication, research assistant, Water Supply and Urban Development Department, World Bank, Washington, D.C.

CONTENTS

1 **INTRODUCTION**   **1**

Examples of Inappropriate Technology, **2**
Purpose and Organization, **4**
Summary, **6**

2 **BASIC CONSIDERATIONS**   **9**

General Design Guides for Practical Water Treatment, **9**
Water Quality Criteria, **11**
Choice of Source, **16**
Choice of Treatment Processes, **17**
Construction Materials and Practices, **22**

   Concrete, **22**
   Structural Design Considerations, **26**
   Alternative Construction Materials for Small Plants, **27**
   Valves and Gates, **28**

3 **PRETREATMENT**   **30**

Plain Sedimentation, **31**
Storage, **37**
Roughing Filtration, **39**

   Vertical flow roughing filters, **40**
   Horizontal flow roughing filters, **41**

Grit Removal, **45**
Control of Microorganisms, **50**

4 CHEMICALS AND CHEMICAL FEEDING   51

The Jar Test, **52**
Primary Coagulants, **53**

>   Alum salts, **55**
>   Ferric salts, **56**

Alkalies for pH Control, **57**
Natural Coagulant Aids, **58**

>   Adsorbents-weighting agents, **59**
>   Natural polyelectrolytes, **59**

Disinfection, **66**

>   Gaseous chlorine, **68**
>   Hypochlorite compounds, **69**
>   On-site manufacture of disinfectants, **69**

Chemical Feeding, **73**

>   Chlorine feeding, **74**
>   Solution-feed chlorinators, **74**
>   Direct-gas feed chlorinators, **76**
>   Solution-type feeders, **79**
>   Constant-rate feeders, **79**
>   Proportional feeders, **81**
>   Saturated solution feeders, **84**
>   Dry-chemical feeders, **85**
>   Eductors, **86**

5 HYDRAULIC RAPID MIXING   91

Design Criteria, **91**
Rapid Mixing Devices, **92**

>   Hydraulic jump mixers, **92**
>   Parshall flume, **93**
>   Palmer-Bowlus flume, **96**
>   Weirs, **97**
>   Baffled mixing chambers, **100**
>   Hydraulic energy dissipators, **101**
>   Turbulent pipe flow mixers, **101**

Application of Coagulants in Open Channels, **102**
Flow-Measurement Systems in Open Channels, **104**

6 HYDRAULIC FLOCCULATION 105

Design Criteria, **106**
Baffled-channel Flocculators, **109**
Hydraulic Jet-action Flocculators, **116**
Gravel-bed Flocculators, **120**
Surface-contact Flocculators, **124**

7 SEDIMENTATION 126

Horizontal-flow Sedimentation, **127**

   Design criteria, **128**
   Inlet arrangement, **133**
   Outlet arrangement, **135**
   Manual sludge removal, **137**

Inclined-Plate and Tube Settling, **139**
Upflow Sedimentation, **144**

8 FILTRATION 146

Rapid Filtration, **147**

   Dual-media filters, **151**
   Filter bottom and underdrains, **156**
   Backwashing arrangements, **159**
   Auxiliary-scour wash systems, **165**
   Filter Control Systems, **165**
   Constant-rate Filtration, **168**
   Declining-rate Filtration, **171**
   Design and operation of interfilterwashing units, **173**
   Direct filtration, **179**

Upflow-Downflow Filtration, **183**
Slow Sand Filtration, **190**

   Design of slow sand filters, **193**
   Dynamic filtration, **199**
   Information sources on slow sand filtration, **202**

9 MODULAR AND PACKAGE DESIGNS FOR STANDARDIZED WATER TREATMENT PLANTS   203

Package Water Treatment Plants, **205**
Modular Water Treatment Plants, **209**

10 COSTS OF WATER TREATMENT PLANTS IN DEVELOPING COUNTRIES   214

The General Cost Equation, **215**
Construction Costs of Water Treatment Plants, **216**
Operation and Maintenance Costs of Water Treatment Plants, **229**
Conclusion, **231**

11 HUMAN RESOURCES DEVELOPMENT   232

An Overview of Manpower Development in the Developing Countries, **235**
Classifications of Plant Personnel, **239**
Numbers of Plant Personnel, **241**
Training, **244**

APPENDIX A   Common Chemicals Used in Water Treatment in the United States   **246**

APPENDIX B   Hydraulic Calculations for Selected Unit Processes   **252**

APPENDIX C   Checklist for Design of Water Treatment Processes   **261**

APPENDIX D   Simple Procedures for Jar Testing   **265**

APPENDIX E   Simplified Procedures for Water Quality Analysis   **272**

APPENDIX F   Individuals and Institutions Referred to in Text   **277**

REFERENCES   **283**

SELECTED BIBLIOGRAPHY   **291**

AUTHOR INDEX   **293**

SUBJECT INDEX   **295**

chapter one

INTRODUCTION

The design of water supply facilities for communities in developing countries should be based upon the proper application of current technology. The social and economic differences between the developed and the developing world explain why conventional approaches for designing water systems in the industrialized countries are not appropriate in developing countries. In industrialized countries, water projects use capital-intensive designs with a high degree of mechanization and automation in order to reduce the need for labor, which is high in cost. The prevailing economies in developing countries, however, are labor-intensive. This implies that a facility which can be built and operated with local labor will likely be more economical and more easily operated than a facility utilizing extensive technology. An investment of about $600,000 in capital equipment would be warranted to replace an around-the-clock attendant in the United States based upon a total cost of $20,000 per year including fringe benefits, for each of the four persons required to provide continuous attendance, 15-year life of the equipment, and 10% interest (Okun, 1982). On the other hand, in a developing country, $20,000 might be the maximum investment warranted, based upon a wage of $1000 per year, 10-year equipment life, and 20% interest. This difference is exacerbated when transportation costs for imported equipment are considered. Moreover, the importation of mechanized equipment leaves the client in the developing country dependent on the foreign manufacturer for spare parts and maintenance skills which are not available locally.

The widespread, if inappropriate, use of sophisticated technologies in the developing countries can be readily explained:

1. The expatriate engineers employed in developing countries are generally familiar only with the technology espoused in the industrialized countries

and are unfamiliar with the culture and competence of the people in the developing country.

2. The client in the developing country wants to appear up-to-date; therefore, he desires "only the best," which is erroneously translated to mean the latest, or the most complex, technology.

3. The water treatment facilities are often the most expensive and visible of all investments made by communities in developing countries, therefore, clients are more likely to opt for modern, sophisticated designs rather than simple technology.

4. Because professors tend to concentrate on the most advanced technology, their students tend to be ignorant of the technology appropriate to developing countries.

Turnkey projects undertaken by expatriate companies are also a major contributor to treatment plant failures in countries under development. They call for a single organization to take on the responsibility for planning, designing, constructing, and providing equipment and oftentimes the financing for an entire water supply project. This approach gives rise to numerous disadvantages for developing countries, the most important being the propensity for the turnkey contractor to select capital-intensive designs because of their great profitability. The end result is community dependence on the turnkey contractor for spare parts and skilled maintenance, both of which are exceedingly expensive and slow to arrive, so that facilities are often inoperative for years at a time.

EXAMPLES OF INAPPROPRIATE TECHNOLOGY

The aftermath of implementation with inappropriate technology is especially noticeable in the treatment units of water plants: coagulation and rapid mixing, flocculation, sedimentation, filtration, and disinfection. Several examples from the developing world can be cited (Okun, 1982).

In a relatively new water treatment plant serving a capital city in Africa, extensive equipment and instrumentation have been installed. Despite the fact that the plant was only two years old, few of the instruments and none of the recorders were operative and much of the equipment was in poor condition. A representative of the expatriate consulting engineering organization was asked how this plant might have been designed differently had it been designed for his own city. After a moment of thought, he said with a touch of pride. "We did everything for this city that we would have done for ourselves." In his city the personnel were available to assure sound operation. Supporting services, particularly from the manufacturers of the equipment, were readily available by reaching for the phone. In many developing countries, reaching for the phone promises the first frustration.

Prior to World War II, the capital city of an Asian country was amply served with a conventional water treatment plant, including a low-cost horizontal-flow sedimentation basin built of concrete, conforming to the topography and involving no mechanical equipment. On occasion, the tank was dewatered for sludge removal. Following the war, when the population of the city began to explode, a turnkey contractor was invited to increase the capacity of the plant. What was installed was a highly complex upflow unit made of steel with steel distributors and launders, all of which require extensive maintenance. Also, the unit was exceedingly difficult to operate. Here is a situation where local conditions should help dictate the most appropriate design. Upflow clarifiers are generally more economical than horizontal-flow tanks and are widely used in the industrialized world because they take little space, require little manpower for their operation, and can provide excellent solids removal so long as their hydraulic capacity is not exceeded. In developing countries, on the other hand, horizontal-flow sedimentation tanks without mechanical sludge removal are much to be preferred because they require no importation of equipment, and labor for cleaning the tanks is readily available. Space is generally not restricted. Most important, horizontal-flow tanks can be overloaded without serious detrimental impact on the subsequent filters, as most of the solids will still settle out. When upflow units are overloaded, however, sludge escapes from the blanket in large volumes and clogs the filters, interfering with the entire process. And water plants in developing countries are almost always overloaded, because capacity generally lags far behind demand.

On a site for a new water treatment plant in a large city in Asia, large numbers of men and women with baskets on their shoulders were removing earth that had been hand excavated for the construction of the settling tanks. The local Asian contractor had decided correctly that it was more economical to use low-cost labor than to invest in excavating equipment. However, the design for the settling tank to go into that excavation called for the most modern mechanical sludge removal devices.

At the same plant, when visitors asked to see how the filters are washed, three laborers were immediately available to turn the hand wheels on the valves for the filter influent, effluent, and wash water. In a neighboring large city, on the other hand, where labor is just as plentiful and low cost, the operating tables for the filters are all automatic, electrically operated, with push-button standbys, and the option for operation from a central control room.

At a large new plant in Asia, a modern solids contact unit with mechanical sludge removal facilities was found to be out of operation. The mechanical equipment had never been operative. They had been waiting more than a year for a spare part from Europe but in the meanwhile had been putting water through the unit. As it happens, the raw water is of such high quality that in a year's time virtually no sludge had accumulated. The investment in equipment was clearly unnecessary.

At this plant, a sampling table had been installed in the laboratory that would permit pumping samples from any one of 96 points in the plant at the turn of a switch. However, only two samples were being tested each day. Furthermore, long sample lines distort sample quality. Better samples would have been obtained at lower cost and more would have been done for the economy and the people if 96 persons were employed for sample collection.

In an Asian country substantial training programs were instituted as part of capital investment programs in the water sector financed by both international organizations and bilateral donors. Personnel receiving the training soon left the public sector water organization for higher paying jobs in the private sector or the Near East oil rich countries. Thus high turnover rates, continuous training, and less than adequate operation and maintenance were the norm. More appropriate technology does not generate the need for skilled technicians who are so marketable elsewhere.

Accordingly, rather than transferring technology from the industrialized world to the developing world, engineers from the industrialized world might well learn something of the simplified practices that have been found satisfactory in the developing world so that, as they provide assistance to other countries of the developing world, they would be using technologies that are appropriate and that can be easily operated and maintained.

PURPOSE AND ORGANIZATION

In recent years much information has been disseminated on appropriate technology in the water supply and sanitation fields, but most has been directed to facilities for individual households or groups of a few houses. The subject matter of this text is the use of appropriate technologies for *communities* that require public water supplies as opposed to individual on-site facilities. However, it is not concerned generally with water treatment in very large urban centers which have the resources and infrastructure to adopt mechanized water treatment practices. The solutions that are proposed herein address the proper application of water treatment technology in developing countries by advocating the design of treatment plants which are labor intensive, have low capital and recurrent costs and, by using indigenous resources, are tailored to the social and economic milieu of the region.

The text is intended to be an aid to engineers designing new water plants or upgrading old ones in developing countries, as well as to government officials in these countries who need information concerning appropriate and economical water treatment. Moreover, this text should enable planners and policy makers to take an initial step toward the development of simple design criteria and standard design manuals that are tailored for local conditions.

The intention of this text is not to repeat information that is already well-documented in standard engineering works, but rather to focus on technologies that are not readily available in books or journals and moreover are not generally used in conventional water treatment practice. Of course, some types of conventional technologies in the industrialized countries are applicable in the developing countries;

where such technologies are mentioned in this text, references are made to appropriate sources for additional information. The selected bibliography at the end of the text should be particularly valuable to users of this manual. It lists, in part, relevant books pertaining to water treatment in developing countries that have been published by the International Reference Center for Community Water Supply (IRC), the World Health Organization (WHO), the Pan American Health Organization (PAHO-CEPIS), and the German Agency for Technical Cooperation (GTZ). These publications can be readily obtained from the appropriate agency.

Chapter Two discusses the basic considerations that must be addressed prior to the actual design of water treatment plants, and Chapters Three through Eight present appropriate treatment requirements and processes for plants that are to be designed for communities in developing countries. A presentation of standardized designs, particularly those pertaining to package and modular-designed plants, is presented in Chapter Nine. Chapter Ten reviews cost data for water treatment plants constructed in developing countries, which may be useful for planning purposes. Chapter Eleven examines the human resources needed to operate and maintain water treatment plants in developing countries and considers the requirements for the training of these personnel. The more valuable and proven technologies are summarized at the end of this chapter.

Material on chemicals, hydraulic calculations, jar testing, and simple methods for water analysis, together with a checklist for design and a listing of names and addresses of individuals and institutions referred to in the text are included in appendixes. A selected bibliography and references conclude the text.

The metric system of measurement, in units familiar to those working in the water supply field, is used predominantly in this text. Common conversion factors for units used in the United States are given in Table 1.1.

Table 1.1. Conversion to U.S. Customary Units

To Convert From	Units	Multiply By	To Obtain	Units
1. Flow				
cubic meters per day	m³/day	2.64×10^{-4}	million gallons per day	MGD
cubic meters per second	m³/sec	22.8	million gallons per day	MGD
2. Surface loading rate (or settling velocity)				
meters per day	m/day	9.91	gallons per square foot per day	g/ft²/day
3. Filtration rate				
meters per hour	m/hr	9.91	gallons per square foot per hour	g/ft²/hr
4. Water demand				
liters per capita per day	lpcd	0.26	gallons per capita per day	gpcd
5. Power				
watt	W	1.34×10^{-3}	horsepower (water)	hp
6. Pressure				
pascal	Pa	1.45×10^{-4}	pound-force per square inch	psi

Unless otherwise stated, costs in this text have been adjusted to March 1982 U.S. dollars using the Engineering News Record (ENR) index; currency conversions have been made using July 1982 exchange rates. The procedure that has been utilized for adjusting costs is outlined in Chapter Ten.

SUMMARY

The following technologies are judged to be of merit in considering options for surface water treatment in communities in developing countries. Planners, managers, and engineers would do well to see that these technologies are among those that are evaluated before selection of design approaches.

1.   Pretreatment.   Pretreatment refers to the "roughing" treatment processes such as plain sedimentation, storage, and roughing filtration, which are designed to remove the larger-sized and settleable material before the water reaches the initial treatment units. Appropriate pretreatment during periods of excessive turbidity can reduce the load on subsequent treatment units and yield substantial savings in overall operating costs, especially for chemicals.

2.   Chemicals.   The chemicals commonly necessary in water treatment include a coagulant, generally alum; disinfectants, generally chlorine or hypochlorites; and, when necessary, alkalies, generally lime, for pH control. Coagulant aids may be used to improve treatment and/or reduce coagulant consumption, with natural aids preferred over synthetic types.

3.   Chemical Feeders.   Feeders should be simple in design and easy to operate. Hypochlorite and coagulant solutions may be fed by simple solution-type feeders that can be constructed locally. Chlorine gas controllers are more complex than solution-type feeders; hence their use is limited to larger plants where skilled supervision is available. The use of saturated solution feeders makes it possible to use inexpensive chemical compounds of low purity (e.g., lime or alum lumps) which may be available locally.

4.   Hydraulic Rapid Mixers.   Rapid mix units are located at the beginning of the plant and are designed to generate intense turbulence in the incoming raw water. Hydraulic rapid mixers, such as hydraulic jumps, flumes, or weirs can usually achieve sufficient turbulence without the need for mechanical equipment and are easily constructed, operated, and maintained with local materials and personnel. The coagulant is added to the raw water by means of an above-water perforated trough or pipe diffuser and placed immediately upstream of the zone of maximum turbulence.

5.   Hydraulic Flocculators.   Flocculation follows directly after the rapid mix process and provides gentle and continuous agitation, during which suspended particles in the water coalesce into larger masses so that they may be removed from the water in subsequent treatment processes, particularly by sedimentation. Hydraulic flocculators, such as baffled-channel, gravel-bed, and heliocoidal-flow types do not require mechanical equipment or a continuous power supply, and can be built with local materials and labor at relatively low cost.

6. *Horizontal-Flow Settling Basins.* The sedimentation process is responsible for the settling and removal of suspended material from water. Horizontal-flow basins with manual sludge removal require no importation of equipment, and labor for cleaning the basins is readily available. Equally important, horizontal-flow basins can be moderately overloaded without deleterious effects on subsequent filtration, as most of the suspended solids will still settle out. Upflow units tend to lose most of their floc when overloaded. Inclined plate or tube settlers may be installed in existing sedimentation basins to expand capacity and/or improve plant effluent quality.

7. *Rapid Filters.* Filtration is a physical, chemical, and (in some instances) biological process for separating suspended and colloidal impurities from water by passage through porous media. A rapid filter consists of a layer of graded sand, or in some instances, a layer of coarse filter media placed on top of a layer of sand, through which water is filtered downward at relatively high rates. The filter is cleaned by backwashing with filtered water.

(a) *Interfilter-Washing Units.* Interfilter-washing filtration units are easier to build, operate, and maintain than conventional rapid filters. Only two valves are needed for filter control, the entire system may be designed with concrete channels or box conduits, and it is possible to eliminate completely the elaborate piping, valves, and controlling systems that are common to conventional filtration schemes. The filter units must be substantially deeper than those employed in conventional designs because of the head required for backwashing.

(b) *Direct Filtration.* The direct filtration process subjects the water to rapid mixing of coagulants, and sometimes flocculation, followed directly by filtration. Direct filtration is generally practicable only for raw waters that are low in turbidity, but is a comparatively low-cost option when it is feasible, particularly in reducing the use of costly coagulants.

(c) *Upflow-Downflow Filters.* In this type of system an upflow roughing filter replaces the conventional arrangement for mixing, flocculation, and sedimentation used in rapid filtration plants. The downflow filter is a conventional rapid filter. This design can result in reduced construction and operational costs for small treatment plants.

8. *Slow Sand Filters.* A slow sand filter consists of a layer of sand through which water is filtered at a relatively low rate; the filter is cleaned by periodically scraping a thin layer of dirty sand from the surface at intervals of several weeks to months. Slow sand filters are effective in removing organic matter and microorganisms from raw waters of relatively low turbidity, resulting in savings in disinfection. In addition, the cost of construction for slow sand filters in developing countries is low, imports of material and equipment are negligible, and they are easily constructed, operated, and maintained.

9. *Modular Water Treatment Plants.* Modular plants are compact treatment units, preferably made of concrete or masonry in developing countries, and assembled

either partly or entirely on-site without large or complicated equipment. Modular designs that are standardized reduce the type and number of plant devices, thereby facilitating a more efficient system of procurement of spare parts, training of operators, and ease of repairs. To further shorten the time span for project implementation, plants may comprise modular units that are prefabricated, and transportable to construction sites for final assembly.

chapter two

BASIC CONSIDERATIONS

This chapter considers the principal factors upon which the appropriate selection of water treatment schemes is based. General design criteria are established for the implementation of water supply projects that reflect the prevailing social, economic, and technical conditions encountered in developing countries. Following this, the remaining sections of the chapter consider several important preliminary factors such as water quality criteria, choice of source, and choice of treatment processes, which should be investigated thoroughly before embarking on the design of treatment units. Appropriate construction materials and practices are also reviewed. The individual unit processes are considered in subsequent chapters.

The selection of plant capacity, which is dependent upon many factors including population, design period, storage facilities, the distance between source and plant, and financial resources are beyond the scope of this volume. Selection of the design period alone is no simple matter, depending as it does on rate of population growth, interest rates (which are a function of financial resources), ease of expansion of the facilities, and the useful life of component structures and equipment (Fair, Geyer, and Okun, 1971). Many text and reference books deal at considerable length with these issues. This text is directed at the design of the treatment facilities after design capacity has been established.

GENERAL DESIGN GUIDES FOR PRACTICAL WATER TREATMENT

Design practice in any locality, whether it be in a developed or a developing country, should strive to optimize the total investment of available capital, material, and human resources, recognizing the limited resources of each that may exist. Inasmuch as socioeconomic and technical conditions differ sharply between industrial and developing countries, a different set of design criteria would govern the implementation of water supply projects in each area.

In the industrialized countries the prevailing capital-intensive economy has called upon the water supply industry to fulfill the following general conditions: (1) a high degree of automation in order to reduce labor costs that are substantially higher than those found in developing countries; (2) extensive utilization of equipment and instrumentation that are easily procured from and serviced by a variety of proprietors; and (3) preference for mechanical solutions rather than hydraulic ones. Treatment plants that have been designed under these conditions have performed reasonably well in the industrialized countries for decades, although in some instances, particularly in small communities, sophisticated plants that employ highly mechanized labor-saving equipment often produce no real savings. Moreover, the reliability of the supply may not be increased, especially if adequate maintenance of such equipment cannot be assured. A common and unfortunate occurrence is the exportation of such design criteria, together with the equipment, to developing countries where they are entirely inappropriate, except perhaps in very large urban centers where technical resources, support services, and qualified personnel are available.

Among the reasons why conventional technologies (such as those found in treatment plants in the United States and other industrialized countries) are inappropriate in most of Asia, Africa, and Latin America is that the ability of the consumers in the developing world to pay for water is small, from one-fifth to one twenty-fifth of that in the United States; so that plants constructed with expensive, imported technologies are not economically feasible (Wagner, 1982a). Moreover, even where capital costs are subsidized, operation and maintenance costs, which are borne by the host country, increase proportionately with the complexity and sophistication of the treatment plant, resulting in higher water services charges for the consumer.

Second, there is a shortage of skilled personnel to operate and maintain treatment plants in the developing world; the limited numbers of qualified individuals are often attracted to higher-paying industry. On the other hand, there is an abundance of unskilled labor, which makes labor-intensive technologies more attractive.

Third, the water utilities that must administer water systems in developing countries are generally weak and suffer from excessive staff turnover.

Accordingly, the following guides are recommended for the design and construction of water treatment plants in developing countries (Arboleda, 1976; Wagner, 1982a):

1. To the extent possible, the utilization of mechanical equipment should be limited to that produced locally;
2. Hydraulically based devices that use gravity to do such work as mixing, flocculation, and filter rate control are preferred over mechanized equipment.
3. Head loss should be conserved where possible.
4. Mechanization and automation are appropriate only where operations are not readily done manually, or where they greatly improve reliability.

5. Indigenous materials and manufacture should be used to reduce costs and to bolster the local economy and expand industrial development.

6. For a variety of reasons (e.g., little or no fire demand, few water-using amenities, little lawn-watering), design estimates for per capita consumption and peak demands in the developing world should generally be much lower than those used in the United States. On the other hand, water may be used for animals and home-grown crops, which increase demand. Also the demand will be greatly influenced by the method of distribution. Occasional standposts, where consumers carry water to their homes, yard taps, and house connections, stimulate different demands.

7. Design period for construction should be shorter to reduce the financial burden on the present population; designs should be for 5 to 10 years rather than 15 to 20 years.

8. The plant must be designed to treat the raw water available. Because all waters are different, specific treatment objectives must be determined before initiating the design of plants.

9. The organization operating and maintaining the facility should have the capacity to recruit, train, and retain the various levels of personnel required for continuous operation.

The selection of water treatment methods that conform to the above-mentioned criteria does not require the creation of new technologies, but rather the innovative application of proven technologies. In some cases it may be appropriate to use methods that were abandoned in the industrialized countries decades ago, such as weir or hydraulic jump rapid mixers, baffled channel flocculators, and solution-type feeders. Such simple technologies are readily adaptable to tailor-made treatment plant designs, which are likely to provide more reliable service at lower cost to the community than those plants that are comprised largely of "shelf items" ordered from manufacturers abroad.

WATER QUALITY CRITERIA

With the virtual disappearance of waterborne infectious diseases in the industrialized countries, more attention is being directed in those countries toward the public health effects of chronic diseases resulting from the presence of low concentrations of organic chemicals such as, for example, the chlorinated hydrocarbons (e.g., trihalomethanes) in drinking water supplies. The chronic effects of such chemicals require many decades of exposure before their impact can be discerned; thus they are not likely to be of importance where life span is short and the relatively high incidence of waterborne infectious diseases such as typhoid and paratyphoid fevers, bacillary dysentery, cholera, and amoebic dysentery exact their toll, particularly as reflected in high infant mortality. Therefore, as enteric diseases are the predominant health hazard arising from drinking water in developing countries, standards for water quality should concentrate on microbiological quality. Furthermore, the removal of many chemical constituents from drinking water requires sophisticated treatment processes that are beyond the technical and financial capabilities of most communities in the industrialized countries.

In places where health-endangering chemicals are present in the water supply source, such as excessive nitrates (which can cause infant cyanosis) or excessive fluorides (which can cause bone diseases), it is preferable to change the source, if at all possible, rather than to provide sophisticated treatment.

A safe and potable drinking water should conform to the following water quality characteristics. It should be:

1. Free from pathogenic organisms.
2. Low in concentrations of compounds that are acutely toxic or that have serious long-term effects, such as lead.
3. Clear.
4. Not saline (salty).
5. Free of compounds that cause an offensive taste or odor.
6. Noncorrosive, nor should it cause encrustation of piping or staining of clothes.

In order to assure that such levels of water quality are maintained, developing countries should establish national standards for water quality, preferably adapted from the new guidelines issued recently in three separate volumes by the World Health Organization (WHO, 1984 a,b,c). WHO guideline values are listed in Table 2.1. In brief, the volumes contain the following information (WHO, 1984c):

Volume I—Recommendations *presents recommended guideline values per se, together with essential information required to understand the basis for the recommended guideline values as well as information on monitoring requirements, and where possible suggestions regarding remedial measures to ensure compliance with the guideline values. It provides guidelines with respect to microbiological, biological, chemical, organoleptic and radiological quality of drinking water.*

Volume II—Health Criteria and Other Supporting Information *sets out the health criteria for those drinking water pollutants and other constituents which were examined with a view to recommending guideline values, as well as provides selectivity information regarding detection of some contaminants in water and measures for their control. It contains a review of the toxicological, epidemiological and clinical evidence which was available and used in deriving the guideline values which are recommended.*

Volume III—Drinking Water Quality Control in Small Community Supplies *deals specifically with the problem of small communities predominantly those in rural areas. It contains information on techniques for the assessment and*

Table 2.1.   WHO Guidelines for Drinking Water Quality

Parameter	Unit	Guideline Value
Microbiological Quality		
Faecal coliforms	number/100 ml	zero[a]
Coliform organisms	number/100 ml	zero[a]
Inorganic Constituents		
Arsenic	mg/l	0.05
Cadmium	mg/l	0.005
Chromium	mg/l	0.05
Cyanide	mg/l	0.1
Fluoride	mg/l	1.5
Lead	mg/l	0.05
Mercury	mg/l	0.001
Nitrate	mg/l(N)	10
Selenium	mg/l	0.01
Aesthetic Quality		
Aluminum	mg/l	0.2
Chloride	mg/l	250
Color	True color unit (TCU)	15
Copper	mg/l	1.0
Hardness	mg/l (as $CaCO_3$)	500
Iron	mg/l	0.3
Manganese	mg/l	0.3
pH		6.5 to 8.5
Sodium	mg/l	200
Solids (total dissolved)	mg/l	1000
Sulphate	mg/l	400
Taste and odor		Inoffensive to most consumers
Turbidity	NTU	5
Zinc	mg/l	5.0

Source: WHO, 1984a.
[a]Treated water entering the distribution system.

control of contamination of such supplies. It covers simple methods for sampling and analysis, sanitary surveys and other means of investigating and controlling drinking water quality in these areas.

In developing national standards, it will be necessary to consider a variety of local, geographic, socioeconomic, dietary, and industrial conditions. This may lead to national standards that differ appreciably from the guideline values. Table 2.2 presents, for illustrative purposes, drinking water chemical and physical standards recommended by WHO and adopted in the USA and several developing countries. Similarly microbiological water quality standards are presented in Table 2.3.

For small community water supplies only a limited selection of parameters could

Table 2.2. Comparison of Chemical and Physical Drinking Water Standards Recommended by the WHO, the United States, and Several Developing Countries

Chemical and Physical Parameters	WHO Guideline Values[a]	United States (1977)	China[b] (1976)	India (1973)	India Recommended (1975)	Korea	Philippines (1963)	Qatar	Tanzania (Temporary) (1974)	Thailand
Total hardness (mg/l as CaCO₃)	500			600	600	300			600	300
Turbidity (NTU)	5	1 to 5	5			2		5	30	5
Color (TCU)	15	15	15			2		20	50	20
Iron, as Fe (mg/l)	0.3	0.3	0.3	1	1	0.3	1	0.3	1	0.5
Manganese, as Mn (mg/l)	0.3	0.05	0.1	0.5	0.5	0.3	0.5	0.3	0.5	0.3
pH	8.5		6.5 to 8.5	6.5 to 9.2	6.5 to 9.2		6.5 to 9.2		6.5 to 9.2	6.5 to 8.5
Nitrate, as NO₃ (mg/l)	45	45		50	45	45	50		100	45
Sulphate, as SO₄ (mg/l)	400			400	400	200	400	250	600	250
Fluoride, as F (mg/l)	1.5	1.4 to 2.4	0.5 to 1.0	2.0	1.5	1.0	1 to 1.5	1.6	8.0	1 to 1.5
Chloride, as Cl (mg/l)	250	250		1000	1000	150	600	250	800	330
Arsenic, as As (mg/l)	0.05	0.05	0.04	0.2	0.05	0.05	0.2		0.05	0.05
Cadmium, as Cd (mg/l)	0.005	0.01	0.01		0.01		0.01		0.05	
Chromium (mg/l)	0.05	0.05	0.05	0.05	0.05	0.05	0.05	0.05	0.05	0.05
Cyanide, as Cn (mg/l)	0.1	0.01	0.05	0.01	0.05		0.01		0.2	0.2
Copper, as Cu (mg/l)	1.0	1.0	1.0	3.0	1.5	1.0	1.5	0.3	3.0	1.0
Lead, as Pb (mg/l)	0.05	0.05	0.1	0.1	0.1	0.1	0.1	0.1	0.1	0.05
Mercury, as Hg (mg/l)	0.001	0.002	0.001		0.001					
Selenium, Se (mg/l)	0.01	0.01	0.01	0.05	0.01		0.05		0.05	0.01

Source: Adapted from World Bank, 1977.
[a]Data extracted from WHO, 1984a.
[b]Data extracted from IDRC, 1981.

Table 2.3. Comparison of Microbiological Drinking Water Standards Recommended by the WHO, the United States, and Several Developing Countries

WHO guideline values[a]	1. Water entering distribution system; chlorinated or otherwise disinfected samples—0/100 mg/l; nondisinfected supplies E. coli 0/100 ml; coliform 3/100 ml occasionally.
	2. Water in distribution system: 95% of samples in a year—0/100 ml coliform; E. coli—0/100 ml in all samples; no sample greater than 3 coliform/100 ml; coliform not detectable in 100 ml of any two successive samples.
	3. Individual or small community supplies: less than 10/100 ml coliform; 0/100 ml E. coli in all samples.
United States	Number of coliform bacteria as determined by membrane filter test shall not exceed one per 100 ml as the arithmetic mean of all samples examined per month. When 10 ml fermentation tubes are used, coliform bacteria shall not be present in more than 10% of the portions in any month. When 100 ml tubes are used, coliform shall not be present in more than 60% of the portions in any month.
China[b]	Total colony count not more than 100/ml; E. coli not more than 3/ml.
India (1973)	Coliform = 0 to 1.0/100 ml permissive; 10 to 100/100 ml excessive but tolerated in absence of alternative, better source; 8 to 10/100 ml acceptable only if not in successive samples; 10% of monthly samples can exceed 1/100 ml.
India recommended (1975)	E. coli = 0/100 ml. Coliform = 10/100 ml in any sample, but not detectable in 100 ml of any two consecutive samples or more than 50% of samples collected for the year.
Philippines (1963)	Coliform—not more than 10% of 10 ml portions examined shall be positive in any month. Three or more positive 10 ml portions shall not be allowed in two consecutive samples; in more than one sample per month when less than 20 samples examined; or in more than 5% of the samples when 20 are examined per month.
Qatar	Coliforms 0/100 ml if present in two successive 100 ml samples, give grounds for rejection of supply.
Tanzania (temporary 1974)	Nonchlorinated pipe supplies: 0/100 ml coliform—classified as excellent; 1 to 3/100 ml coliform—classified as satisfactory; 4 to 10/100 ml coliform—classified as suspicious; 10/100 ml coliform—classified as unsatisfactory; one or more E. coli/100 ml classified as unsatisfactory. Other supplies: WHO standards to be aimed at.
Thailand	Coliform = 2.2/100 ml. E. coli = 0/100 ml.

Source: Adapted from World Bank, 1977.
[a]Data extracted from WHO, 1984a.
[b]Data extracted from IDRC, 1981.

possibly be used to survey and measure the water quality for public supplies. Volume III of the WHO guidelines suggests that the main emphasis be placed on microbiological quality, followed by aesthetic considerations, such as turbidity, color, and taste and odor. The selected guideline values often have to be considered as long-term goals rather than rigid standards that have to be complied with at all times and in all supply systems.

CHOICE OF SOURCE

The selection of the source determines the adequacy, reliability, and quality of the water supply. The raw water quality dictates the treatment requirements. For example, most groundwaters that are free from objectionable mineralization are both safe and potable, and may be used without treatment, provided the wells or springs are properly located and protected. Surface waters, on the other hand, are exposed to direct pollution, and treatment is usually a prerequisite for their development as a drinking water supply. The location of the source also defines the energy requirements for raw water pumping, which can directly affect recurrent operational costs.

Whenever possible, the raw water source of highest quality economically available should be selected, provided that its capacity is adequate to furnish the water supply needs of the community. The careful selection of the source, and its protection, are the most important measures for preventing the spread of waterborne enteric diseases in developing countries. Dependence upon treatment alone to assure safe drinking water in developing countries is inappropriate, because of inadequate resources, as illustrated by their poor record in operating and maintaining water treatment plants, particularly with respect to adequate disinfection before the treated water enters the distribution system (NEERI, 1971). The American Society of Civil Engineers (1969) has characterized water sources for potable supplies according to water quality, using parameters of biochemical oxygen demand (BOD), coliform, pH, chlorides, and fluorides, as shown in Table 2.4. Turbidity is not included because it is easily removed in treatment.

Other contaminants, such as heavy metals, may be of concern, but their possible presence can be determined by a "sanitary survey" which identifies the major sources of pollution on a watershed. Evidence of industrial development on a watershed warrants investigation to determine whether hazardous chemicals are being discharged. Urban runoff or industrial pollution may well identify a degraded source even where BOD and coliform levels are not high.

Groundwater is the preferred choice for community water supplies, because it generally does not require extensive treatment and operation is limited to pumping and possibly chlorination. When not available from a natural source, groundwater can often be obtained by artificial recharge. In the event that no suitable aquifers are available, relatively clear waters from lakes or streams are preferred, because these can be treated by slow sand filtration. If river waters are heavily silted, pretreatment may be provided by plain sedimentation or roughing filters prior to slow

Table 2.4.   Quality of Raw Water Sources

	Excellent Source	Good Source	Poor Source	Rejectable Source
Average BOD (5 days) (mg/l)	0.75 to 1.5	1.5 to 2.5	2.5 to 4	>4
Average coliform, most probable number (MPN) per 100 ml	50 to 100	100 to 5000	5000 to 20,000	>20,000
pH	6 to 8.5	5 to 6	3.8 to 5	<3.8
		8.5 to 9	9 to 10.3	>10.3
Chlorides (mg/l)	<50	50 to 250	250 to 600	>600
Fluorides (mg/l)	<1.5	1.5 to 3	>3	—

Source: Adapted from ASCE, 1969.

sand filtration. Only as a last resort should sources be developed that require chemical coagulation, rapid filtration, and disinfection. Even then, only simple, practical technologies such as gravity chemical feed with solutions, hydraulic rapid mixing and flocculation, horizontal-flow sedimentation, and manually operated filters should be used.

A sanitary survey of potential drinking water sources for a community is an essential step in source selection. The survey should be conducted in sufficient detail to determine: (1) the suitability of each source, based upon its adequacy, reliability, and its actual and potential for contamination; and (2) the treatment required before the water can be considered acceptable. Physical, bacteriological, and chemical analyses can, in addition, be helpful in providing useful information about the source and the conditions under which it will be developed. Guidelines for sanitary surveys are given in the WHO monograph *Surveillance of Drinking Water Quality* (1976a).

CHOICE OF TREATMENT PROCESSES

The broad choices available in water treatment make it possible to produce virtually any desired quality of finished water from any but the most polluted sources; therefore economic and operational considerations become the limiting constraints in selection of treatment units. A treatment plant may consist of many processes, including pretreatment, chemical coagulation, rapid mixing, flocculation, sedimentation, filtration, and disinfection, which are arranged, in general, as shown in Figure 2.1. However, water quality varies from place to place and, in any one place, from season to season, and the resources for construction and operation vary from place to place, so the treatment selected must be based on the particular situation.

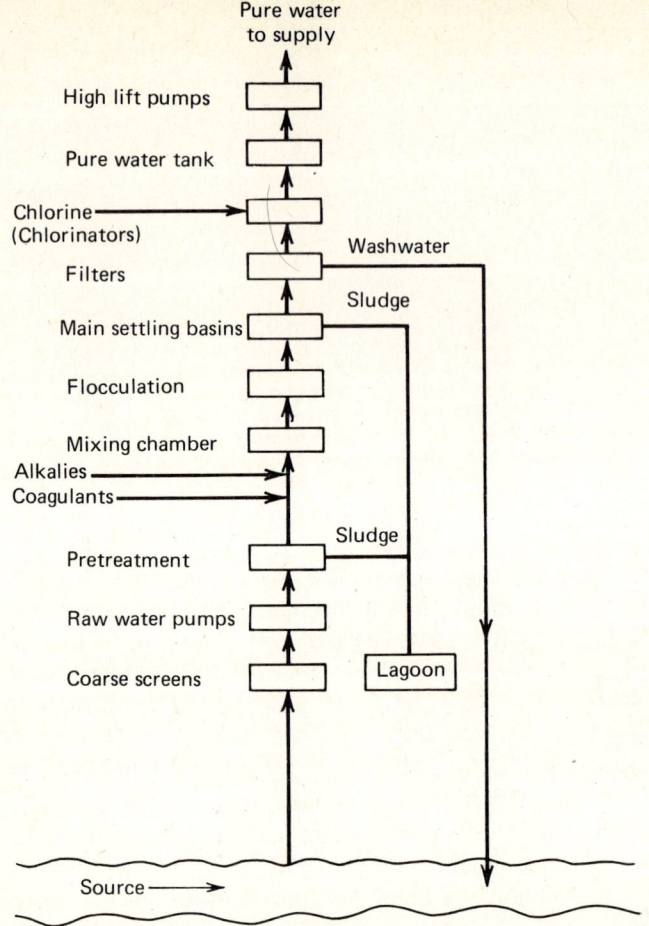

Figure 2.1. Flow diagram showing possible treatment stages in a conventional rapid filtration plant. *Source: Adapted from Smethurst, 1979.*

The primary factors influencing the selection of treatment processes are:

1. Treated water specifications.
2. Raw water quality and its variations.
3. Local constraints.
4. Relative costs of different treatment processes.

These factors are discussed below.

Finished water requirements and raw water quality generally exert the greatest influence on process selection. Finished water specifications, as prescribed by the WHO, are presented in Tables 2.1 and 2.2; whereas Table 2.5 indicates the treatment necessary for raw waters of a variety of bacteriological, physical, and chemical characteristics.

Local constraints that govern the implementation of water supply projects in developing countries, as discussed previously, are quite different from those in the industrialized countries. Considerations that local engineers or water supply planners must evaluate include:

1. Limitations of capital.
2. Availability of skilled and unskilled labor.
3. Availability of major equipment items, construction materials, and water treatment chemicals.
4. Applicability of local codes, drinking water standards, and specifications for materials.
5. Influence of local traditions, customs, and cultural standards.
6. Influence of national sanitation and pollution policies.

The selection of appropriate treatment processes is facilitated by field and laboratory investigations. A sanitary survey that identifies sources of pollution and can help characterize raw water quality during dry and wet seasons is essential. Raw water analyses are helpful but unless taken at all seasons, may be misleading because seasonal variations in raw water quality are often extreme in countries with well-defined rainy seasons. In some instances, it is possible to select designs based on experience in other plants treating water of similar quality, especially if this water derives from similar catchment areas in the same geographical region. For chemical coagulation, laboratory jar tests can be used to assess the optimum pH, the type and range of dose of primary coagulant, and the suitability of using coagulant aids. Pilot plant studies are useful for evaluating design parameters for filtration processes and to a lesser extent sedimentation processes, but should be conducted over a sufficient period so that adequate information is generated for the entire range of expected operating conditions (although such studies need not be run every day or every week).

Sedimentation cannot be reproduced accurately on a small scale because of the effects of density currents and wind action on full scale settling tanks. Filtration, on the other hand, scales up readily and pilot plant results can be used to determine filter run lengths, filtered water quality, and type, depth, and size of the filter media. Moreover, such studies are helpful in deciding whether direct filtration is feasible or if conventional treatment must be used (see Chapter Eight, under "Direct Filtration"). Pilot plant filter testing and jar testing and evaluation are covered fully by Camp (1970) and Hudson (1981). Simple procedures for conducting such tests for sedimentation, slow sand filtration, and rapid filtration are covered in the IRC publication *Small Community Water Supplies* (1981b).

Records of all such studies, sanitary surveys, raw water analyses, jar tests, and pilot plant investigations, should be kept, because the accumulation of experience in a region can be the best guide to the planning of new water treatment schemes.

Table 2.5. Classification of Raw Waters with Regard to Treatment Processes

Classes	Average Coliform, Most Probable Number (MPN) per 100 ml	Turbidity (NTU)	Color (Platinum-Cobalt Scale)	Iron (mg/l)	Total Solids (mg/l)	Chlorides (mg/l)	Hardness (mg/l)	Plankton and Algal Growth
I	<1	<25	<50	<1.0	<1500	<600	<250	insignificant
II	<2 <50	<25	<50	<1.0	<1500	<600	<250	insignificant
III	<2 <50	<25	<50	<1.0	<1,500	<600	<250	excessive
IV	<50	<25	<50	>1.0	<1,500	<600	<250	insignificant
V	<50	<25	<50	<1.0	<1,500	<600	>250	insignificant
VI	<1,000	<50	<70	<2.5	<1,500	<600	<250	insignificant
VII	<5000	<75	—	<2.5	<1,500	<600	<250	insignificant
VIII	<20,000	<250	—	<2.5	<1,500	<600	<250	insignificant
IX	<20,000	<250	—	>2.5	<1,500	<600	<250	insignificant
X	<20,000	>250	—	<2.5	<1,500	<600	<250	insignificant
XI	<20,000	<250	—	<2.5	<1,500	<600	>250	insignificant

Classes	Minimum Treatment Possible	Example of Source
I	None	protected spring
II	Chlorination	spring
III	Chemical pretreatment and chlorination	impounded reservoir
IV	Iron removal and chlorination	groundwater
V	Hardness reduction and chlorination	groundwater
VI	Slow sand filtration and chlorination	mountain stream
VII	Pretreatment-slow sand filtration-chlorination; upflow-downflow filtration and chlorination	clear water from lakes or reservoirs
VIII	Coagulation-sedimentation-filtration-chlorination	river
IX	Upflow-downflow filtration and chlorination	
X	Aeration-coagulation-sedimentation-filtration-chlorination	river or lake low in oxygen
XI	Pretreatment-coagulation-sedimentation-filtration-chlorination	very turbid river
	Coagulation-sedimentation-filtration-hardness reduction-chlorination	river

Source: Adapted from Azevedo-Netto.

CONSTRUCTION MATERIALS AND PRACTICES

To the extent possible, indigenous materials and local construction practices should be adopted for the construction of water treatment plants in developing countries. For large structures, standard construction practices as employed in the industrialized countries (but tailored to local conditions) are warranted; for small structures alternative methods (e.g., using ferrocement or masonry in place of reinforced concrete) should be investigated to lower construction costs. In earthquake regions special measures, sometimes incorporated into codes of practices, may need to be taken to limit damage.

Guidelines for the construction of treatment plants of reinforced concrete are described in detail by the American Concrete Institute (Wilder, 1971), Portland Cement Association (1963, 1969), and the British Standards Institution, Council for Codes of Practice (1966). Hudson (1981) covers jointing and sealing of water-holding structures and methods for connecting pipes to concrete structures. The intent of this section is to outline some important criteria for constructing concrete water treatment plants, and to introduce alternative constructive materials that may be appropriate for smaller plants. Much of this section is drawn from the WHO publication *Community Wastewater Collection and Disposal* by Okun and Ponghis (1975).

Concrete

Portland cement, with or without additives to improve quality, is used almost exclusively in concrete construction projects; pozzolanic materials either alone or preferably mixed with cement may be used for small projects. A variety of admixtures, available commercially or from indigenous mineral and vegetable sources can improve watertightness, workability, and prevent or minimize shrinkage of concrete. In tropical climates, a major problem is shrinkage, which takes place during the curing of the concrete, and can lead to cracking and leaking.

The importance of a properly designed concrete mix, properly handled, placed, and cured, cannot be overemphasized. Basins, filters, storage reservoirs, and other water-holding structures of concrete must fulfill the following requirements. They must: (1) be extremely dense and impermeable (i.e., watertight), to prevent bacterial and chemical pollution of the water supply; (2) be resistant to prevailing environmental conditions to prevent atmospheric, groundwater, and frost damage; (3) be resistant to water treatment chemicals; and (4) provide a surface which is smooth and well-formed so that resistance to flow will be minimized.

Strength, durability, and watertightness will be reasonably assured by using water–cement ratios as low as possible, consistent with obtaining satisfactory workability and good compaction, as indicated in Figure 2.2; graphs *a* and *c*, and Table 2.6. A slow, moist curing period also improves the performance of concrete, as shown in graph *b*. In hot, dry climates, wood forms remaining in place do not permit adequate curing. They should be removed or loosened so that the concrete surfaces can be kept moist.

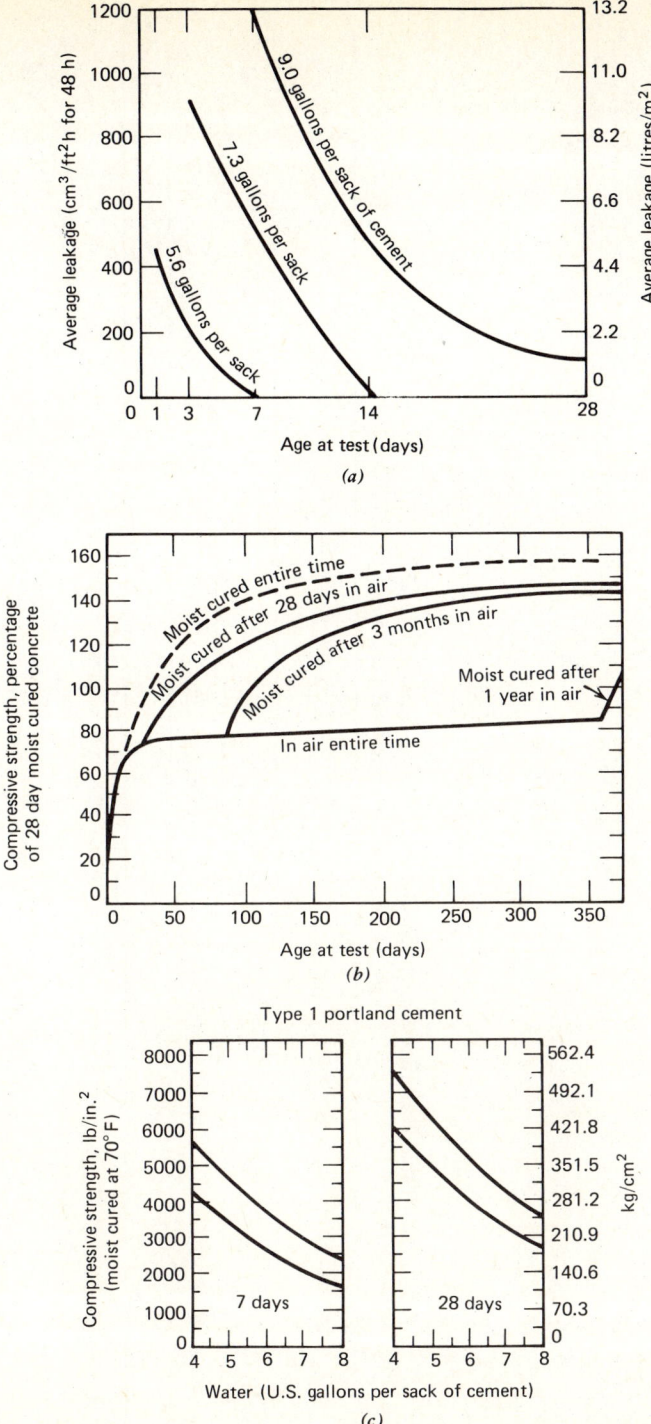

Figure 2.2. Factors affecting concrete quality. *Source:* Okun and Ponghis, 1975.

Table 2.6. Recommended Maximum Water-Cement Ratios for Plastic Concrete Mixes for Various Types of Structures and Exposure Conditions[a]

	Mild: Temperature Rarely Below Freezing, Infrequent Snow or Frost. Air Entrainment Recommended			
	In Air, or Continuously Submerged in Fresh Water		In Fresh Water Within the Range of Fluctuating Water Level or Exposed to Spray	
	(Liters)	(U.S. Gal.)	(Liters)	(U.S. Gal.)
Irrigation canal, storm-water conveyance or flood-control canal lining	30	7.0	27	6.5
Water treatment units; tank walls, tank floor slabs or roofs	25	6.0	25	6.0
Concrete deposited by tremie in water	21[b]	5.0[b]	21[b]	5.0[b]

Type of concrete or structure				
Concrete that will be protected from weather by enclosure or backfill, but may be exposed to freezing and thawing for several months a year before such protection	As appropriate[c]		Not appropriate	
Concrete protected from weather: interior of buildings, concrete below grade	As appropriate[c]		As appropriate[c]	
Thin sections: railings, curbs, sills, ledges, piles	25	6.0	25	6.0
Moderate sections: retaining walls, piers, abutments, beams, girders	As appropriate[c]		25	6.0
Mass portions of concrete (exterior), minimum thickness of these portions, 2 ft (0.6 m)	As appropriate[c]		25	6.0

Source: Adapted from Okun and Ponghis, 1975. Information supplied by Portland Cement Association, USA.

[a]Figures in liters are per 50-kg sack, those in U.S. gal. per 100-lb sack.

[b]Plus air entrainment for workability.

[c]Water-cement ratios should be selected on basis of strength and workability requirements.

Structural Design Considerations

The level of the groundwater table is a basic design consideration in the design and construction of water treatment structures. Sidewalls and bottoms of tanks must withstand combined water and soil pressures, and the tank itself must resist flotation. Of course, the filter unit poses less of a problem, inasmuch as filter media placed within the filter box are normally heavy enough to withstand uplift pressures. For the basins of the plant uplift may be controlled by: (1) designing a heavy structure of concrete to withstand uplift (not economically viable in most situations); (2) relief valves that bleed water into the tanks when permissible groundwater levels are exceeded; or (3) installing drains under or around the tank to lower the groundwater table (these drains must have a free fall to an open outlet). When none of these methods is feasible, a simple alarm system, consisting of a tube with a float, can be installed to alert the operator when a dangerous groundwater elevation is reached, so that he can immediately start filling the tank to balance the uplift force.

Proper drainage is important for durability. Structures should drain free of puddles when not in service. Tops of walls, flat slabs, and all exposed surfaces should be designed to have sufficient slope to avoid standing water.

Construction and expansion joints are points at which deterioration may begin in hydraulic structures. The durability of concrete at such joints (e.g., in tanks) is affected by the quality of the concrete immediately below the joint and by the care taken in preparing the joint surface before the wall is poured. If a layer of concrete is obviously of poor quality, the top surface should be cleaned just before the concrete becomes thoroughly hard. The new concrete should always be placed on a layer of cement mortar spread evenly over the joint surface. When possible, the mortar should be scrubbed into the surface of the joint with wire brooms.

Keying or notching the joint is a common construction practice. Premolded sealing strips, to serve as water-stops, are attached to the key before fresh concrete is poured, as shown in Figure 2.3. When used in vertical joints, the premolded

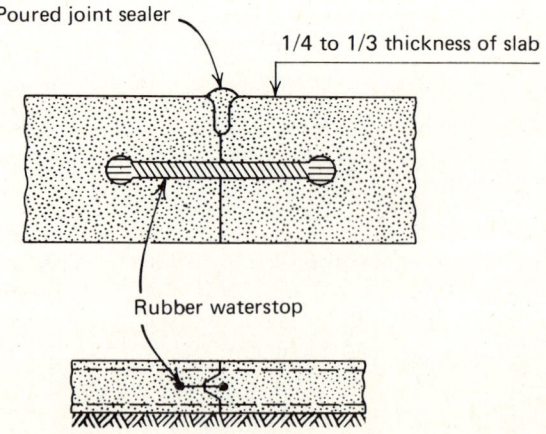

Figure 2.3.   Watertight construction joints. *Source:* Okun and Ponghis, 1975.

strip must be inserted to a sufficient depth in the joint opening to permit application of caulking compound or cement grout.

Joints in concrete water-retaining structures must often be provided with a means of preventing water pressure from passing through them. Metal and rubber water-stops are used for this purpose. Because of their flexibility, durability, and resilience, rubber water-stops will maintain a watertight seal under severe service conditions and high hydrostatic heads. They may be made of synthetic rubber, neoprene, or natural rubber. A joint sealer is used to seal the outside of the joint space. Hot-poured rubber-asphalt compounds and cold-poured joint sealers are commonly used. Joint sealers require periodic inspection and replacement. A detailed discussion on jointing and sealing of water-holding structures is given by Hudson (1981).

Alternative Construction Materials For Small Plants

Portland cement is in short supply in many developing countries, and local supplies often have to be augmented by imports, at high-cost in foreign exchange and transportation. Accordingly, for small water treatment plants that do not require the high strength of reinforced concrete, alternative construction materials should be investigated to reduce the quantity of portland cement required.

Pozzolanic materials (volcanic ash, fly-ash, raw or heat-treated shales and clays) mixed with lime have been used widely in the past in many of the situations where today Portland cement is used. In fact, over 2000 years ago the Romans used this type of cement extensively and many of their structures stand to this day (e.g., the Pantheon), attesting to the strength and durability of the material. Two distinct advantages of lime pozzolan cements for developing countries are (1) the general availability of the raw materials needed, and (2) the relative ease of converting the raw materials to the finished product. Lime is a readily available chemical and the pozzolan, or silica-bearing ingredient is potentially available from a wide variety of mineral and vegetable sources (e.g., volcanic deposits, and ash resulting from the combustion of certain woods and other plant materials). Tests conducted at Pennsylvania State University (Cady and Groney, 1976) on lime-pozzolan cement blocks, using rice husk ash as the pozzolan, yielded a maximum compressive strength of 76.5 kg/cm^2 for samples containing a 2:1 lime–pozzolan ratio, which is comparable to the strength of masonry cement. Extreme care must be exercised in the use of lime-pozzolan mixtures to make concrete, because they are generally much weaker than normal portland cement. Where such materials are being considered in an area, their strength and methods of pouring and curing (which may take longer) should be determined by a professional.

Ferrocement—a thin wall cement mortar reinforced with wire mesh and/or steel rods—is a labor-intensive technology using locally available materials, and hence is particularly suited for community projects done on a self-help basis. The reinforcement is built up of vertically and horizontally placed rods of small diameter (5–6 mm) imbedded in chicken wire. The mortar is spread onto the reinforcement with a trowel. The thickness of the finished walls generally varies from 60 to 120 mm. The properties of ferrocement are well-established and it has been applied successfully to a variety of water-holding structures in developing countries. A useful handbook on ferrocement has been produced by the Asian Institute of Technology (Paul and Pama, 1978). The IRC manual on slow sand filtration (1978) discusses a slow sand filter constructed of ferrocement, including construction drawings.

Ferrocement was used for the construction of a small rapid filtration plant (400 m³/day) in Kuta Ampel, Indonesia (Hofman and van Kervooden, 1982). The plant consisted of three circular units: (1) an upflow clarifier with a separate flocculation compartment (6 m tank diameter; 40 mm wall thickness); (2) a dual-media filter (1.6 m tank diameter; 25 mm wall thickness); and (3) a filtered-water storage reservoir (5.5 m³ capacity). The entire plant was constructed by a team comprising one skilled supervisor and six unskilled laborers working over a period of three months. The total construction cost of the plant was $8600 ($21.50/m³/day; 1982 US$).

Stone masonry has been used for the construction of small water treatment units throughout the world, where masonry is readily available. Such a plant in India is shown in Figure 8.28. The design of slow sand filters constructed of masonry is discussed in the IRC manual (1978). High quality masonry and mortar (consisting of portland or lime-pozzolan cement, depending on availability) are necessary to assure a watertight structure. Normal wall thickness ranges from 0.3 to 0.4 m, and the diameter of circular basins should range from 5 to 10 m. When constructing masonry walls for water-holding structures, vertical joints should never be placed above each other; and the blocks should not be split into blocks smaller than half the standard size locally available.

Valves And Gates

Gate valves are commonly used in pipes as, when open, they offer no obstruction to the flow. Valves with rising stems and outside yokes are preferred in treatment plants, because the height of the stem shows whether the valve is open or closed and, if open, the degree of opening.

Hand-operated valves are far less costly than hydraulically, pneumatically, or electrically controlled valves; these are seldom economically justified in developing countries where manpower is plentiful and wages are low. For larger valves (i.e., 400 mm and above), gears may be provided to make them easier to operate. Where remote control is absolutely necessary, mechanically activated valves can be used.

Butterfly valves are used in pipelines to control flow, particularly as rate controllers in rapid filters, inasmuch as the flow rate is closely related to the percentage of opening of the movable disk within the valve. They are simple and fast to operate, and relatively easy to maintain.

Sluice and slide gates are used on conduits that do not operate under pressure. Slide gates of steel or aluminum are particularly suited for small channels where they are not operated frequently. For larger conduits, reinforced plates may be used but, because of their weight, fixed or movable A-frames may be used to lift them. Guides for slide gates are preferably cast in place after precoating with bituminous compounds to protect the metal from the chemical action of the concrete.

Float valves are used to maintain a constant-head for chemical solution feeding and for maintaining levels in water treatment units.

Mud valves are used for drains at the bottom of tanks and channels.

Telescopic valves, consisting of brass or bronze sleeves that slide up and down in an iron pipe, are used in smaller plants for filter control (see Figure 8.36).

chapter three

PRETREATMENT

Rivers throughout the world exhibit wide fluctuations in flow and turbidity, with high turbidities resulting from silt carriage in rainy seasons. Silt concentrations at 1000s of mg/l are not uncommon during floods. Appropriate pretreatment during periods of excessive turbidity may reduce the load on subsequent treatment units and yield substantial savings on overall operating costs, especially for chemicals. In this text, pretreatment refers to the "roughing" treatment processes designed to remove the larger-sized and settleable material from raw water before the water reaches the initial treatment units; that is, chemical coagulation and mixing in rapid filtration plants or slow sand filtration. Pretreatment is only justified for treating waters from turbid rivers or streams; lakes, surface reservoirs, and other quiescent bodies of water inherently provide natural settling of the heavier suspended material. Furthermore, the seasonal variations of raw water quality in the rivers may take pretreatment necessary only in a part of the year, such as during seasonal flooding. For other times of the year the pretreatment units can be bypassed.

Proper location and design of intakes can minimize the requirements for pretreatment and protect treatment units. The heavier material tends to move along the bottom during periods of high flow; accordingly intake pipes should be located well above the bottom. In streams that vary significantly in level it may be necessary to have intakes at different elevations in the stream with the lower intake being used for dry periods and the higher intake being used for wet periods. A box intake structure, with the inlet facing downstream, can be installed with the use of stop logs or planks to permit skimming water from the upper levels of a stream regardless of its depth. Bars can be placed over intakes to exclude debris. Sometimes they are mounted in frames which are duplicated so that one frame can be lifted for cleaning or repair without allowing unscreened water to the plant.

Table 3.1. Conventional Methods of Pretreatment

Pretreatment	Turbidity Range[a] (NTU)[b]
Plain sedimentation	20 to 100
Storage	>1000
Roughing filtration	20 to 150

Source: Adapted from Huisman and Wood, 1974.
[a]For pretreatment prior to slow sand filtration.
[b]The nephelometric turbidity unit (NTU), the Jackson turbidity unit (JTU), and the formazin turbidity unit (FTU) are all numerically the same, and are interchangeable for all practical waterworks purposes.

Table 3.1 indicates the usual conventional methods of pretreatment (plain sedimentation, storage, and roughing filtration) for given turbidities. The turbidity ranges in the table are suggested by Huisman and Wood (1974) for pretreatment prior to slow sand filtration, and are used here to serve as guidelines. Each of these methods is discussed below.

For slow sand filtration, pretreatment is essential if the raw water has a turbidity of more than 50 NTU for periods longer than a few weeks (Huisman and Wood, 1974). The best purification occurs when the average turbidity of the water on top of the slow sand filters is 10 NTU or less. In general, slow sand filters, even when preceded by some type of pretreatment, should not be used for treating turbid river waters. The filter beds are likely to clog much too quickly with the finer particulate matter that cannot be readily removed in pretreatment units.

For rapid filtration plants, pretreatment can improve the performance of the unit processes for the following reasons: (1) better operation of the unit processes is likely because raw water quality is less variable; (2) less voluminous sludge is produced, and therefore less cleaning is needed for the main sedimentation basins; and (3) because a large portion of the suspended solids is removed in the pretreatment step, fewer chemicals are used in subsequent treatment.

The selection of the most suitable type of pretreatment for a particular design should be made on the basis of field investigations, in which samples are taken from all regimes of the river to determine variations in raw water characteristics.

PLAIN SEDIMENTATION

The process of plain sedimentation allows for the removal of suspended solids in the raw water by gravity and the natural aggregation of the particles in a basin, without the use of coagulants. The efficiency of this process, as measured by turbidity removal, is largely dependent on the size of the suspended particles and their settling rate. Table 3.2 shows particle diameters and settling ranges for suspended materials found in water. It is obvious from the data in Table 3.2 that plain sedimentation

Table 3.2.   Effect of Decreasing Size of Spheres on Settling Rate

Diameter of Particle (mm)	Order of Size	Total Surface Area[a]	Time Required to Settle[b]
10	Gravel	3.14 cm^2	0.3 sec
1	Coarse sand	31.4 cm^2	3 sec
0.1	Fine sand	314 cm^2	38 sec
0.01	Silt	0.314 m^2	33 min
0.001	Bacteria	3.14 m^2	55 hr
0.0001	Colloidal particles	31.4 m^2	230 days
0.00001	Colloidal particles	0.283 ha	6.3 yr
0.000001	Colloidal particles	2.83 ha	63 yr minimum

Source: Adapted from AWWA (1971), p. 70.
[a]Area for particles of indicated size produced from a particle 10 mm in diameter with a specific gravity of 2.65.
[b]Calculations based on sphere with a specific gravity of 2.65 to settle 30 cm.

would serve no practical purpose for the removal of material smaller than 0.01 mm.

Plain sedimentation is quite effective in tropical developing countries for the following reasons: (1) the turbidity in rivers can be attributed largely to soil erosion, the silt being settleable; and (2) the higher temperatures in these countries improve the sedimentation process by lowering the viscosity of the water. Experience has shown that waters of high turbidity are clarified more effectively than waters of low turbidity. Plain sedimentation (or presettling) basins can be used as pretreatment units for both rapid and slow sand filtration plants. In the latter case, however, its use is limited to where it is possible to reduce the raw water turbidity to 30 NTU or less to avoid too frequent clogging of the sand bed. The economic and technical feasibility of achieving such a limit using plain sedimentation may be determined from settling tests of the raw water (Camp, 1946; IRC, 1981b).

The design of plain sedimentation basins is similar to that of conventional settling basins except that the detention times are generally shorter and the surface loadings are higher. The minimum depth of the basin is also somewhat less, because the sludge storage requirements are not as great as for conventional basins that follow coagulation and flocculation. The basin may operate on an intermittent basis, being held empty until needed. Practical experience by Smethurst (1979) in Baghdad and elsewhere confirms that effluents with considerably less than 1000 mg/l of suspended solids can be withdrawn from presettling tanks of one hour detention, even though the incoming water might have suspended solids of 10,000 mg/l or more. The design criteria for rectangular plain sedimentation basins are summarized in Table 3.3. The values listed are generalized, and serve only as guidelines. Chapter Seven contains additional information on sedimentation basin design and construction.

The construction of plain sedimentation basins can be quite simple. Three types of such basins are shown in Figures 3.1 to 3.3. The first type is constructed with wooden sheet piles, the second type is dug out of the earth with sloping sides, and the third type consists of a triangular shape with variable depth for achieving a

Table 3.3. Design Criteria for Plain Sedimentation Basins

Parameter	Range of Values
Detention time (hr)	0.5 to 3
Surface loading (m/day)	20 to 80
Depth of the basin (m)	1.5 to 2.5
Length/width ratio	4:1 to 6:1
Length/depth ratio	5:1 to 20:1

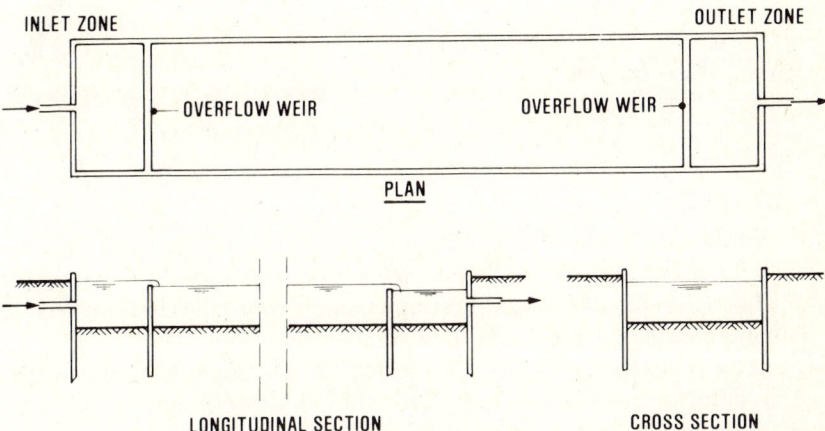

Figure 3.1. Presettling basin constructed with wooden sheet piles. *Source:* IRC, 1981b.

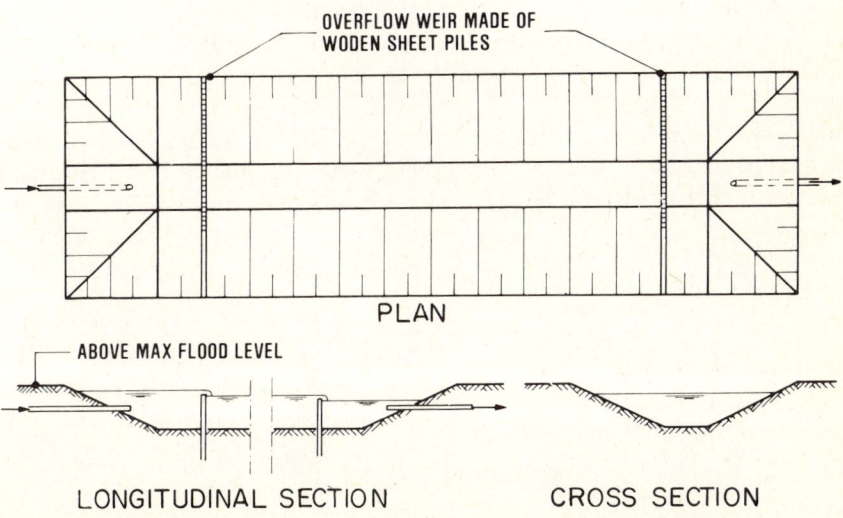

Figure 3.2 Dug basin as a presettling tank. *Source:* IRC, 1981b.

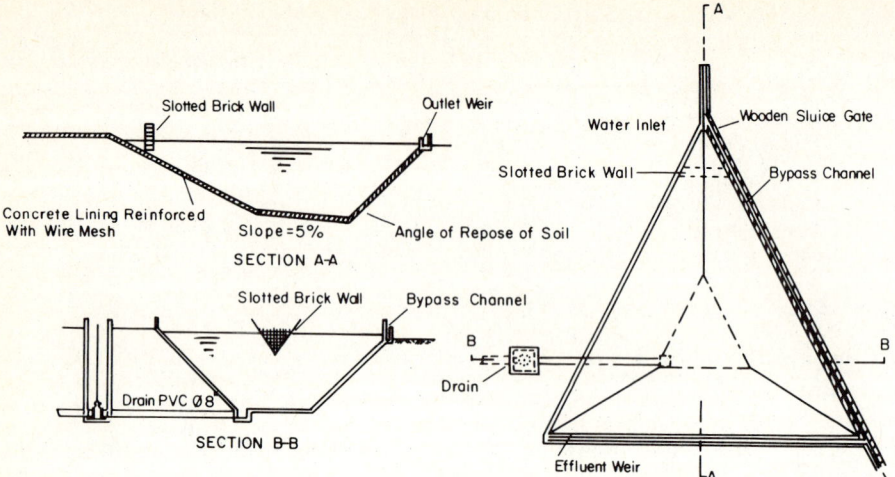

Figure 3.3 Triangular presettling basin with variable depth. *Source:* CEPIS, 1982, vol. 2. plan no. 4.

uniform distribution of water both at the inlet and in the settling unit. The design in Figure 3.3 has a slotted brick wall at the inlet side to distribute the flow uniformly, a bypass channel to be used when presedimentation is not necessary or when the unit must be cleaned, and a drain for hydraulic discharge of sludge. Also, the sidewalls of the basin are gently sloped so that a thin cement lining reinforced with wire mesh (ferrocement) can be used in place of reinforced concrete, thereby substantially reducing construction costs.

Earthen basins may have to be lined with plastic or an impermeable layer of clay, masonry, or concrete in places where seepage occurs, and protected from flooding. The addition of overflow weirs or baffles across the width of the basin can improve the uniform distribution of velocities and mitigate short-circuiting. In areas where floating algal growths present a problem, outlet orifices may be placed behind a deflecting baffle and some distance below the water surface as shown in Figure 3.4.

Under conditions of continuous treatment, at least two settling basins should be built to allow one to be shut down periodically for cleaning. It is not essential in pretreatment that the basins be designed to handle the full plant capacity at all

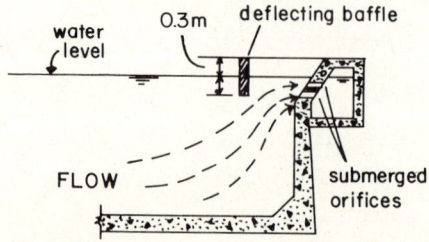

Figure 3.4. Submerged orifice basin outlet system. *Source:* Arboleda, 1973.

times. For example, a plant with two presettling basins can be designed so that each basin carries three-quarters of the design flow. This will limit the overload on the operating basin to about 33%, when one unit is shut down temporarily for cleaning. When possible, cleaning and sludge removal should be done during periods of low turbidity in the raw water when the shutdown of one basin will not overburden the main treatment units. The process of manual sludge removal is laborious, but normally is not necessary more than once a year, depending on basin size. Fire hoses or fixed nozzles can ease the cleaning process (see Chapter 7, under "Manual Sludge Removal").

Settling tests, using cylinders 2 to 5 m high and 20 cm in diameter; particle-size distribution tests, using a hydrometer analysis; and graphical analyses of the data generated from these tests, are reported in a study from Mosul, Iraq, involving the design of plain sedimentation basins for several rapid filtration plants that draw water from the Tigris River (Ahmad, Wais, and Agha, 1982). Four sets of experiments were conducted for water having turbidities of 500, 1200, 1800, and 2500 NTU. The results, presented in Table 3.4, indicate that much of the turbidity was removed within 3 to 4 hr, whereas a turbidity of 50 NTU was reached in 24 hr in all cases. On the basis of these pilot studies, an overflow rate of 12 m/day, detention time between 2.25 to 4 hr, and basin depth of 2 m were selected for the design of the plain sedimentation basins. The procedures that were followed in this study may serve as a general guide for the design of presedimentation basins, although it should be recognized that the particular results obtained in this study cannot be applied directly to other situations.

The use of tube settlers for upgrading the pretreatment of water in existing plain sedimentation basins or for reducing the size of new basins was studied by Ahmad and Wais (1980). It was found experimentally that tube settlers were effective in removing turbidity from water containing silt and clay particles before coagulation. The best tube inclination for turbidity removal, determined by experiment, was 40° from the horizontal. At a surface loading rate of 59 m/day the turbidity removal was as high as 85% in a 2.6 cm diameter tube. A modular design employing a presettling basin with inclined-plate settlers prior to slow sand filtration is shown

Table 3.4. **Turbidity Removal with Different Settling Times (Mosul, Iraq)**

Initial Turbidity (NTU)	Turbidity Remaining (NTU)	
	After 2 hr	After 3 hr
500	145	90
1200	620	120
1800	450	90
2500	610	120

Source: Ahmad, Wais, and Agha, 1982, p. 442.
Note: Turbidity removals are very much a function of particulate size distribution, but these data suggest the order of magnitude of removals.

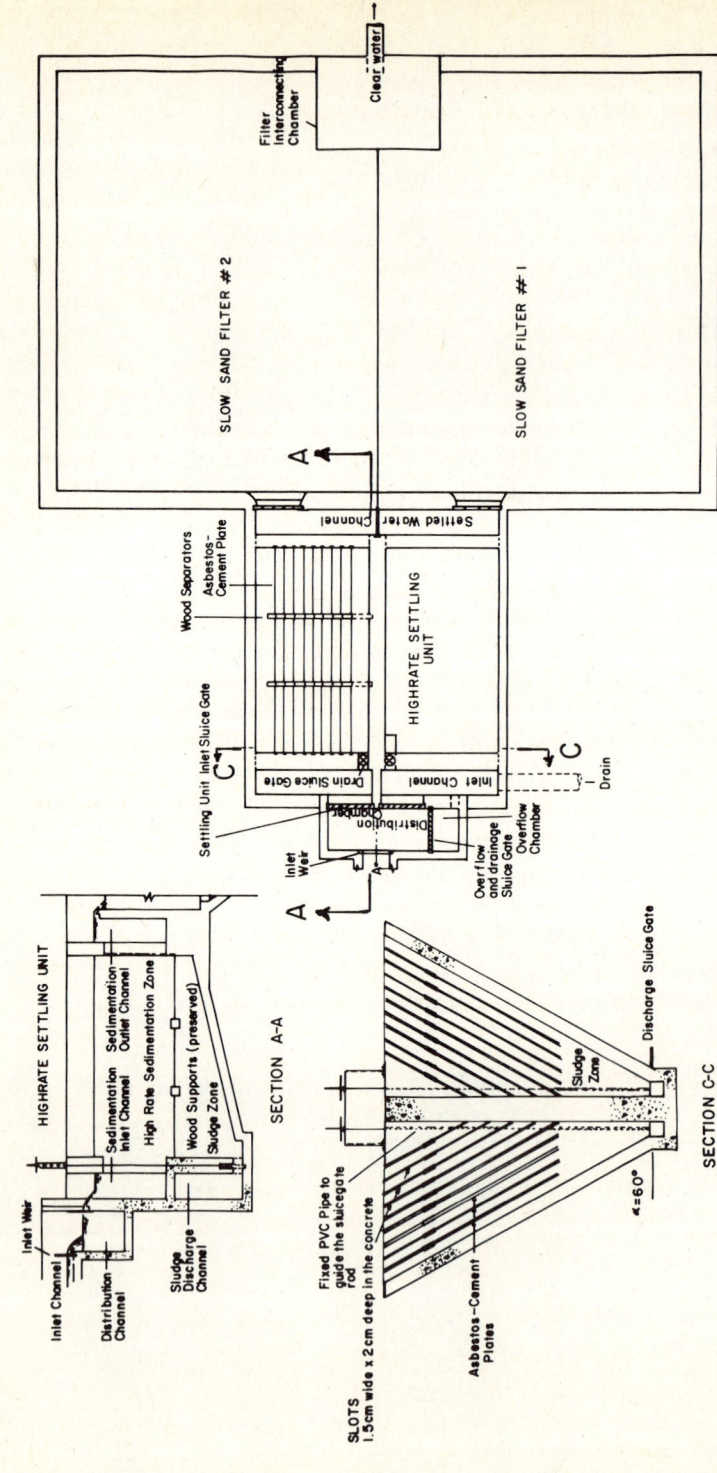

Figure 3.5.   High-rate plain sedimentation with inclined-plate settlers before slow sand filtration. *Source:* Adapted from CEPIS, 1982, vol. 2, plan nos. 12, 13.

36

in Figure 3.5 (CEPIS, 1982). The settling unit is operated at a high rate of 60 m/day and has a detention time of 25 minutes. This design is able to treat waters with turbidities up to 500 NTU, but only when the turbidity results from particles larger than 1 micron.

STORAGE

Storage basins or reservoirs can be used for presedimentation. The detention time is generally much greater than that for conventional sedimentation basins, ranging from about one week to a few months. For extremely turbid rivers or streams (average annual turbidity over 1000 NTU), storage provides the best pretreatment. Storage serves several purposes in water treatment: (1) it reduces the turbidity by natural sedimentation; (2) it attenuates sudden fluctuations in raw water quality; (3) it improves the quality of water by reducing the number of pathogenic bacteria (if the storage site is protected); (4) it improves the reliability of the water supply because it can be drawn upon during periods of short supply of raw water; and (5) it can be drawn upon during short periods of exceedingly high turbidity, during which river water is not fed into the storage basin.

The design of storage basins is not subject to well-defined criteria, but should take into consideration local conditions, especially land availability for the construction of the basin. Storage basins may be shaped into ponds or lagoons formed from the natural topography, or constructed from manmade earth dams. The capacity of a storage basin should allow for losses due to evaporation and seepage, especially in arid regions. In places where seepage is a problem, the bottom of the basin should be covered with some type of impermeable layer, such as clay or concrete. In some instances, it may be desirable to restrict public access to the storage basin to maintain the quality of the water. A simple method for protecting earthen storage basins is to plant a natural barrier of heavy vegetation (such as thorn bushes) around the periphery of the basin to conceal it and break wind effects as well as to thwart potential dumping.

The costs of constructing storage pretreatment facilities are often quite high because they usually require flat and valuable riverside land. The maximum economical water depth, because of the need for earthen dikes, is about 15 m. The cost of such facilities is about $2.00 per cubic meter of water stored (1979 U.S.$; Smethurst, 1979).

Smethurst (1979) has compiled extensive plant-operating data on the quality of water before and after storage for several water supply facilities in England that use storage for pretreatment. These data are tabulated in Tables 3.5 and 3.6. A remarkable improvement in bacteriological quality as well as significant reductions in turbidity due to storage is evident from these tables. Dr. A. V. Houston's studies on bacterial die-away in London's storage basins along the Thames almost a century ago were the basis for water treatment until chlorination was introduced.

Table 3.5. Quality of Water Before and After Storage for Water Supplies in England

	River Thames at Teddington[a]		River Thames at Oxford[b]		River Great Ouse at Diddington[c]	
	Raw River	Stored Water	Raw River	Stored Water	Raw River	Stored Water
Color, Hazen	830	450	19	9	30	5
Turbidity (NTU)	35	5.3	14	3.2	10	1.5
Presumptive coli, MPN per 1000 ml	—	—	60,000	200	6500	20
Presumptive coli, % of samples:						
% of 100 cm^3	99.9	57.2	—	—	—	—
% of 10 cm^3	97.7	32.5	—	—	—	—
% of 1 cm^3	83.1	13.4	—	—	—	—
% of 0.1 cm^3	48.3	3.3	—	—	—	—
E. coli, MPN per 100 ml	—	—	20,000	100	1700	10
Colony counts per ml:						
3 days at 20°C	4465	208	—	—	50000	580
2 days at 37°C	280	44	—	—	15000	140

Source: Adapted from Smethurst, 1979.
[a]Time of storage unavailable.
[b]Storage of 140 days.
[c]Storage of 200 days.

Table 3.6. Change in Water Quality Due to Storage for Water Supplies in England

	Reduction Due to Storage, %	
	River Severn at Hampton Loade[a]	River Derwent at Draycott[b]
Color	28	67
Turbidity	70	51
Presumptive coli	95	99
E. coli	94	>99
Colony counts per 1 ml:		
3 days at 20°C	88	—
2 days at 37°C	89	—

Source: Adapted from Smethurst, 1979.
[a]Storage of 12 to 15 days.
[b]Storage greater than 15 days.

ROUGHING FILTRATION

Particles removed in filters are much smaller than the pore spaces in the media, so the process of filtration is not straining. The principal processes are sedimentation in the pore spaces, adhesion to the media particles and, in slow sand filters, biochemical degradation of particles that are captured.

Roughing filters allow deep penetration of suspended materials into a filter bed, and they have a large silt storage capacity. The solid materials retained by the filters are removed by flushing, or if necessary, by excavating the filter media, washing it, and replacing it. Roughing filtration uses much larger media than either slow or rapid filtration, as indicated in the following comparison:

Slow Sand Filters	0.15 to 0.35 mm diameter
Rapid Sand Filters	0.4 to 0.7 mm diameter
Roughing Filters	>2.0 mm diameter

The rate of filtration, however, can be as low as those used for slow sand filters or higher than those used for rapid filters, depending on the type of filter, the nature of the turbidity, and the desired degree of turbidity removal.

Roughing filters are often used before slow sand filters because of their effectiveness in removing suspended solids. Field studies in Tanzania have shown that, in many cases, neither presedimentation nor storage is as effective as roughing filters for pretreating raw water to the physical standards required by slow-sand filters (Wegelin, 1982). Roughing filters are limited, however, to average annual raw water turbidities of 20 to 150 NTU, so as to prevent too frequent clogging and to ensure their continuous operation for an extended period of time.

There are basically two types of roughing filters, which are differentiated by their direction of flow: vertical flow (VF) and horizontal flow (HF). Structural constraints limit the depth of the filter bed in VF filters, but higher filtration rates and backwashing of the filter media are possible. On the other hand, HF filters enjoy practically unlimited filter length, but normally are subject to lower filtration rates, and they generally require manual cleaning of the filter media.

Vertical Flow Roughing Filters

VF roughing filters are further subdivided into upflow and downflow units. A typical arrangement for a VF upflow filter is shown in Figure 3.6. Several gravel layers, tapering from a coarse gravel layer (10 to 15 mm) located directly above the underdrain system, to successively fine gravel layers (7 to 10 mm and 4 to 7 mm) permit deep penetration of suspended solids into the filter bed. Filtration rates in gravel upflow filters are relatively high (up to 20 m/hr) because of the large pore spaces in the filter bed that are not likely to clog rapidly. Low backwashing rates are used because no attempt is made to expand the bed; but longer time periods for adequate cleaning of the gravel are usually necessary (about 20 to 30 minutes). Filter underdrains can be fabricated locally, using either a "teepee" type of design or a main and lateral system (both are described in Chapter Eight, under "Filter Bottom and Underdrains"). Upflow filters are used predominantly in upflow-downflow type filtration to replace the unit process of flocculation and sedimentation found in conventional rapid filtration plants (see Chapter Eight, under "Upflow-Downflow Filtration"). They are similar in design and construction to gravel bed flocculators (see Chapter Six, under "Gravel-Bed Flocculators").

VF downflow roughing filters using shredded coconut fibers for the filter medium have been said to be successful in Thailand (Frankel, 1974), and have been installed in over 100 rural villages in Southeast Asia (Frankel, 1981). The raw coconut husks are found throughout Southeast Asia and have little market value; hence they provide a low-cost filter medium for treatment plants in that part of the world.

Shredded coconut fiber may be prepared manually by soaking the husk for 2 to 3 days in water, then shredding the husk by pulling off the individual fibers and removing the solid particles which bind the fibers. Shredded coconut fibers may

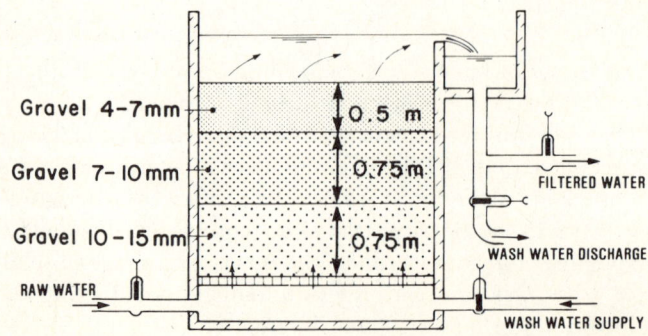

Figure 3.6. Gravel upflow roughing filter. *Source:* Adapted from IRC, 1982b.

also be purchased directly from upholstery stores or coir (coconut fiber) factories. The shredded fiber should be immersed in water for about three days, until the fiber does not impart any more color to the water (Frankel, 1977). The depth of the coconut fiber in the filter box is usually 60 to 80 cm. There are no backwashing arrangements for cleaning the coconut fibers, because the fibers do not readily relinquish entrapped particles due to their fibrous nature. Instead, water is drained from the filter box and the dirty fibers are removed and discarded. Coconut fiber stock, which has been properly cleaned, is then packed into the filter. The filter medium generally must be replaced every three or four months. The availability of the raw coconut husks at low cost, as well as the elimination of backwash pumps and ancillary equipment, combine to make this manual filter bed regeneration process economical in areas where coconut trees are common. The use of such indigenous materials for filter media is also a practical alternative to conventional filter design (see Chapter Eight, under "Dual-Media filters").

Several small filter plants ranging in capacity from 24 to 360 m^3/day were constructed from 1972 to 1976 in the Lower Mekong River Basin countries (Thailand, Viet Nam, Cambodia) and in the Philippines (Frankel, 1981). Two-stage filtration, using shredded coconut fibers and burnt rice husks for the roughing and polishing filter, respectively, was typical for all filter plants. The filtration systems generally produced a clear effluent (less than 5 NTU) when treating raw water with a turbidity less than 150 NTU. The units were designed at a filtration rate of 1.25 to 1.5 m/hr, which is about 10 times higher than that used for conventional slow sand filters. Bacterial removals averaged 60 to 90% without the use of any disinfectant. The media generally required changing once every 3 to 5 months, depending on the level of turbidity in the raw water.

Horizontal Flow Roughing Filters

Horizontal flow (HF) roughing filters have a large silt storage capacity because of their coarse filter media and long filter length. Filter operation commonly extends over a period of years before the filter must be removed from service and cleaned. HF roughing filters have been operating successfully ahead of slow sand filtration at several water treatment plants in Europe (Kuntschik, 1976). On the basis of several pilot projects conducted on HF filters which have substantiated their effectiveness in pretreating turbid river waters in Thailand, Tanzania, and Honduras, their use in small treatment plants in developing countries holds much promise (Thanh, 1978; Wegelin, 1982; CEPIS, 1982).

The main features of an HF roughing filter are shown in Figure 3.7. For overall efficiency it is best to use a graded gravel scheme for the filter medium. The HF filter is usually divided into several zones, each with its own uniform grain size, tapering from large sizes in the initial zone to small sizes in the final zone. In this way, penetration of suspended solids will more easily take place over the entire filter bed and result in longer filter runs. The following design guidelines have been suggested by Wegelin (1982), and are based on extensive field testing of HF roughing filters ahead of slow sand filters in Tanzania:

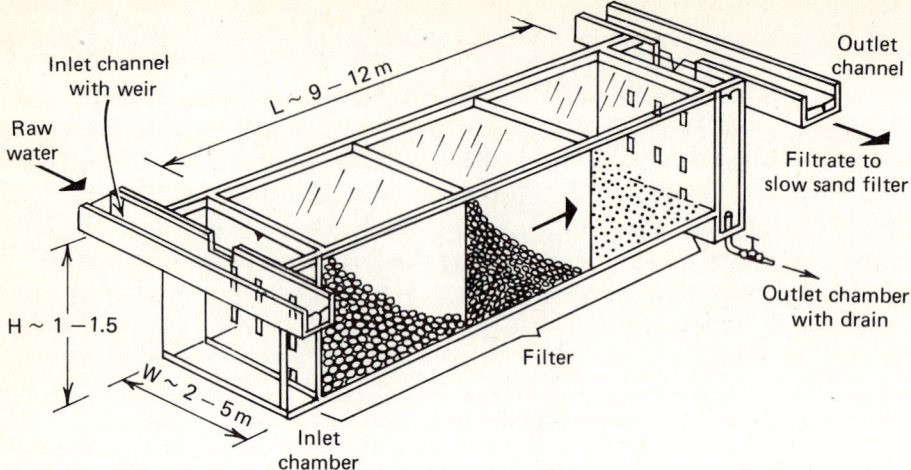

Figure 3.7. Basic features of a horizontal-flow roughing filter *Source:* Wegelin, 1982.

1. The acceptable range for the filtration rate is 0.5 to 4.0 m/hr (face velocity), but an upper limit of 2 m/hr should be observed for waters with very high suspended solids load and/or colloidals.

2. The filter grains to be used should have two to three zones with sizes ranging from 4 to 40 mm. The sequence of arrangement in the longitudinal direction should be from coarse to fine.

3. Because the first zone of the filter bed stores a higher percentage of suspended solids than the others, the length of the coarse zone provided should be greater than that of the finer zones in order to provide a large silt storage volume. Thus, the following range of lengths of individual zones should be provided:

First, coarse fraction:	4.5 to 6.0 m
Middle, medium fraction:	3.0 to 4.0 m
Last, fine fraction:	1.5 to 2.0 m

    As a result, the total length of filter should be 9.0 to 12.0 m.

4. For a filter with side walls that are above the ground surface, the height should be below 1.0 to 1.5 m to allow for easy manual digging out of gravel and refilling it after cleaning.

5. The free water table should be covered by a 10 to 20 cm thick gravel layer in order to prevent plant and algal growth. Hence, the top level of the filter medium should be 30 to 40 cm above the crest level of the outlet weir.

6. The filter floor should slope in the direction of flow (about 1:100).

7. The outlet should be provided with a V-notch weir to facilitate discharge measurements.

The filter length is the most critical dimension in the design of HF roughing filters and should be selected after considering an appropriate balance between construction costs and the frequent cleanings required when filter lengths are short. Consideration of such a balance led to the construction of several HF gravel filters in the city of Dortmund, West Germany, each of a length of about 50 to 70 m to pretreat raw water from the Ruhr River (Kuntschik, 1976). The total operating period for these extremely long filters is about 5 years, after which the gravel has to be removed, cleaned, and replaced. The high cost of labor in countries such as West Germany dictates the design of long filters to minimize the frequency of cleaning, which is relatively expensive. For developing countries, however, such great lengths usually are not warranted; instead, filter lengths between 4 and 15 m seem reasonable, because the cleaning of the filter can be accomplished at much lower cost. Of more concern, however, for design purposes, is the availability of the filter media. The gravel or crushed stone that is required for HF filters must be of reasonably uniform size, which may be difficult to obtain in large enough quantities if large filter beds have to be filled.

An HF filter has been operating successfully prior to slow sand filtration for the village of Jedee-Thong, Thailand (Thanh, 1978). The filter incorporates six gravel zones in a filter box with a volume of 6 × 2 × 1 m, and is designed for a capacity and face velocity of 76.5 m³/day and 4 m/hr, respectively (see Figure 3.8). The characteristics of the filter media are presented in Table 3.7. The filter box is constructed from bricks covered with a layer of fine mortar. The six compartments are separated by removable strong wire mesh, which allows for easy cleaning and changing of the media. The filtering area is preceded and followed by chambers without gravel, the effluent chamber serving as a wet well for the pumps. Thanh (1978) reported a removal efficiency of 60 to 70% for this filter for raw water turbidities ranging from 30 to 100 NTU.

HF roughing filters may also be constructed adjacent to a stream bed so as to allow raw water to flow through a porous-stone wall and into a gravel bed (see Figure 3.9). The drain system is made of a perforated PVC pipe that leads to a

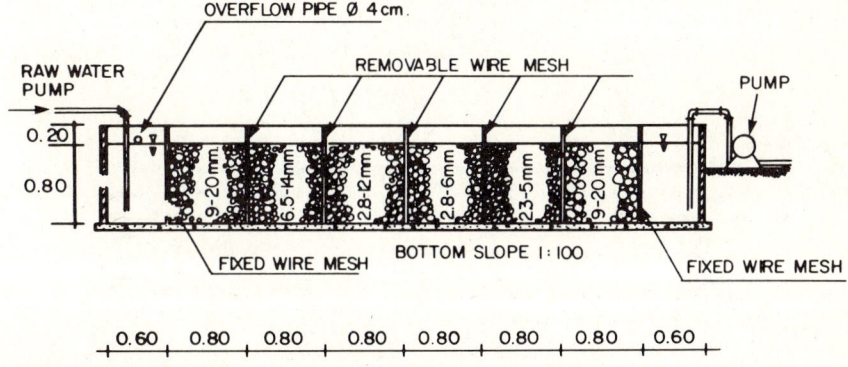

Figure 3.8. Horizontal-flow roughing filter used ahead of slow sand filtration in Jedee-Thong, Thailand. *Source:* Thanh, 1978. (Dimensions are in meters.)

Table 3.7.   Characteristics of Filter Media for Horizontal-Flow Roughing Filters in
Jedee-Thong, Thailand

Size Range (mm)	Effective Size (P_{10})	Uniformity Coefficient (P_{60}/P_{10})
9 to 20	15	1.38
6.5 to 14	11	1.5
2.8 to 12	6.1	1.47
2.8 to 6	3.8	1.36
2.3 to 5	2.6	1.27
9 to 20	15	1.38

Source: Thanh, 1978, p. 17.

junction box. To avoid the infiltration of surface runoff, an impermeable layer of
clay or a polyethylene liner can be placed over the gravel bed. This particular design
has a capacity ranging from 85 to 860 m³/day, and is intended to operate at a
filtration rate of 0.5 m/hr. It can treat waters of turbidities less than 150 NTU prior
to slow sand filtration. The length of the filter is variable, depending on the design

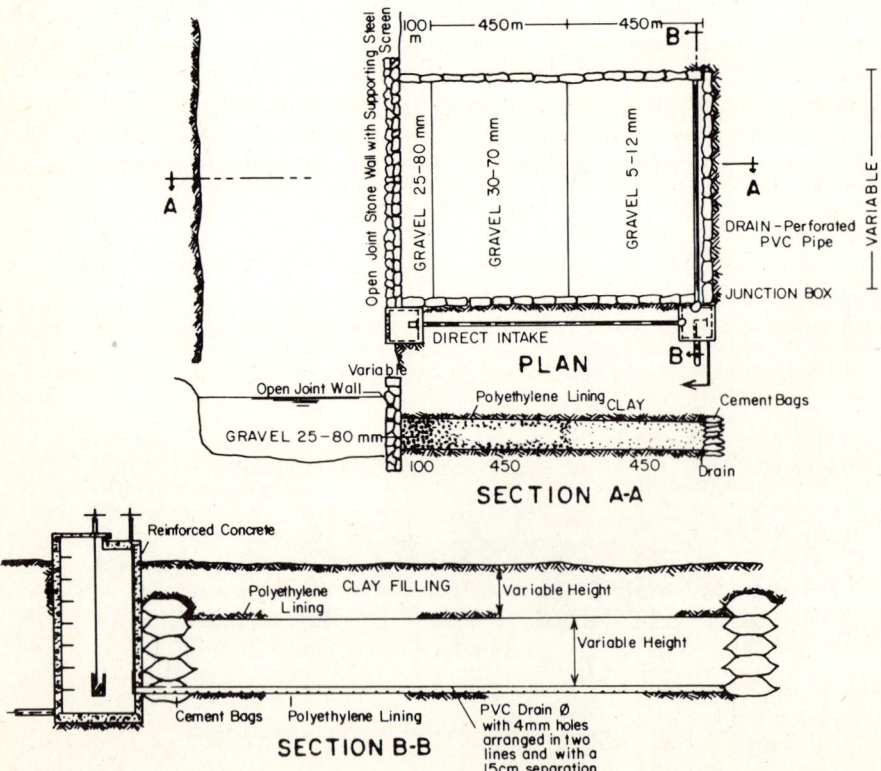

Figure 3.9.   Horizontal-flow roughing filter constructed adjacent to a stream bed. *Source:* CEPIS,
1982, vol. 2, no. 3.

capacity. The state agency SANAA, in Honduras, is currently field testing such roughing filters as part of their rural water supply program (CEPIS, 1982).

GRIT REMOVAL

Grit consists of particles of sand, gravel, or other mineral matter, and is often found suspended in fast-flowing streams and rivers. Without grit removal, facilities that draw water from these sources often find their pumps, transmission mains, and tanks fouled with grit, which can cause excessive wear on equipment and interfere with plant operation. Although many mechanically operated degritting devices are available, an appropriate type for developing countries is a grit chamber cleaned by hand. Sand separators, which are proprietary items, may be suitable, if purveyors of the equipment are available.

Grit Chambers

Grit chambers are generally designed to maintain a constant velocity of flow, usually 0.3 m/sec, and a time of passage through the chamber of about 30 sec to 1 min; they are therefore approximately 10 to 20 m in length. These criteria permit most of the grit to settle to the floor of the chamber, with lighter organic solids being kept in suspension. Constant velocity in a grit chamber is achieved by one of several methods (Okun and Ponghis, 1975):

1. Multiple chambers are suitable for large plants where one or two units may be used during low flow periods and additional chambers cut in, as the flow increases.

2. Proportional weirs at the end of rectangular channels. Inasmuch as the head on such a weir is proportional to the flow ($Q = kh$), in a rectangular channel the cross-sectional area will then be proportional to the flow and the velocity will be constant. Figure 3.10 shows a three-channel grit chamber with a proportional weir.

3. A rectangular discharge section, such as a weir or a Parshall flume, may be used to maintain a constant velocity if the grit channel itself is made parabolic. An approximately parabolic chamber can be designed by sloping the sides of the channel to a narrow width at the bottom. This also makes grit removal easier.

An alternative, for smaller units, is to set the flume lower than the floor of the grit chamber, so that the grit chamber can be made with vertical sides and the entire bottom used for grit storage. A hydraulic profile should then be plotted for the range of flows expected. The advantage of a Parshall flume for a control section is that the head loss is low and the flume can be used to measure flow into the plant (see Chapter Five, under "Parshall Flume" and "Flow Measurement Devices in Open Channels"). Figure 3.11 shows a typical arrangement for a grit chamber incorporating a Parshall flume.

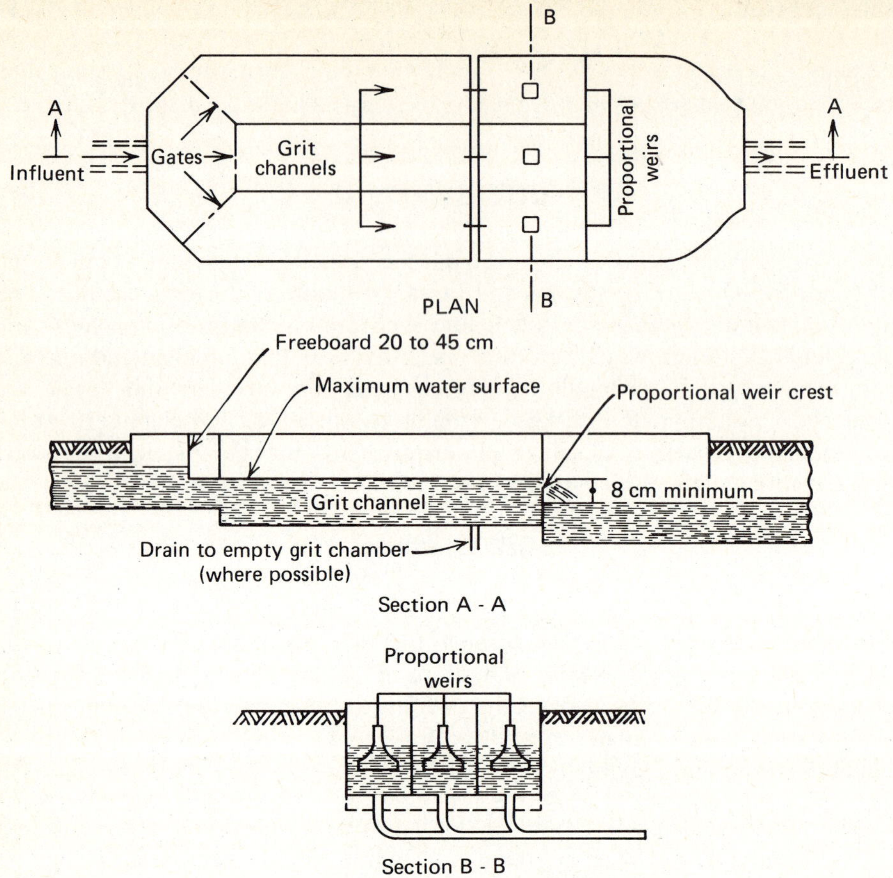

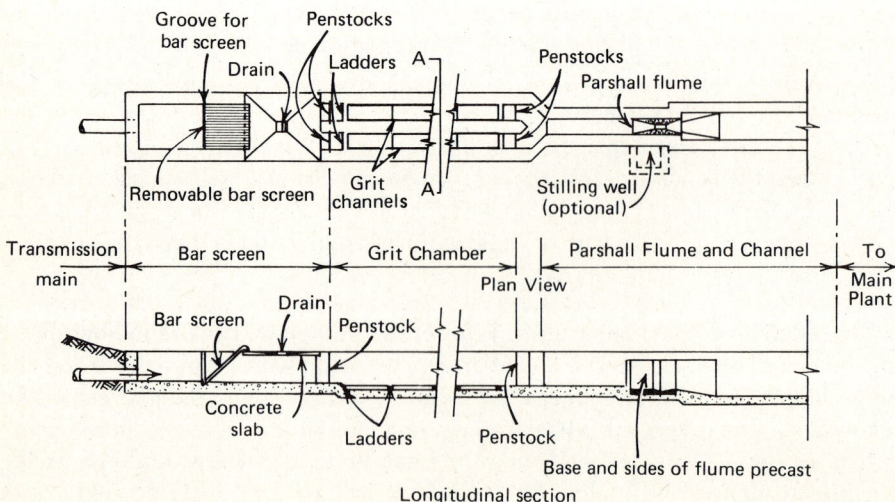

Figure 3.10. Grit chamber with proportional weir. *Source:* Okun and Ponghis, 1975.

Figure 3.11. Velocity control grit chamber with Parshall flume. *Source:* Adapted from Marais, 1969.

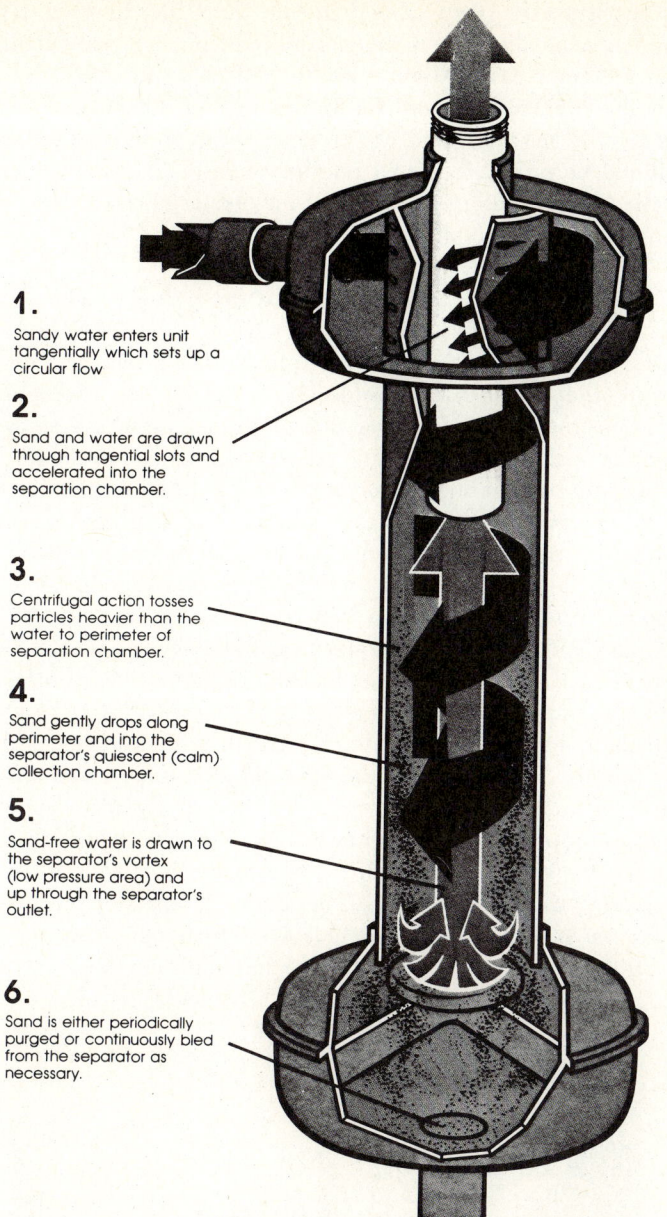

1.

Sandy water enters unit tangentially which sets up a circular flow

2.

Sand and water are drawn through tangential slots and accelerated into the separation chamber.

3.

Centrifugal action tosses particles heavier than the water to perimeter of separation chamber.

4.

Sand gently drops along perimeter and into the separator's quiescent (calm) collection chamber.

5.

Sand-free water is drawn to the separator's vortex (low pressure area) and up through the separator's outlet.

6.

Sand is either periodically purged or continuously bled from the separator as necessary.

Figure 3.12. How a sand separator works. *Source:* Claude Laval Corporation.

In small installations it is advisable to provide from three to seven days of storage in the channel, depending upon the reliability of plant personnel to empty the chamber on a regular basis (Marais, 1969). Storage is provided by lowering the floor level of the grit channel, as shown in Figure 3.11. Crossbars (or ladders) may be added across the grit storage area to enhance the deposition of grit. At least two channels should be provided so that one may be isolated, drained, and cleaned while the other is in operation. Although it may be disposed of safely in land fills, grit may also be used in the plant for walks and driveways.

Sand Separators

A hydraulically-operated sand separator is a simple and effective device for removing sand, grit, and other abrasive solids before they enter the raw water pump, thereby reducing wear on impellers and bearings and prolonging pump efficiency. It operates on the principles of centrifugal force and gravity, as described in Figure 3.12. There are no moving parts to wear out and replace, and operation is limited to the periodic purging of solids from the collection chamber, which can be done manually. A steady pressure loss is maintained in the system, ranging from 30 to 85 kPa (4.5 to 12 psi). The manufacturer of the system, Claude Laval Corporation (see Appendix F for the company's address), claims that it can remove up to 98% of all particles as small as 74 microns. Standard sizes are available for flowrates ranging from 24 m³/day to 63,200 m³/day. For large flowrates, it is possible to manifold several separator units in parallel to reduce costs.

Sand separators can be installed directly onto the suction side of turbine and submersible pumps, as shown in Figure 3.13. This arrangement includes a coarse inlet screen to prevent internal plugging and a vertical discharge outlet to accommodate typical pump suction connections. The solids that accumulate in the collection chamber are purged hydraulically by an eductor. If conditions are not suitable for its installation on the suction side of the pump, a separator can be installed on the discharge side to keep transmission mains and treatment units free of solids, in effect, serving the same function as a conventional grit chamber.

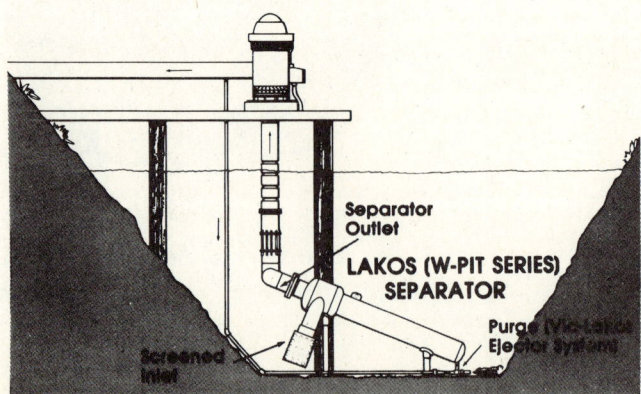

Figure 3.13.   A sand separator installed on the suction side of a vertical-turbine pump. *Source:* Claude Laval Corporation.

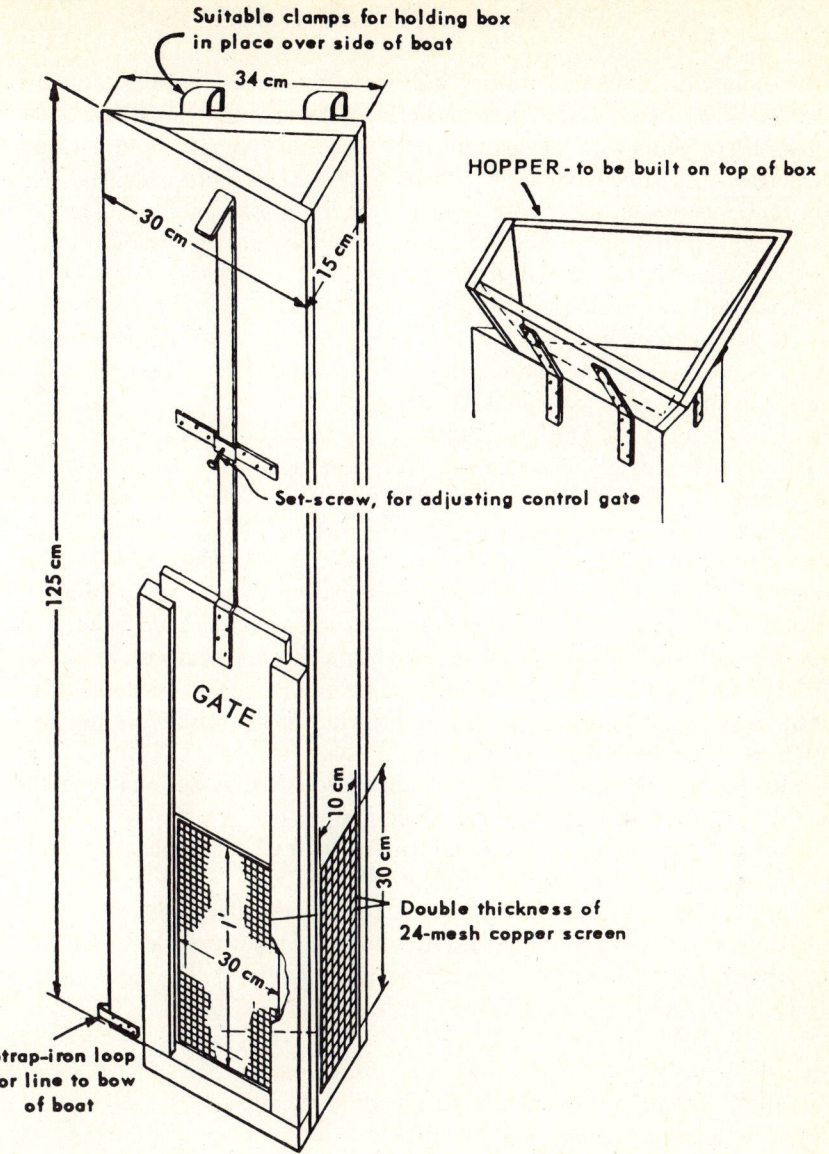

Figure 3.14. Box for controlled distribution of copper sulfate solution in lakes or reservoirs. *Source:* Cox, 1964.

CONTROL OF MICROORGANISMS

Microorganisms that cause tastes and odors are usually harmless. They are visible under a microscope and include bacteria, protozoa, algae, small worms, crustacea, and others. Storage of water in an open reservoir provides an opportunity for microorganisms, particularly algae, to grow and develop. The greater the concentration of nutrients in the water, the larger will be the growth of algae. The potential for algal growth in a particular area can be found by observing ponds or lakes in that area. In general, because algae require sunlight, growths are not likely to be heavy where waters are turbid as a result of silt.

In the industrialized countries, microstrainers are sometimes used for the removal of algae and other microorganisms from drinking water supplies. Such screening mechanisms are foreign exchange items.

It is better to limit the growth of those organisms by covering reservoirs to exclude light. Where this is not possible, an algicide such as copper sulphate or chlorine can be used, if available at reasonable cost. When algicide treatment is deemed appropriate, it is desirable to apply it at the first signs of algal growth, so as to reduce the amount of decaying matter that can cause taste and odor problems. The required dose is a function of the types of organisms and their relative numbers and, therefore, microscopic examination of samples of the impounded water is desirable. A dose of 0.3 mg/l is usually sufficient when copper sulphate is used, except when alkalinity is high (Cox, 1964). When alkalinity is above 50 mg/l, the dosage should be doubled, because Cu is precipitated as $CuCO_3$. Experience at a specific location will provide the best guide to dosage.

In the treatment of reservoirs, copper sulphate may be applied by two simple methods; namely, the burlap bag method and wooden box method. The first method is accomplished by hanging burlap bags containing a weighted quantity of copper sulfate crystals from the sides or stern of small rowboats. The boat is usually propelled in parallel courses about 8 to 15 m apart, and then at right angles so as to treat the entire body of water thoroughly. The rapidity with which the chemical goes into solution may be controlled by regulating the fabric of the bag used, varying the velocity of the boat, using crystals of large or small size, or by combinations of these variables. Boats equipped with outboard motors greatly improve the mixing of the chemicals in the water due to the action of the propeller. The second method is simply an improvement on the first method, whereby the burlap bags are replaced by permanent boxes attached to both sides of a small boat. The design for a copper sulphate distribution box is shown in Figure 3.14. The rate of solution of the crystals can be controlled by changing the position of the control gate. Cox (1964) and Salvato (1982) provide further information on the use of copper sulphate in water treatment including dosage, toxicity, and frequency and method of application.

chapter four

CHEMICALS AND CHEMICAL FEEDING

The chemicals necessary in water treatment plants include a coagulant, generally alum; disinfectants, generally chlorine; and, when necessary, alkalies, generally lime, for pH control. Coagulant aids may also be used to improve the coagulation-flocculation process or reduce coagulant consumption. Fortunately, alum, chlorine, and lime are the most readily available water treatment chemicals in developing countries, albeit expensive when imported. Where iron salts are available as a by-product locally, it is an economical coagulant. Other water treatment chemicals widely used in the industrialized countries to provide fluoridation, taste and odor removal, and stability and corrosion control are not recommended for communities in developing countries, except perhaps in the major cities where the chemicals and skilled supervision are available.

Improper selection, handling, and feeding of chemicals can be detrimental to water treatment plant performance, and have been the bane of many such plants in developing countries. A survey of plants conducted in India (NEERI, 1971) revealed that about 80% of the plants were dosing alum in an unscientific and primitive way (by dumping blocks of alum into the raw water channels), because alum equipment was out of order. Similarly, in 50% of the plants studied, the chlorine dosing equipment was out of order, and chlorination was done by bubbling chlorine gas directly into the filtered water channel. Bacteriologically safe water was not being produced in most cases, and no plant had the capacity to switch over to breakpoint or superchlorination under emergency conditions. To avoid problems in small communities in developing countries, alternative chemicals that are easily handled and applied should be explored; for example, hypochlorite compounds in place of chlorine gas. Similarly, chemical feeders should be simple in design, easy to operate, and preferably gravity feed. Whenever possible, the local manufacture of these items is preferable to importing them.

The chemical storage facilities and solution tanks should be located high enough, for example, on the second floor of the chemical building, so that chemicals can be fed by gravity. Trucks can unload materials on the second floor by means of ramps, or they can be carried by manual labor. Otherwise, when storage facilities and solution tanks are placed on the first floor, the chemical solution must be pumped to an elevated feeder, from which the solution then flows by gravity. The requirement for pumps usually involves importation of equipment and spare parts, with resulting down time when the pumps fail.

This chapter begins with a brief discussion of the jar test, which is the standard laboratory procedure for selecting chemicals and optimal doses, followed by sections on the primary coagulants, alkalies, natural coagulant aids, disinfection, and chemical feeding. The table of Chemicals Used in Water Treatment (found in Appendix A) summarizes the characteristics of the more common chemicals used in water treatment.

THE JAR TEST

The required chemical dosage for a particular raw water is virtually impossible to determine analytically because of the complex interrelationships that exist between these chemicals and the constituents of the water being treated, as well as such factors as pH, temperature, and the intensity and duration of mixing. Consequently, a laboratory procedure known as the jar test is used to determine the most effective and economical dose of coagulant for a particular mixing intensity and duration. Table 4.1 indicates the advantages of using proper alum dosages for coagulation. Jar test procedures are presented in Appendix D.

Laboratory stirring equipment, as shown in Figure 4.1, is used in jar testing to provide uniform mixing for a number of samples simultaneously and can be adjusted to match plant scale velocity gradients for rapid mixing and flocculation. (Formulas

Table 4.1.   Effects of Alum Dosage on Treatment Criteria

	Under Dose	Proper Dose	Over Dose
Turbidity removal	Poor	Good	Fair
Color removal	Poor	Good	Fair
Algae removal	Poor	Good	Fair
Length of filter runs	Medium	Long	Short
Residual aluminum	High	Low	High
Dollar value	Wasted	Good	Poor

Source: Ontario Ministry of the Environment, 1976.

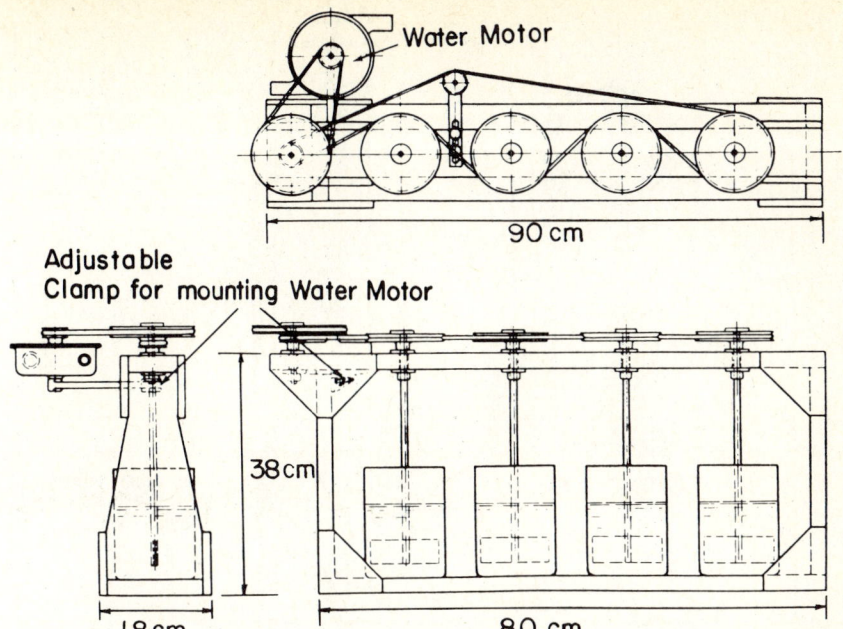

Figure 4.1. Laboratory stirring equipment for jar testing. *Source:* Adapted from Cox, 1968.

for calculating plant scale velocity gradients are given in Chapter 6, under "Design Criteria.") These units may be purchased from laboratory supply houses or manufactured locally. The multiple stirrer shown in Figure 4.1 is powered by a water motor (a jet stream of water provides the motive force to turn an impeller), but units operated by a hand crank or a small variable-speed electric motor with speed reducing systems may also be used. Metal rods with stirring paddles are attached to pulleys suspended directly over the beakers, which rotate at the same rate due to the common drive. The rods have a handle attached to the upper end so that they may be lifted vertically while the pulleys are turning.

Standard 2-liter laboratory beakers are commonly used for jar testing, both with and without stators. The effect of the stators is to inhibit vortex formations at high stirring velocities and to increase the value of G at all velocities. The calibration curves presented in Figure 4.2 were developed for 2 liters of water in a 2-liter beaker.

PRIMARY COAGULANTS

Metal coagulants based on aluminum, and to a lesser extent on iron, are used almost exclusively in the coagulation process. The choice of coagulant should be determined under laboratory conditions (e.g., jar testing) with the final choice influenced by economic considerations.

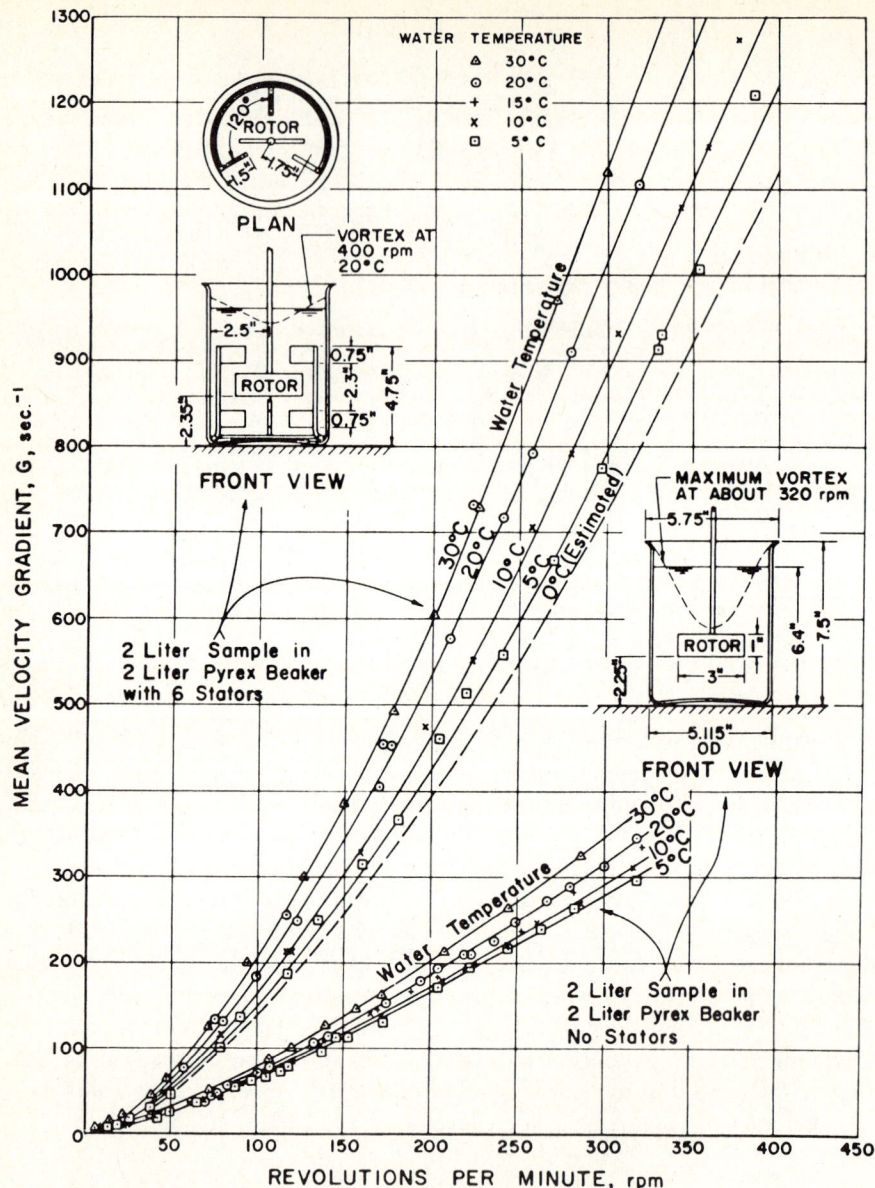

Figure 4.2.   Velocity gradient calibration curves for water., *Source:* Camp, 1970.

Alum Salts

Aluminum (aluminum sulfate; $Al_2(SO_4)_3 \cdot 14H_2O$) is available commercially in industrialized countries in lump, ground, or liquid form. Dry alum (density about 480 kg/m^3) is measured by volume or weight and is normally dissolved in water prior to its introduction into the rapid mixer. The content of water-soluble alumina in dry alum is 11 to 17%. Figure 4.3 shows 100-kg bags of lumped alum manufactured in Kenya. This form of alum is easily handled and stored in the treatment plant. Low-grade lumped alum has been used effectively in saturated solution feeders in Latin America and Indonesia (see section on "Saturated Solution Feeders"). Liquid alum may be obtained economically if alum-producing industries are nearby (e.g., to serve large paper mills). It is usually less expensive than dry alum when obtained at the source, but the shipping weight is double or more (e.g., 1270 kg/m^3 for 7.2% Al_2O_3-content grade) and requires special shipping containers because of its corrosiveness. The water-soluble alumina content in liquid alum is 5.8 to 8.5%.

Alum is a relatively inexpensive coagulant if local production is possible. In some developing countries, however, alum must be imported at substantially increased costs. Countries in West Africa, for example, import most of their alum from Europe, paying as much as U.S. $700 per ton (Wagner and Hudson, 1982). This compares with a price of about U.S. $100 per ton for commercially produced alum in the United States. Accordingly, treatment plants in those areas should be designed so that alum consumption is minimized. The dosage of alum may be reduced in some instances by (1) pretreating excessively turbid river waters, (2) direct filtration of low turbidity (<50 NTU) waters, (3) the use of coagulant aids, and (4) optimum pH adjustment.

Figure 4.3. 100-kg bags of lumped alum stored at a treatment plant in Nairobi, Kenya. *Source:* Schulz.

The correct alum dosage is determined initially from jar tests of the raw water, and then modified by actual plant operation experience. Optimal floc formation using alum occurs when the pH value of the water is between 6.0 and 8.0. If insufficient alkalinity is present to react with the alum, an alkali such as lime must be added. The reactions of alum with the alkalinity in the water are impossible to determine accurately, but the following quantities serve as a useful guide (American Water Works Association, 1971):

1 mg/l of alum reacts with:

0.50 mg/l natural alkalinity, expressed as $CaCO_3$

0.33 mg/l 85% quicklime as CaO

0.39 mg/l 95% hydrated lime as $Ca(OH)_2$

0.54 mg/l soda ash as Na_2CO_3

These approximate amounts of alkali, when added to water, will maintain the water's alkalinity when 1 mg/l of alum is added. For example, if 1 mg/l of alum is added to raw water, the alkalinity will drop by 0.50 mg/l; however, if 0.39 mg/l of hydrated lime is added with 1 mg/l of alum, the alkalinity of the raw water will remain the same.

A suitable method for feeding alum in developing countries is via solution-type chemical feeders. The alum is dissolved in water at a concentration of 3 to 7% (5% is most commonly used) in tanks, and then fed to the raw water. The highest concentration that can be practically achieved in alum solutions is 12 to 15% by weight. Such saturated solutions are used in alum saturated solution feeders.

Ferric Salts

Four types of ferric salts are used as coagulants: (1) ferrous sulfate (copperas), (2) chlorinated copperas, (3) ferric sulfate, and (4) ferric chloride. The physical and chemical characteristics of each are summarized in Appendix A and covered more fully in standard references (American Water Works Association, 1971; Cox, 1964). In general, they give similar results when their doses are compared in terms of iron content.

A number of practical differences between alum and ferric coagulants have been noted by Cox (1964):

1. Ferric hydroxide is insoluble over a wider range of pH values than aluminum hydroxide. This is illustrated in the pH zone-coagulation relationship shown in Figure 4.4 for aluminum sulfate and ferric sulfate. The two curves indicate that for alum the pH zone for optimal coagulation is relatively narrow (6.5 to 7.5), whereas for ferric sulfate it is much broader, ranging from 5.5 to 9.0.

2. Ferric hydroxide is formed at low pH values, so that coagulation is possible with ferric sulfate at pH values as low as 4.0 and with ferric chloride at pH values as low as 5.0.

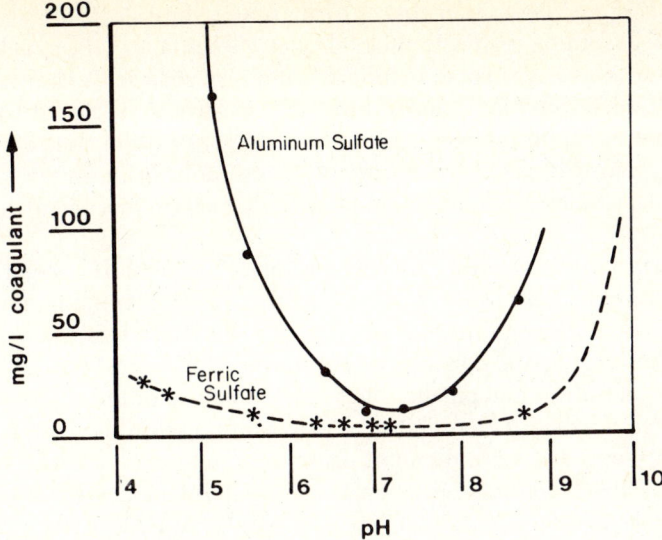

Figure 4.4. pH zone-coagulation relationship. Coagulation of 50 mg/1 kaolin with aluminum sulfate and ferric sulfate. Comparison of pH zones of coagulation of clay turbidity by aluminum sulfate, and ferric sulfate. Points on the curves represent coagulant dosage required to reduce clay turbidity to one-half its original value. *Source:* AWWA, 1971.

3. The floc formed with ferric coagulants is heavier than alum floc.
4. The ferric hydroxide floc does not redissolve at high pH values.
5. Ferric coagulants may be used in color removal at the high pH values required for the removal of iron and manganese and in the softening of water.

Iron coagulants are now recovered from steel mill waste pickling liquor and have become increasingly competitive in recent years, often yielding superior results at lower cost than alum. Although ferric salts are not as widely available as alum (which may prevent their widespread adoption as coagulants in developing countries), the possibility of using ferric salts should be investigated, especially when raw water pH is about 8.0 or higher.

ALKALIES FOR PH CONTROL

If the natural alkalinity of the raw water is insufficient to react with the dosage of coagulant that is added (i.e., the pH drops below the range for optimum coagulation), then alkalies should be added. Three types of alkalies are suitable for use in developing countries: (1) soda ash (sodium carbonate, Na_2CO_3), (2) quicklime (calcium oxide, CaO), and (3) hydrated lime (calcium hydroxide, $Ca(OH)_2$). Caustic soda (NaOH), a chemical often used in the industrialized countries, is not generally recommended because of its high cost and the extreme care that must be exercised in its handling.

Soda ash is a white powder that is easily soluble in water, so that there is less difficulty in its feeding than with other alkalies. Nevertheless, it can cake in storage bins because of its hygroscopic nature. Solutions of soda ash will not clog dosing orifices or feeding lines and, unlike lime, do not have to be stirred continuously. Furthermore, soda ash provides CO_3 ions to aid in corrosion control. The cost of soda ash is normally about three times that of lime, but in countries that mine soda ash, it provides a relatively trouble-free and economic means for alkalinity control.

Lime is produced in either unslaked (quicklime) or slaked (hydrated) form. Quicklime must be slaked (i.e., hydrated, by using a small amount of water) before it can be used. Hydrated lime, however, does not have to be slaked, does not deteriorate when stored, and contains fewer impurities than most quicklimes, making the clogging of orifices and pipelines less of a problem. For these reasons, hydrated lime is preferred when it is available economically. Quicklime is sometimes used at softening plants and large filtration plants because of its lower cost.

Lime is relatively insoluble in water and in most instances is fed as a suspension. Solution tanks and feed lines that store and convey lime-water suspensions clog easily and must be routinely cleaned. Provision for easy cleaning is essential in the design. The use of rubber hose for lime-water suspension feed lines will facilitate cleaning. Lime cake on the interior of the hose is easily broken away by hammering or pounding the hose and then flushing the hose. Open channels are also useful. In any event, the feeders should be as close to the point of application as possible. An alternative to feeding lime-water suspensions is the use of lime saturation towers that yield saturated solutions of lime, and have reduced maintenance problems in Brazil (Arboleda, 1973). Lime saturators are discussed in some detail below (see section on "Saturated Solution Feeders").

NATURAL COAGULANT AIDS

A great variety of both natural and synthetic materials is available to aid in the clarification of water. The correct application of these coagulant aids may improve the settling characteristics and toughness of the floc which in turn permit shorter sedimentation periods and higher rates of filtration. More important, however, such aids may significantly reduce the required dosage of the primary coagulant (e.g., alum), which is beneficial to those developing countries that must import coagulants.

A number of synthetic chemicals (e.g., cationic, anionic, and ampholytic polyelectrolytes) that can successfully cope with certain types of coagulation-flocculation problems (especially those arising from seasonal changes in water quality and ambient temperature) have been developed by chemical manufacturers in the United States and Europe. In general, however, the use of these chemicals in developing countries is inappropriate, because of the need for importation, careful monitoring and regulation, and their high cost. Continued supply may be questionable. A reasonable alternative, then, is natural coagulant aids that are available at low cost in most developing countries.

Natural coagulant aids fall into two categories, namely, (1) adsorbents-weighting agents, and (2) natural polyelectrolytes, both of which are discussed below.

Adsorbents-Weighting Agents

Bentonitic clays, fuller's earth, and other adsorptive clays are used to assist in the coagulation of waters containing high color or low turbidity. They supply additional suspended matter to the water upon which flocs can form. These floc particles are then able to settle rapidly due to the high specific gravity of the clay. Some clays swell when added to water, and can produce a floc when used alone or with a limited dosage of alum. Practical experience has shown that doses of clay ranging from about 10 to 50 mg/l result in good floc formation, improved removal of color and organic matter, and a broadening of the pH range for effective coagulation (American Water Works Association, 1971). For low turbidity raw waters (less than 10 NTU), the addition of adsorptive clays may often reduce the dosage of alum required.

Powdered calcium carbonate (limestone) is also effective as a weighting agent and, in addition, supplies alkalinity to the water upon dissolving. It is a common construction material (known as whiting in the building industry) and is easily stored, handled, and applied. A dosage of about 20 mg/l is sometimes used to treat low turbidity waters (Cox, 1964).

Natural Polyelectrolytes

Polyelectrolytes are either derived from natural sources or synthesized by chemical manufacturers. In both instances, their structure consists of repeating units of small molecular weight, chemically combined to form a large molecule of colloidal size, each carrying electrical charges or ionizable groups. Polyelectrolytes are often classified by the type of charge they carry. Thus, polymers possessing negative charges are *anionic*, those possessing positive charges are *cationic*, and those that carry both positive and negative charges are *ampholytic*. A wide variety of ampholytic polymers is derived from natural sources.

The application of synthetic polyelectrolytes as coagulant aids in water treatment is appropriate only in the industrialized countries or in the largest cities in countries such as Brazil, Argentina, Colombia, and India, which have reasonably developed water supply infrastructures that are able to regulate and monitor the manufacture and dosage of those chemicals. For example, a polyacrylamide gel manufactured in India under the trade name Polymix, has been used in conjunction with alum in the Delhi Water Works for over a year. The cost of this coagulant aid is presently $1.10 (1982 US$) per kilogram. It was found that the maximum dose required did not exceed 0.3 mg/l of Polymix and 10 mg/l of alum, even during the rainy season when turbidity levels are at their highest. An overall savings of 30% in the cost of chemicals used for coagulation was achieved.

A report published by the IRC (1973) summarizes the health aspects of using synthetic polyelectrolytes in water treatment; this report outlines procedures for their control that have been adopted and have proven effective in the United States and England. Nevertheless, in most developing countries, natural coagulant aids are preferable, because they do not require such regulatory control and are usually less costly.

Background

Interestingly, natural polyelectrolytes have been used for many centuries in developing countries for clarifying water. Sanskrit writings from India reported that seeds of the nirmali tree *(Strychnos potatorum)*, illustrated in Figure 4.5, were used to clarify turbid river water 4000 years ago. In Peru, water has been traditionally clarified with the mucilaginous sap of "tuna" leaves obtained from certain species of cacti (Kirchmer, Arboleda, and Castro, 1975). Jahn (1979) reports that in several countries in Africa (Chad, Nigeria, Sudan, and Tunisia) indigenous plants are added to drinking water by rural villagers to remove turbidity or unpleasant tastes and odors. Thus, the clarifying powers of natural polyelectrolytes are known to the rural inhabitants of numerous developing countries. At the same time, though, these substances have also been proven effective as coagulant aids in community water treatment, based on practical experience with such aids in Great Britain and research undertaken in several developing countries.

Figure 4.5.   Fruit of *strychnos potatorum*. The fruit (1) is a shiny black berry the size of a cherry. Its two stones (2-5) are the "clearing nuts" that are effective as coagulants. *Source:* Adapted from Jahn, 1981.

The British were among the first to use natural polyelectrolytes as coagulant aids in urban water supplies (*Manual of British Water Engineering Practice*, 1969; Packham, 1967). Sodium alginate, a natural polymer extracted from brown seaweed, has been employed by a number of water authorities at doses of 0.4 to 0.5 mg/l as an aid to alum, particularly during periods of low temperatures. Sodium alginates are widely used as thickening and stabilizing agents in the food, textile printing, and paper industries. Other natural polymers that have been used successfully in England are hydroxyethyl cellulose (HEC) and Wisprofloc, a derivative of potato starch. Starch products, cellulose derivatives, and alginates are all used in food processing.

Experience In India

The National Environmental Engineering Research Institute (NEERI) in India has completed studies on several plant species to determine their effectiveness as coagulant aids (NEERI, 1976; Tripathi et al., 1976). Seeds from the following plants were studied:

1. Nirmali tree; *Strychnos potatorum*.
2. Tamarind tree; *Tamerindus indica*.
3. Guar plant; *Cyamopsis psoraloides*.
4. Red Sorella plant; *Hibiscus sabdariffa*.
5. Fenugreek; *Trigonella foenum*.
6. Lentils; *Lens esculenta*.

Laboratory, pilot-plant, and full-scale plant studies were conducted using raw water turbidities that ranged from 50 to 7500 NTU. The following conclusions were reached:

1. The effective dose is 2 to 20 mg/l in the pH range 4 to 9.
2. The aids are uneconomical for water of turbidity below 300 NTU.
3. The aids are effective at high turbidity levels, during which 40 to 54% savings in alum consumption are possible.
4. The aids deteriorate in about three months, even after the addition of preservatives.

Trials with coagulant aid treatment in the water works operation of the Kanhan Plant, Nagpur, India (Jahn, 1981), allowed the evaluation of the economic benefits of this alum-saving method. The results are shown in Table 4.2.

The natural polymers studied in India are all prepared in the same manner:

1. The raw material is cleaned of any fibrous material and pulverized.
2. The powder is sieved to remove the husk and is mixed with soda ash at a ratio of 9 to 1.

Table 4.2.   Cost Savings in 1982 US$ Achieved by the
Use of a Natural Coagulant Aid in the Treatment
Plant at Kanhan Water Works; Nagpur, India[a]

Total quantity of water treated in July (1970)	1.74 million cubic meters
Total amount of alum conserved	56,040 kg
Value of 56,040 kg of alum at $.09 per kg = 56,040 × .09	$5,000
Labor required for extra operation: 9 laborers per day for 31 days	279 days labor
Cost of labor at $1.20 per laborer per day	= $330
Power, water, and depreciation of the machinery used for coagulant aid application at $1340 per million cubic meters	= $2,300
Additional cost incurred by using coagulant aid: $330 + $2,300	= $2,630
Net saving = $5,000 − $2,630	= $2,370
Saving per million cubic meters treated	$1,360

Source: Adapted from Jahn, 1981, p. 182.
[a]1982 ENR index = 3729
 1970 ENR index = 1385
 8.8 rupees = U.S. $1.00.

3.   A volume of 0.5 m^3 of water is added to 1.0 kg of the mixture to form a milky solution.

4.   The solution is heated (but boiling is not necessary).

The dose is calculated as 1 ml of solution equivalent to 1 mg of coagulant aid. A volume up to 1 m^3 of this solution can be prepared for use. The solution may be dispensed by conventional solution-type feeders.

Nirmali Seeds

The effectiveness of crushed nirmali seeds as a coagulant was verified in jar test studies conducted at Johns Hopkins University, Baltimore, Maryland (Kazoyoshi Kawata). The test runs involved six samples of 500 ml of turbid water (4.5 NTU) that were made using a suspension of bentonite clay in water. Crushed nirmali seeds were added to the six samples in doses ranging from 0 to 70 mg/l, and then the samples were subjected to rapid mixing for 1 minute, followed by gentle stirring for 2 to 3 minutes and settling for 15 minutes. The turbidity reductions are shown in Table 4.3 for water temperatures of 20°C and 30°C. A check on alkalinity and pH before and after the additions of the crushed nirmali seeds showed no appreciable change.

Moringa Oleifera Seeds

Laboratory studies conducted in the Sudan (Jahn and Dirar, 1979) revealed that seeds from the moringa oleifera tree act as a primary coagulant, and compare favorably with alum with respect to reaction rates and turbidity reductions in the raw

Table 4.3.   Turbidity Reductions Using Scrapings of Seeds of Strychnos Potatorum
           (Nirmali Seeds) at 20°C and 30°C, and 15 min. Settling Time

	Turbidity (NTU)	
Dose (mg/l)	20°C, 15 min. Settling	30°C, 15 min. Settling
0	4.5	4.5
20	2.9	1.25
40	2.5	1.1
50	2.25	1.5
70	2.25	—

Source: Kazoyoshi Kawata.

water. The results from jar testing showed that alum gave only a further 1% reduction
in turbidity (Figure 4.6).

The efficiency of several plant materials (including moringa oleifera and nirmali
seeds) as natural coagulants in comparison with alum have been studied experimentally by several investigators. Their work is summarized in Table 4.4.

Chitosan

A remarkable cationic polyelectrolyte called chitosan acts faster than any known
coagulant from plant materials (Jahn, 1981). It is derived from chitin, which is the
organic skeletal substance in the shells of crustacea such as shrimp, prawns, and
lobster. The yield of chitosan varies from about 20% of the weight of heavily armored
shells such as the claws of crabs, to about 35 to 40% of the weight of shrimp shells.
Chitosan is produced from the partial deacetylation of chitin in concentrated alkali
solutions at 135 to 150°C. Chitosan has been given interim approval for doses up
to 10 mg/l by the U.S. Environmental Protection Agency for use in drinking water

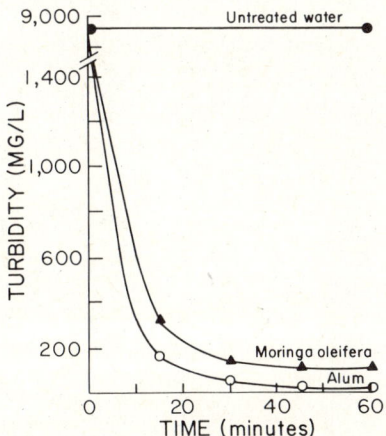

Figure 4.6.   Coagulating properties of moringa oleifera seeds in comparison with alum. *Source:* Adapted
from Jahn, 1979. (Concentration of coagulants: 200 mg/l)

Table 4.4. Plant Materials Tested in Comparison with Alum as Primary Coagulants and Coagulant Aids for Natural Waters

Country	Source	Turbidity (NTU)	Alum Dose (mg/l)	Residual Turbidity[a] (NTU)	Plant Matter as Primary Coagulant (mg/l)	Residual Turbidity[a] (NTU)	Plant Matter as Coagulant Aid (mg/l)	Residual Turbidity[a] (NTU)
				Moringa oleifera (seeds)				
Sudan	Hafir of El Querabin pH 8.5	470	200	11 (1 hr)	200	16 (1 hr)		
Sudan	Nile at Mogren water works pH 8.4	75	40	3 (1 hr)	50	10 (1 hr)		
				Moringa stenopetala (seeds)				
Sudan	Nile at Mogren water works pH 8.4	75	40	3 (1 hr)	5	8 (1 hr)		

Location							
Sudan			*Maerua pseudopetalosa* (root)				
Hafir of El Qerabin pH 8.5	470	200	200	11 (1 hr)	35 (1 hr)		8.5 to 1.5 (0.5 hr)
India			*Strychnos potatorum* (seeds)				
Yamuna River pH 8.2	2200	65	3 to 3.5	40 (2 hrs)	16 to 17 (1 hr)	1, followed after 3 min by 10 to 15 mg/l of alum	
Canal Water	500		2		30 (1 hr)		
Peru			*Opuntia ficus indica* (sap extracts)				
Rio Rimac	28	32		5 (0.5 hr)		0.4 + 0.5 mg/l of alum	14 to 15 (0.5 hr)

Source: Adapted from Jahn, 1981, p. 170.

[a]The settling time following coagulation is indicated directly below the turbidity results.

treatment plants. A patent has been issued for its use as a water treatment coagulant aid (Peniston and Johnson, 1970). In Japan, chitosan has been used as a coagulant aid since about 1950 (Kawamura).

Table 4.5 summarizes the results of laboratory tests on chitosan conducted in India. Jar tests were run for both alum and chitosan using: (1) flash mix for one minute at 100 rpm; (2) flocculation for 9 min at 40 rpm; and (3) settling for 10 min. There was no change in the pH after treatment with chitosan. The results show that the coagulation properties of chitosan surpass those of alum at high turbidities. Moreover, as a coagulant aid, chitosan can effect an even greater reduction in turbidity than using either primary coagulant alone.

A pilot filter study conducted to develop design criteria for a new water treatment plant on the island of Maui, Hawaii showed that chitosan in conjunction with alum coagulation was very effective in reducing turbidity and color (Kawamura). At a filtration rate of 17.9 m/hr, the combination of alum at 8.8 mg/l and chitosan at 0.22 mg/l removed 75% of the influent turbidity (from 19.4 to 4.8 NTU), and 90% of the color (from 90 units to 10 units). Alum at 8 mg/l and chitosan at 0.5 mg/l removed 98% of the influent turbidity and about 90% of the color, yielding an effluent turbidity of 0.25 NTU. On the other hand, 75 mg/l of alum alone removed 97% of the influent turbidity and about 90% of the color, yielding an effluent turbidity of 0.38 NTU.

The local production of chitosan for water treatment applications should be considered in coastal areas with large seafood industries where shells are a waste product, particularly if alum is costly.

A potentially deleterious side-effect of some natural polyelectrolytes is their propensity for increasing the growth of bacteria in the water being treated. Independent studies conducted in India and the Sudan showed that seeds from the nirmali and moringa trees, when used as coagulant aids, initially removed bacteria from the water, but after several hours the bacteria count rose slightly (Jahn, 1979). This phenomenon was attributed to the organic material present in the seeds, which was thought to provide additional substrate for the growth of bacteria. However, proper disinfection of the treated water will kill microbiological organisms, including bacteria.

Other potential problems associated with natural polyelectrolytes are their widespread use as foodstuffs, which may make them difficult to procure without causing local scarcity; their quality tends to deteriorate in time, and, therefore they should not be stored over three months. Chitosan, however, has been shown to be stable indefinitely in dilute solutions (Kawamura); there was no sign of bacterial activity or degradation in a five-year old solution of chitosan.

DISINFECTION

The disinfection of potable water supplies is almost universally accomplished by the use of chlorine gas or chlorine compounds (hypochlorites). Their ability to kill pathogenic organisms and to maintain a residual in the distribution system, as well

Table 4.5.   Efficiency of Chitosan as a Primary Coagulant and a Coagulant Aid

Raw Water Turbidity (NTU)	Alum (mg/l)	Residual Turbidity (NTU)	Chitosan as Primary Coagulant (mg/l)	Residual Turbidity (NTU)	Chitosan as Coagulant Aid (mg/l)	Residual Turbidity (NTU)
3200	300	90	1.00	10	0.15 + 20 mg/l alum	4
1400	100	10	1.00	10	0.1 + 20 mg/l alum	3
500	30	5	0.25	25	0.1 + 5 mg/l alum	5
70	10	14	0.25	18	0.05 + 8 mg/l alum	10

Source: Jahn, 1981, p. 185.

as their wide availability and moderate cost in most regions of the world, make them well-suited for disinfection. Presently, the only viable alternative to chlorination for the disinfection of community water supplies is ozonation, which has been used increasingly in European water supplies. The use of ozone is not generally recommended for developing countries, however, because of the high installation and operating and maintenance costs of ozonation, the need for a continuous supply of power, and the need for importation of the equipment and spare parts.

The period available for the interaction between the disinfectant and constituents in the water, called the contact time, is important in the design of disinfection systems for water treatment. The minimum contact time for chlorination should be 10 to 15 minutes to ensure effective disinfection (Cox, 1964), which can generally be provided in the transmission main before the first consumer. If such contact is not available, a contact chamber should be used. In filter plants, the clear well may be designed to provide the contact through baffles to avoid short-circuiting.

The decision to use either chlorine gas or hypochlorites should be based on several factors: (1) the quantity of water to be treated, (2) the cost and availability of chemicals, (3) the equipment needed for its application, and (4) the skill required for operation and control. Chlorine gas feed equipment is more expensive, more difficult to operate and maintain, and more dangerous than solution-type hypochlorinators; and in most instances has to be imported. On the other hand, chlorine gas is generally much less expensive than hypochlorite containing an equivalent amount of available chlorine, and can be stored for longer periods of time without deteriorating. The convenience and economy of using chlorine gas, then, are counterbalanced by cost and complexity of gas chlorinators as well as safety requirements. Furthermore, chlorine gas often is available only in the capital cities of developing countries; hence, its cost in outlying communities may increase proportionally to the shipping distance from the capital city. A study conducted in Brazil has shown chlorine gas to be more economical than sodium hypochlorite for plants with capacities greater than 500 m^3/day (see Chapter Ten, under "Operation and Maintenance Costs of Water Treatment Plants"). The study compared chemical, transportation, installation, operation, and maintenance costs for the two chemicals. The final choice to use either chlorine gas or hypochlorite compounds can be made only after considering each installation on an individual basis.

A *Handbook of Chlorination* by G. C. White (1972) covers the practical and theoretical aspects of both chlorination and hypochlorination. It is widely recognized as the best single reference on chlorination.

Gaseous Chlorine

Chlorine is a greenish-yellowish gas which, under pressure, is converted to liquid form. Liquid chlorine may be purchased in pressurized steel cylinders that are furnished in various sizes from 40 to 100 kg to 900 kg (about 1 ton). The cylinders are so filled that liquid chlorine fills 80% of the capacity at a temperature of 65°C. The flow of chlorine gas from a container depends on the internal pressure, which in turn depends on the temperature of the liquid chlorine. The normal discharge rate for a 50 kg cylinder at 21°C is about 800 gm/hr against a 240 KPa back pressure; the discharge rate for a 900 kg cylinder is about 7 kg/hr under similar conditions. If chlorine demand requires the use of several cylinders, manifolding the cylinders to one chlorinator is preferable to providing separate feeders for each cylinder.

The chlorine cylinders connected to the manifold should be placed on weighing scales, each scale holding one cylinder, to provide uninterrupted service while empty cylinders are being replaced. At least two cylinders should be connected to the manifold. The weights of the cylinders should be recorded at regular intervals (at least once a day) in order to ascertain the actual quantity of chlorine being used, which serves as a check on the accuracy of the dosing.

Chlorine gas is dangerous and corrosive, and care must be exercised in handling containers. Heat should never be applied to chlorine containers or valves. Containers should be stored in a dry location and protected from heat. In tropical countries it is important that chlorine cylinders and feeding equipment be shielded from the direct rays of the sun. The storage area should be outdoors or, if indoors, well-ventilated, preferably by forced ventilation. The containers should be stored and used in the order in which they are received. Gas masks designed to protect against chlorine fumes should be available. Safety requirements for the handling and feeding of chlorine gas are given in White's *Handbook* (1972).

Hypochlorite Compounds

Calcium and sodium hypochlorites and chlorinated lime (bleaching powder) are the compounds commonly used. Their chemical and physical characteristics are listed in the table of chemicals, found in Appendix A. The available chlorine in these compounds varies from 12 to 15% for sodium hypochlorite, to 33 to 37% for bleaching powders, to 65–70% for calcium hypochlorites. Bleaching powders are relatively unstable; and exposure to air, light, and moisture makes the chlorine content fall rapidly. Calcium hypochlorites are considerably more stable; under normal conditions they will lose 3 to 5% available chlorine per year. The properties, storage, and preparation of hypochlorite solutions are covered in several standard texts (White, 1972; Cox, 1964; AWWA, 1971).

The choice of hypochlorite for a particular installation should be based on cost and availability. Other factors being equal, the chemical of choice would be sodium hypochlorite, which exists only as a solution, is easily fed, and poses no hazard when stored. In remote areas to which chemicals must be transported, granular calcium hypochlorites are more economical because their Cl_2 content is greater.

On-site Manufacture of Disinfectant

In small communities and villages in remote areas of the world, where the proper handling, transportation, and storage of lethal chlorine gas and highly reactive chemical compounds cannot be assured, the local production of chlorine may be more practical than its importation. Chlorine is commercially manufactured by the electrolysis of brine, primarily in diaphragm cells. Chlorine comes off as a hot gas

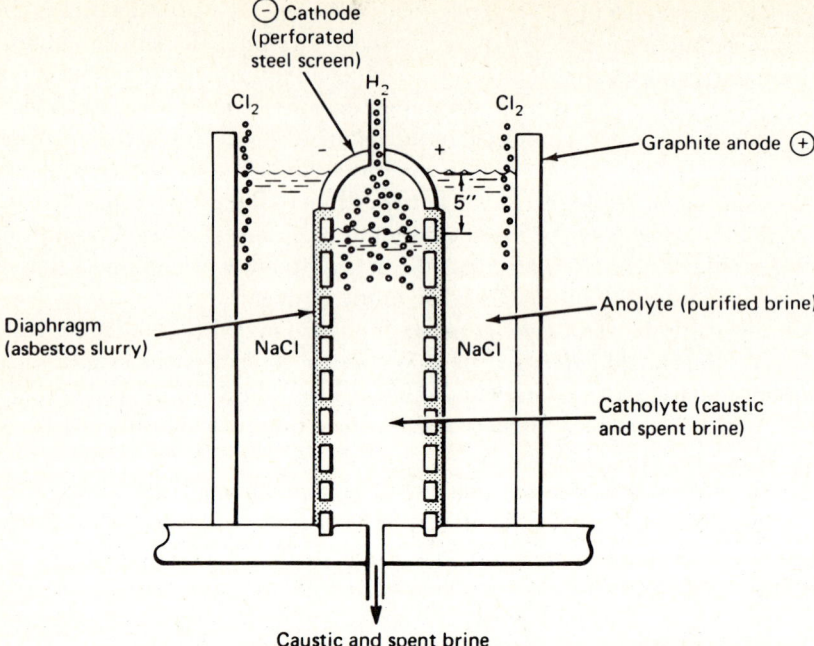

Figure 4.7.   Schematic diaphragm cell for chlorine generation. *Source:* White, 1972.

at the anode, and hydrogen comes off at the cathode together with a co-product, caustic, resulting from the electrolysis of the brine solution (White, 1972). A diaphragm separates the catalyte from the anolyte. Figure 4.7 shows a schematic of a typical diaphragm cell. The raw materials required for on-site hypochlorite generation are brine, water, and electric power. The brine and water should be relatively free from impurities (e.g., Ca, Mg, Fe) to minimize blockage in the cell diaphragm. Also, high ammonia-nitrogen concentrations in the brine water should be avoided, because this can lead to the accumulation of nitrogen trichloride (NCl_3) in the chlorine cylinders; NCl_3 is an extremely volatile substance that can suddenly explode and rupture the cylinders. High ammonia-nitrogen concentrations in the brine can be prevented, where they tend to occur, by super-chlorination of the process water to convert to nitrogen trichloride (NCl_3), followed by aeration to purge NCl_3 before using the water for brine production.

The power and salt requirements for several American proprietary devices for on-site hypochlorite generation are listed in Table 4.6. At power costs of 5 cents per kilowatt hour and salt costs of $10 per ton, the costs for power and salt are 9.6 to 14 cents per pound of sodium hypochlorite. To this, of course, must be added electrode replacement costs and operating labor costs. Total costs are substantially less than purchased hypochlorite and compare favorably with liquid chlorine for smaller installations. For example, the total cost of the "Ionics" hypochlorite generation systems for 1000 lb/day production excluding the capital cost of the generator, is estimated at 10 cents (7.4 operating and 2.6 capital) per pound of available chlorine; whereas the total cost of truck-delivered hypochlorite in the United States is about

Table 4.6.   Power and Salt Requirements for On-Site Hypochlorite Generation as
Reported by U.S. Manufacturers

	Power (kw hr/lb Na)	Salt (lb/lb NaOCl)
Ionics "Colormat"	1.7	2.1
Engelhard "Chloropac"	2.8	(Sea water)
Pepcon "Pep-Clor"	3.5	3.25
Diamond Shamrock "Sanilec"	2.5	3.05

Source: Culp and Culp, 1974, p. 195.

30 to 45 cents per pound of available chlorine and liquid chlorine costs 9 to 15 cents per pound in ton cylinders (Culp and Culp, 1974).

The Intermediate Technology Development Group in England has installed a pilot electrolysis unit for a treatment plant in the town of Beira, Mozambique (Intermediate Technology Services, 1982). Each module is rated at 5 kg/day of chlorine equivalent. The cell is operated by a 0 to 8-V supply from a transformer-rectifier unit connected to a standard 220/250-V power supply. All materials are PVC plastic or titanium, and are corrosion resistant. The manufacturer of the unit, Ecological Engineering Ltd. of the United Kingdom, claims that it can function (1) on untreated water and independently of water pressure, (2) with impure salt, (3) on fluctuating electrical voltage, and (4) with low maintenance costs.

The pilot unit was installed in 1981 at a cost of U.S. $8600; therefore it would have to operate for a period of two years to pay for its foreign exchange costs in terms of calcium hypochlorite or 3½ years in terms of gaseous chlorine (based on 1982 prices in Mozambique). This was a single purchase, and unit costs could be reduced substantially if a countrywide program of on-site hypochlorite generation was initiated. The pilot project has encountered some problems: (1) The production of active chlorine is reduced when the ambient temperature is greater than 30°C (cooling of the unit will increase its efficiency, but this would substantially complicate the unit), and (2) for larger plants the volume of tankage required becomes excessive (2000 liters are required to produce 5 kg/day), although several units could be operated in parallel, if this is considered feasible economically. More information on this particular project may be obtained from the Companhia das Aguas de Beira (Water Authority for Beira, Mozambique).

An alternate approach to conventional electrolytic cells, which are susceptible to clogging when impure brine solutions are used, are nondiaphragm cell systems that produce nascent sodium hypochlorite from a weak brine solution. Unlike diaphragm or membrane cells that separate the products of electrolysis (Cl_2, NaOH, and H_2), chlorine gas is not produced in nondiaphragm cell systems. Instead, the various components are allowed to react within the cell to produce a relatively dilute solution of NaOCl in the 0.5 to 0.8% range (5 to 8 g/l). Also, research conducted by the manufacturer of the system, Sienco, Inc. (1982), has shown that for any given dosage level, nascent hypochlorite is a more effective disinfectant than commercial sodium or calcium hypochlorite. This is attributed to the absence of excessive caustic, which is contained in most commercial products. Chlorine solutions become less bactericidal as the pH increases.

Figure 4.8 shows a nascent sodium hypochlorite generator manufactured by Sienco, Inc. This unit is intended for water treatment in villages and small towns with populations up to approximately 5000 people. One of the features of this system is that pumping is not required. The system utilizes the gas lift and heating effect of the electrolysis to circulate the solution in a continuous closed-loop mode. The unit is capable of producing 350 l of sodium hypochlorite solution in a 12-hr period, at a concentration between 1500 and 2000 mg/l (expressed as equivalent chlorine). The only moving part in the system is the mechanical timer located on the front of the control cabinet. If impurities are present in the brine solution, a slime layer may build up on the inside of the electrolytic cell; this is easily removed by applying an acidic solution provided by the manufacturer. The cost of the unit is about $3000 (1982 US$). This unit has been used in Africa and Asia and is currently being introduced into several Latin American countries.

An alternative to the electrolytic production of chlorine, which functions without electricity, was developed by Stone (1950), while working in the interior of central China. The raw materials employed were common salt, manganese dioxide, sulfuric acid, and low-grade slaked lime; all of which were manufactured or mined within the region. By the direct chemical combination of chlorine gas with the slaked lime, it was possible to make bleaching powder with about 35% available chlorine. The installation is shown in Figure 4.9. The two principal elements are the chlorine gas generator and the absorption chamber.

Figure 4.8. Nascent sodium hypochlorite generator. *Source:* Sienco, Inc., 1982.

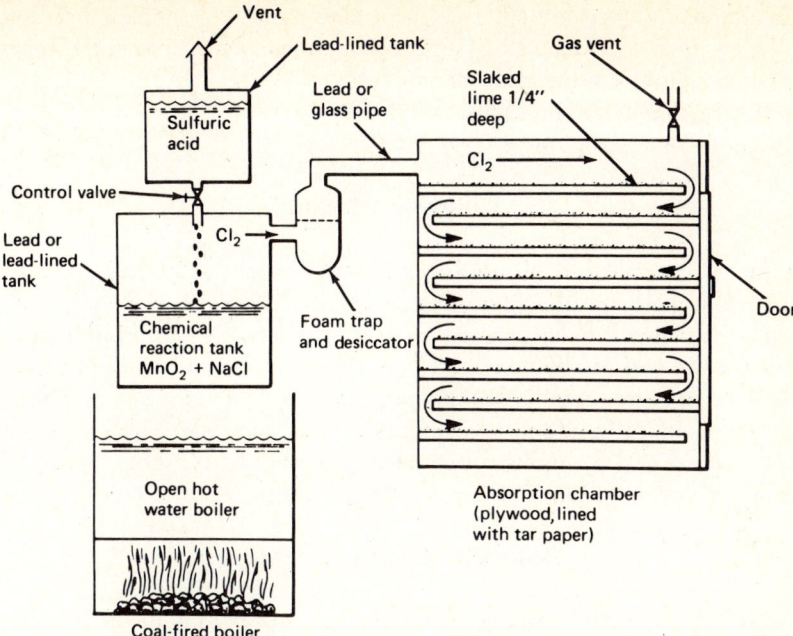

Figure 4.9. Generation of hypochlorite without using electricity. *Source:* White, 1972.

In the generator, the furnace heats the water, which in turn gently warms the chemicals placed in the tank, accelerating the reaction. The amount of chlorine gas that is generated is controlled by the heating operation as well as by the addition of sulfuric acid, whose feed-rate is regulated by a control valve. The chlorine gas is passed through a foam trap and dessicator, where it is dried before going to the absorption chamber. The absorption chamber consists of several trays on which a 0.6 cm layer of slaked lime is placed. The chlorine gas enters at the top of the chamber, and being heavier than air, circulates downward and reacts with the slaked lime to form bleaching powder. It was possible to produce about 14 kg of bleaching powder (at 35 to 39% chlorine strength) after 12 hr of operation. The physical dimensions of this design could easily be enlarged to increase its capacity for larger-sized plants.

The IRC has commenced a multicountry study to promote the technology of on-site generation of hypochlorite for disinfection of drinking water in developing countries. Information on suitable technologies is being assembled from users, equipment suppliers, and funding agencies.

CHEMICAL FEEDING

Chemical feeders should be simple in design and easy to operate. Hypochlorite and coagulant solutions may be fed by simple solution-type feeders that are constructed locally. Dry-chemical feeders are somewhat more complex and have greater maintenance problems. Because chlorine gas feeders are more complex than solution-type feeders, their use is limited to larger plants, where skilled supervision is available and the economy of using chlorine gas counterbalances the disadvantages. Lime may be fed either as a continuously mixed suspension by slurry-type feeders, or as a solution by saturation towers.

Chlorine Feeding

Chlorine gas is applied by two distinct methods: namely, solution-feed and direct-feed. In both instances the chlorine gas is obtained from the evaporation of liquid chlorine maintained under pressure in steel cylinders. Dry chlorine liquid or gas is noncorrosive and can be stored safely in steel cylinders and transmitted safely by black or wrought-iron piping. By contrast, chlorine solution and liquid or gaseous chlorine containing moisture or in humid atmospheres are highly corrosive and must be transmitted by piping constructed of plastic, hard rubber, glass, or other materials of proved resistance (Fair, Geyer and Okun, 1968).

Solution-feed Chlorinators

Solution-feed chlorinators take gaseous chlorine evaporated in the container, meter it, and mix it with water to form a strong chlorine solution. Water supplied by distribution-system pressure or by a special pump is often used to inject the chlorine solution into the water supply. Water under pressure creates a negative pressure, while passing through a venturi tube or orifice, which draws the chlorine gas into the chlorinator. The negative pressure in the chlorinator system prevents leakage of chlorine gas. The water pressure for injection of the chlorine solution should be 170 KPa, or at least three times greater than that of the water being chlorinated (whichever is higher) to ensure effective injection. A schematic for a chlorinator tied to the high-service pumping is shown in Figure 4.10. This is preferable to using separate pumping installations that require electric power. In general, the pressure difference between the discharge side of the high-service pumps, which supplies water to the chlorinator, and the suction side of the pump, where the chlorine-water solution is injected, is more than adequate for proper injection. Also, using the high-service pumps for chlorine injection makes it unlikely that finished water would leave the plant unchlorinated, inasmuch as the stoppage of those pumps due to mechanical failure or power outages would halt both chlorination and the flow of water out of the plant. When adequate contact time (about 15 min) cannot be assured on the discharge side of the high-lift pumps, chlorine should be added ahead of the clear well. When chlorine is injected on the suction side of the pumps, it should be well-mixed and diluted to prevent corrosion of the pump impellers.

An "all-vacuum" system for feeding chlorine is shown in Figure 4.11 (Capital Controls Co. brochure). The operation of the system is relatively simple. Water under pressure (e.g., from the discharge side of a high-service pump or the distribution system) passes through the injector at high velocity causing a negative pressure,

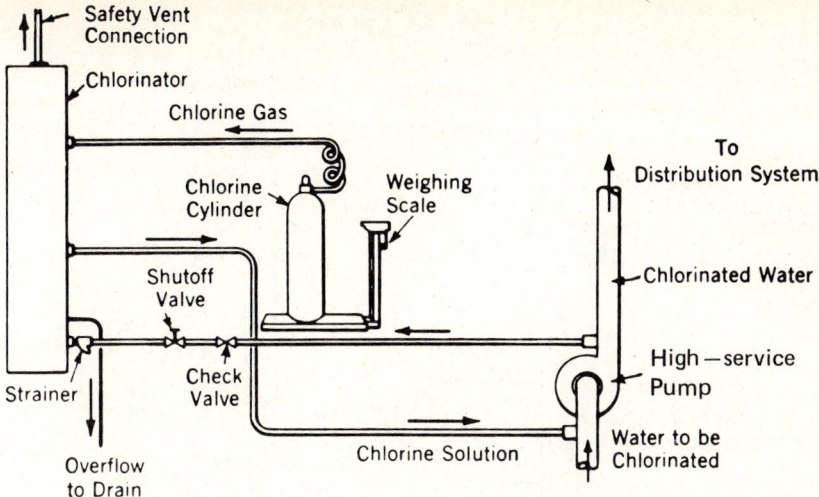

Figure 4.10. Chlorinator installation using pressure from high-service pumping. *Source:* Fair, Geyer, and Okun, 1971.

which opens a diaphragm check valve in the ejector body. When the check valve opens, the negative pressure is transmitted through a plastic tube to the chlorinator, which is mounted on the chlorine cylinder. With negative pressure at the chlorinator, a spring-opposed diaphragm opens the chlorine safety valve at the inlet of the chlorinator. Chlorine at cylinder pressure enters through the inlet valve where the pressure is reduced below atmospheric pressure and regulated by the diaphragm. The gaseous chlorine then passes under negative pressure through the flow meter and rate control valve to the ejector, where it is thoroughly mixed with and dissolved in the water, then carried as a solution to the point of desired chlorination through the ejector outlet.

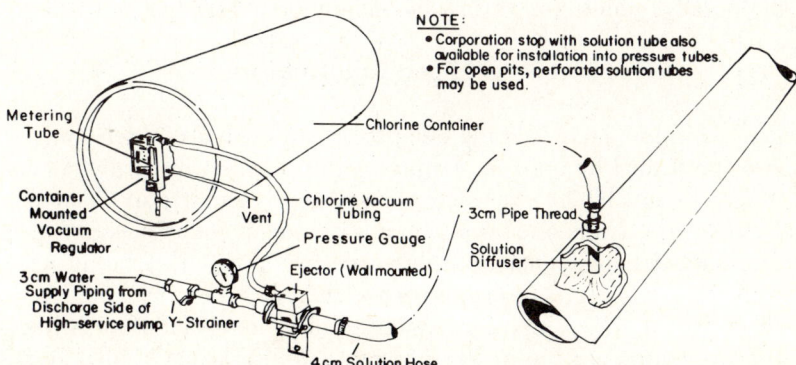

Figure 4.11. "All-vacuum" system for feeding chlorine. The regulator can also be mounted atop small steel chlorine cylinders. *Source:* Adapted from Capital Controls Co. brochure.

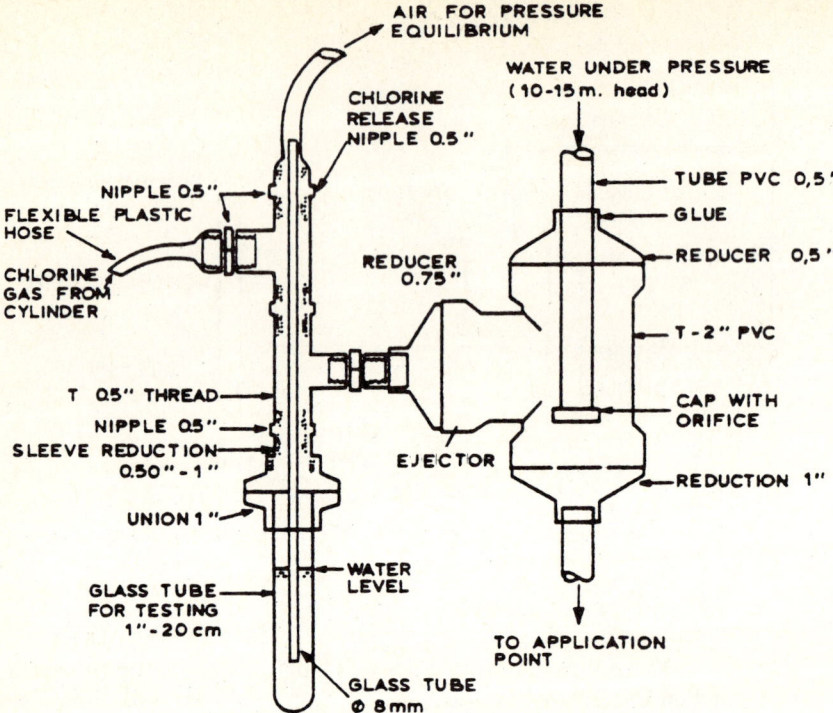

Figure 4.12. Low-cost chlorinator fabricated in Brazil. *Source:* IRC, 1977a.

A locally constructed chlorinator, built from plastic and glass for about $15 (1982 US$), has been used in Brazil (IRC, 1977a), and is shown in Figure 4.12. The chlorine gas is introduced into the water by the negative pressure created in the ejector. The depth of water in the glass test tube is a measure of that negative pressure. The dose can be regulated by the auxiliary valve on the cylinder or by reducing the water flow in the ejector. Unlike the "all-vacuum" system described above, the negative pressure in this system cannot be maintained constant.

Direct-gas Feed Chlorinators

Direct-gas feed chlorinators feed the chlorine gas through a diffuser directly into the water to be treated by utilizing the pressure of the chlorine in the container. This unit may be suitable for use where chlorination is required in the absence of either electricity to run a booster pump, or a water supply under pressure adequate to provide a sufficient differential for a solution-feed chlorinator. It should be noted, however, that the feeder lines in such units operate under pressure; therefore, when accidentally ruptured, they can pose a considerable danger to plant operators. Because of the corrosive nature of chlorine gas in water, a diffuser made of silver or porous stone is needed. Direct-feed units are more effective in warm climates (above 10°C). At lower temperatures, chlorine combines with water to form chlorine hydrate

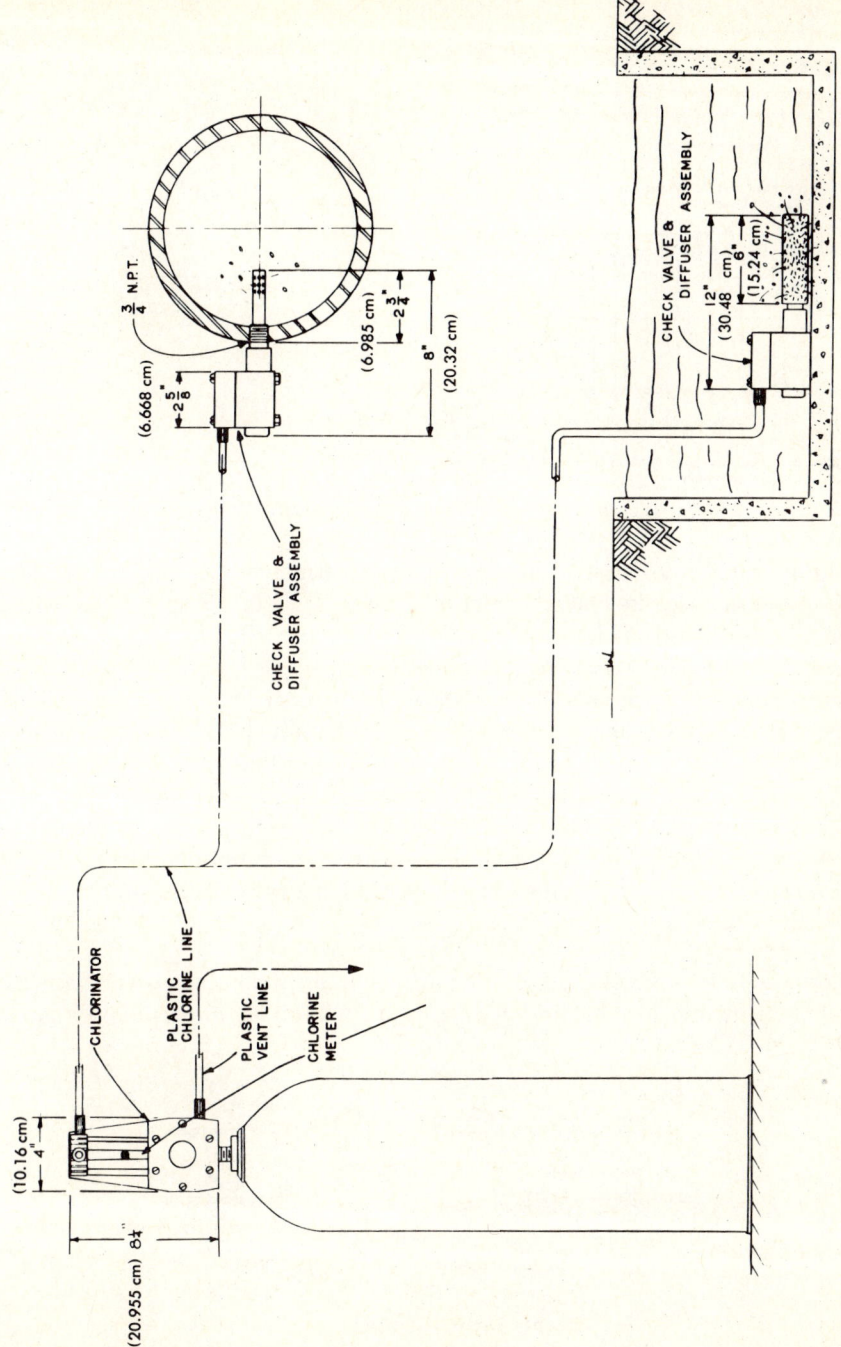

— DIMENSIONS & TYPICAL INSTALLATIONS —

Figure 4.13. Direct-feed gas chlorinator. *Source:* Capital Controls Co. brochure.

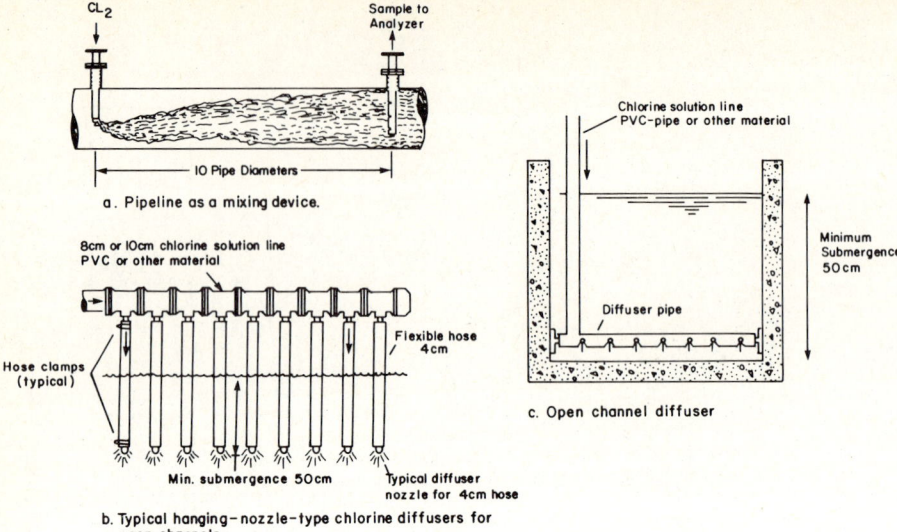

Figure 4.14.   Typical designs for chlorine diffusers. *Source:* White, 1972.

(commonly called chlorine ice); this hydrate may obstruct feeding equipment. The maximum capacity of individual direct-feed units is about 35 kg per day when applied to a pipeline or main, or 135 kg per day when applied to an open tank or channel (American Society of Civil Engineers, 1969).

A simple direct-feed gas chlorinator is shown in Figure 4.13. Chlorine at cylinder pressure (520 to 620 KPa at room temperature) enters through the inlet valve, where the pressure is reduced to 140 KPa and is controlled at that pressure. Pressure at the point of injection should be 70 KPa or less. After passing through the inlet valve, the gas flows through the flow meter and rate control valve to the check valve, where pressure pushes the valve open. After passing through the valve, the gas is diffused into the water being treated. An emergency vent valve in the chlorinator body prevents excess pressure.

Chlorine solutions are discharged into either pressure conduits or open channels. Typical chlorine diffusers that are used for each situation are illustrated in Figures

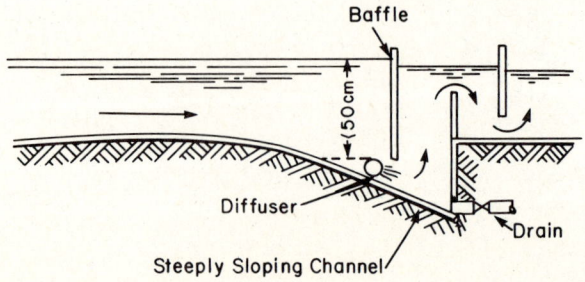

Figure 4.15.   Baffled mixing chamber for chlorine application in open channels. *Source:* Azevedo-Netto.

4.14 and 4.15. They are fabricated from plastic pipe or similar corrosion-resistant material. The type used in pipelines should always be designed to introduce the chlorine solution into the center of the pipe. For channels, the chlorine solution should not be applied at a depth less than 50 cm, because it will not be completely absorbed. Hydraulic jumps or baffled channels are suitable mixing devices for open channels. The chlorine diffuser should be placed immediately upstream of the point of turbulence, as shown in Figure 4.15.

Solution-type Feeders

Solution-type feeders apply to both hypochlorite and coagulant solutions. Some general requirements for the design of these feeders are:

1. A minimum of two chemical feeders must be provided at each application point (or a total of three if two chemicals are to be fed) so that feeding is not interrupted when one unit is out of service.
2. The combined capacity of the chemical feeders must be greater than the maximum dosage required, but not so large as to be inaccurate during periods of low flow.
3. Concrete tanks must be adequately protected against corrosion by a layer of bitumastic enamel, plastic, rubber, or similar-functioning substance.
4. Adequately sized drains are needed to facilitate cleaning the tanks and flushing out accumulated sediments.
5. There must be provision for either hand-operated or motor-driven paddles for mixing the chemical solutions.
6. Chemical feed lines made from rubber or plastic hose are needed, supported at short intervals or placed inside a tile pipe for protection, with provision for easy cleaning. Open channels are best for conveying lime suspensions.
7. The location of chemical feeders must be as close as possible to the point of application of the chemicals, unless eductors are used (see section on "Eductors").

Solution-type chemical feeders may be classified into three groups: (1) constant-rate, (2) proportional, and (3) saturated-solution. Some designs that are practical for developing countries are described briefly here; more detailed information and additional designs are given elsewhere (AID UNC/IPSED a, b, c, 1966; Arboleda, 1973).

Constant-rate Feeders

The constant-head solution feeder has been widely used for chemical feeding in water treatment plants. A typical arrangement for feeding alum solutions is shown in Figure 4.16. The feeding system consists essentially of (1) a solution tank for dissolving and/or mixing the chemicals in water to produce solutions of known strength, (2) a constant-head box with a float-valve inlet, and (3) a dosing tap or

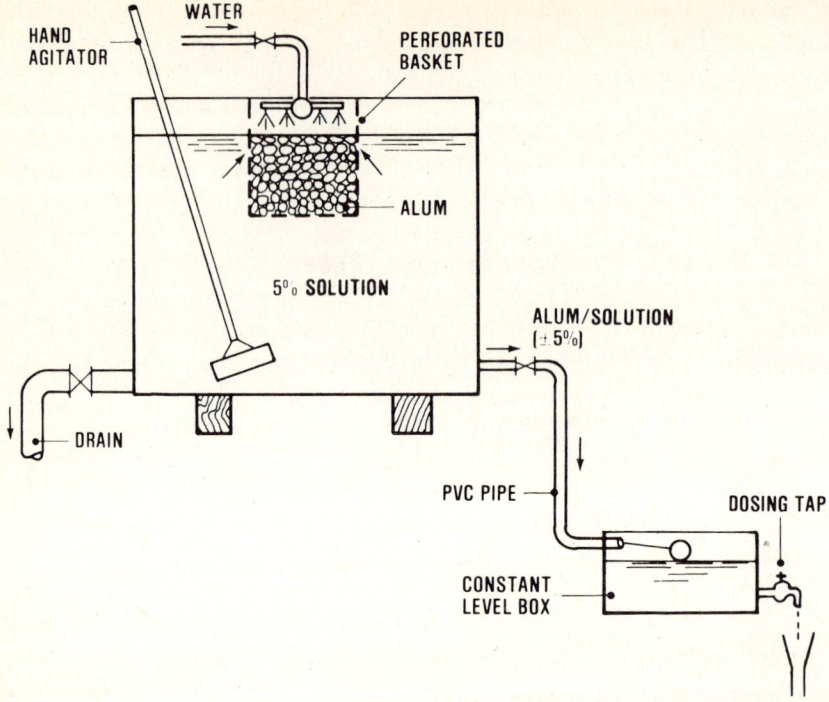

Figure 4.16. Constant-head solution feeder for alum dosing., *Source:* IRC, 1981b.

adjustable orifice attached to the outlet side of the constant-head box for adjusting the flow of chemical solution. All exposed parts should be protected against corrosion, including the float which can be made of hard rubber, glass, or ceramic. Two solution tanks can be connected to a single constant-head box to assure uninterrupted operation while one tank is being filled.

A simple chemical feeder that functions without a constant-head box is a constant-head orifice feeder in which the discharge orifice is easily adjustable (Figures 4.17 and 4.18). The rate of feed is controlled by maintaining a constant head on the discharge orifice by floating it at a constant depth below the liquid surface. The designs differ in the method by which the constant head is maintained and the type of discharge orifice employed. Each type may be easily converted to an alum doser or hypochlorinator simply by adding or removing the alum tray and perforated-water pipe in the tank (shown in the alum doser in Figure 4.17).

These designs have distinct advantages over conventional constant-head tanks in that they do not require float valves, which are difficult to keep clean and are subject to corrosion. Moreover, they have been proven from experience to be more accurate in controlling the rate of flow than either control valves or petcocks. The feed rates should be calibrated for each unit after installation, and a chart provided for ready reference.

Another type of constant-rate feeder is the rotating dipper, which consists of a

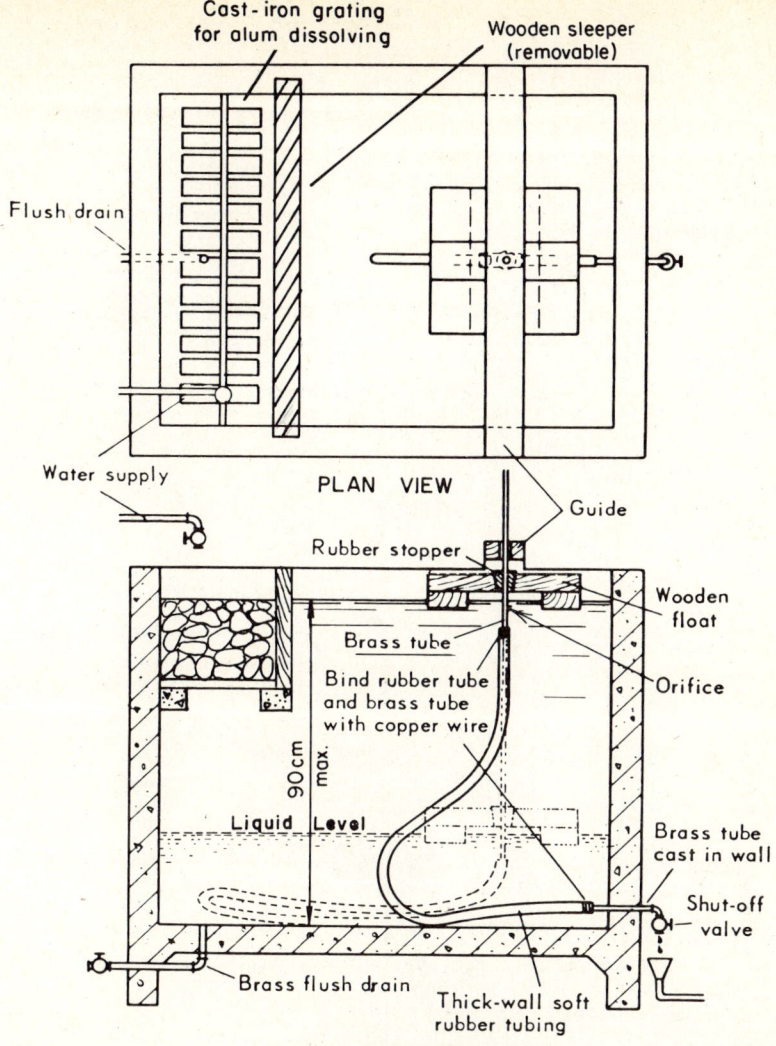

Cast-iron grating
for alum dissolving

Wooden sleeper
(removable)

Flush drain

Water supply

PLAN VIEW

Guide

Rubber stopper

Wooden
float

Brass tube

Orifice

Bind rubber tube
and brass tube
with copper wire

90 cm
max.

Liquid Level

Brass tube
cast in wall

Shut-off
valve

Brass flush drain

Thick-wall soft
rubber tubing

ELEVATION SECTION

Figure 4.17. Floating-arm type alum solution feeder. *Source:* AID, UNC/IPSED, 1966a.

tank in which calibrated cups revolve into the solution and remove adjustable volumes of solution. Such a unit may be powered by small variable-speed motors, variable-speed drives, or by water flow as described below. Rotating dippers are especially suitable for feeding lime suspensions, which must be agitated continuously, and hence cannot be fed by constant-head type solution feeders.

Proportional Feeders

A water-powered version of the rotating dipper that feeds chemicals proportional to the raw water inflow was developed in Swaziland (AID UNC/IPSED Series Item

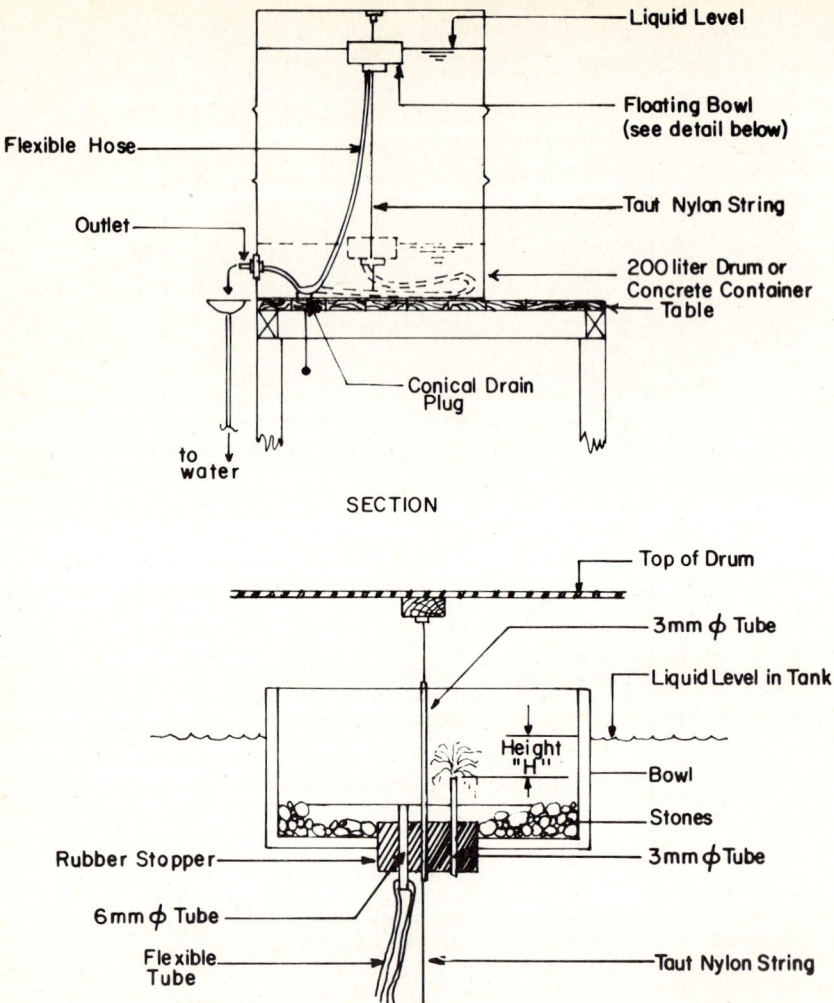

Liquid Level

Floating Bowl
(see detail below)

Flexible Hose

Taut Nylon String

Outlet

200 liter Drum or
Concrete Container
Table

Conical Drain
Plug

to
water

SECTION

Top of Drum

3mm φ Tube

Liquid Level in Tank

Height
"H"

Bowl

Stones

Rubber Stopper

3mm φ Tube

6mm φ Tube

Flexible
Tube

Taut Nylon String

FLOATING BOWL DETAILED SECTION

Figure 4.18. Floating-bowl type hypochlorinator. *Source:* Adapted from AID UNC/IPSED, 1966b.

No. 5, 1966) and is illustrated in Figure 4.19. The flow in the influent channel drives a paddle wheel at a speed proportional to the flow rate. The wheel drives a shaft, which in turn rotates the dosing cups that are attached. The dosing cups are located in two solution tanks positioned on either side of the raw water channel. The cups sweep through the chemical solution in the tank, and each cup empties a fixed volume of solution into a funnel, which conveys the solution into the channel immediately upstream of the paddle through a rubber tube.

A chemical dosing unit that operates hydraulically from distribution-system pressure was developed by Wallace and Tiernan, Ltd. (Richmond, 1981). The assembly

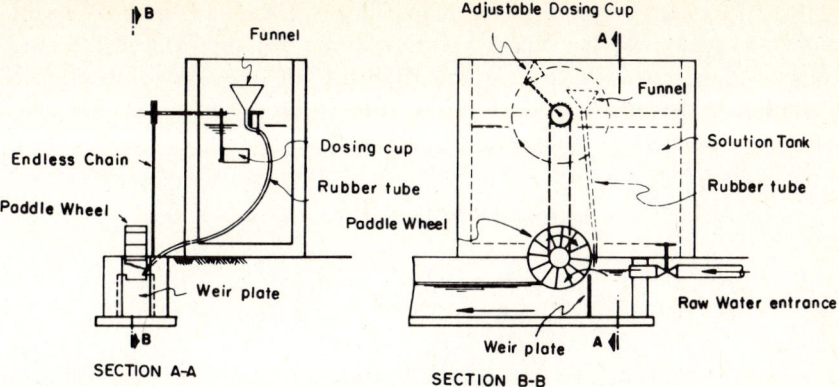

Figure 4.19. Water-powered proportional chemical feeder. *Source*: Arboleda, 1973.

is shown for two operating cycles in Figure 4.20. It consists of three basic components: (1) a water tank of approximately 150-l capacity, (2) a hydraulically powered adjustable dosing head with an integral 10-l capacity chemical reservoir, and (3) a quick-acting, high-capacity syphon.

The principle of operation is simple: Water enters the 150-l tank at any rate between 2.3 m³/day and 100 m³/day. As the level in the water tank rises, a ball float lifts a small container or weir chamber, full of sodium hypochlorite, to a point where a fixed plunger enters the chamber, to a preset variable depth; in so doing, it displaces a controlled amount of the contents. The displaced hypochlorite is directed into the main tank by means of a special pivoted tube that moves clear when the chamber descends into the hypochlorite reservoir, where it is replenished in readiness to the next cycle. As the water in the main tank reaches an appropriate level, the syphon actuates and discharges the contents of the tank into a service reservoir or tank immediately below the unit. At maximum dosage and water flow-

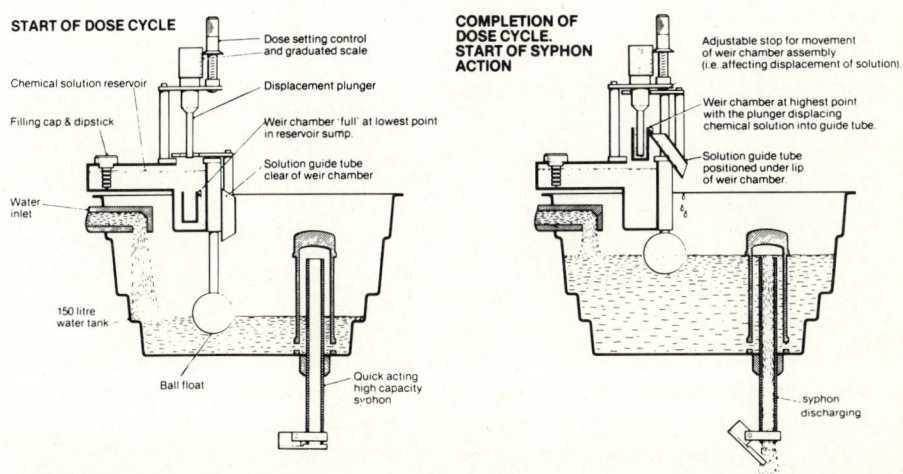

Figure 4.20. Hydraulically-operated chemical solution feeder. *Source*: Richmond, 1981.

rate, the 10-l capacity chemical reservoir will last for about three days without replenishment. Because the unit is based on cyclic operation within the range of the machine, the chemical dose is proportioned to the incoming flow of water.

Chemical feeders that use eductors (described in the last section of this chapter) can also provide proportional feeding.

Saturated Solution Feeders

The use of saturated solution feeders makes it possible to use inexpensive chemical compounds of relatively low purity, such as alum cake, which may be available locally.

A saturated alum solution can be obtained by means of a wooden vat about 4 m square by about 1 m deep (arranged in pairs). The solution is made by dumping lump alum into the vat or a tray in the vat (Figure 4.17) and then filling the vat with water, occasionally stirring the contents by hand using poles or paddles until the alum is dissolved or until no further alum will become dissolved, indicating a saturated solution. By using the solution from one tank while mixing is taking place in the other tank (or vat) a standby saturated solution is always available. This method is successfully used in a 85,000 m³/day plant in Surabaya, Indonesia (Burlingame). The sizes of the dissolving vats or tanks are such that the contents of each will provide about a week's supply of alum. Provision needs to be made for separating and removing insoluble components that may collect in the vat.

In Brazil, a saturation tower for feeding lime solutions has minimized some of

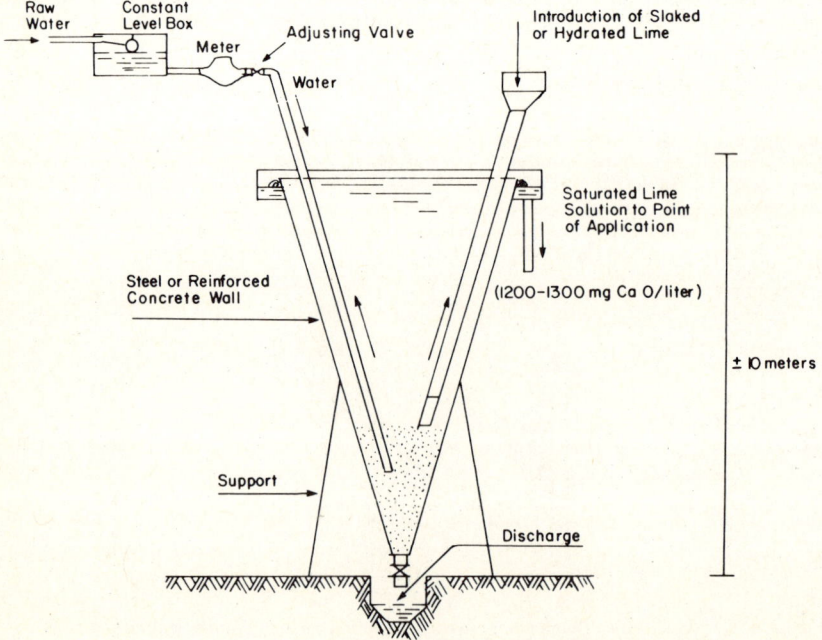

Figure 4.21.   Lime saturation tower. *Source:* Adapted from IRC, 1977a.

the maintenance problems associated with conventional lime-suspension feeders (e.g., clogging pipes and orifices). Although only slightly soluble, a lime solution with concentrations of 1200 to 1300 mg/l can be maintained if cold water is used. Lime is also more soluble if the lime particles are small. Figure 4.21 illustrates the lime saturator. Cold water flows from a constant-level box where it is metered; it is then carried to a conical tank where it is saturated with lime. The saturated solution is removed by a collection trough located along the upper perimeter of the tank, then fed directly to the point of application. The inert material and deposits of calcium carbonate are extracted through a bottom drain. The tank is generally constructed from steel that is not corroded by lime, and therefore needs no protective coating.

Dry-chemical Feeders

Dry-chemical feeders usually consist of conical hoppers fitted at the bottom with either a volumetrically or gravimetrically operated device for dispensing the dry chemicals. Volumetric devices plow, push, or shake the chemical into a receiving chamber, where it is mixed with water and conveyed to the point of application. Dry-chemical feeders enjoy some advantages over solution-type feeders, including (1) more accurate feeding, (2) longer unattended operation, and (3) eliminating the need for making up solutions or slurries. However, they are generally more complex and more difficult to maintain than solution-type feeders, particularly in humid climates where unprotected metal parts corrode easily and hygroscopic chemicals (e.g., alum) can clog the inside of hoppers or jam the devices that control the chemical dosage.

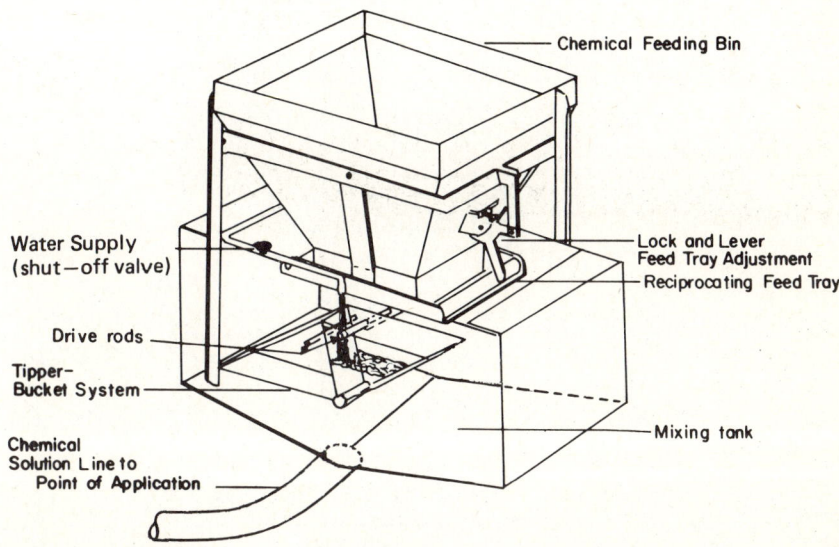

Figure 4.22. Hydraulically-operated dry-chemical feeder. *Source:* Adapted from Moore Fluid Equipment brochure, South Africa.

An unusual dry feeder, which does not require electrical power, was developed by the Moore Fluid Equipment Company of South Africa, and is illustrated in Figure 4.22. It operates on the tipping bucket principle, utilizing water as the drive. The water operates a tipping bucket, which in turn reciprocates a stainless steel tray below a feed hopper, causing chemical to be discharged and to fall into a chamber beneath, where water from the tipping bucket is discharged. This creates turbulence and forms a solution or slurry of the chemical being fed, which is then fed by gravity to the point of application. Two methods are provided for adjusting chemical dosage: controlling the flow of the operating water and adjusting the position of the tray below the feed bin, which is effected by a lever and lock system. The unit also incorporates a bin agitator, consisting of a rubber hammer that strikes the chemical storage bin with every stroke of the feeder. The feeding range for this unit varies between 0.5 to 15 kg/hr of hydrated lime or 1 to 30 kg/hr of granulated aluminum sulphate.

Eductors

Eductors are relatively simple, inexpensive devices for conveying particulates, vapors, and liquids in pressure conduits. They are operated by flow through a nozzle, as shown in Figure 4.23. The motive or operating liquid entrains suction liquid, the two are thoroughly mixed in the venturi, and the mixture is conveyed to the point of application. Typical units are designed for 3 l of suction fluid for each liter of motive fluid. Capacities of 1-in. water jet eductors for various suction lifts, discharge pressures, and available operating water pressures are presented in Table 4.7; ratio factors to determine capacities for sizes other than 1 in. are presented in Table 4.8.

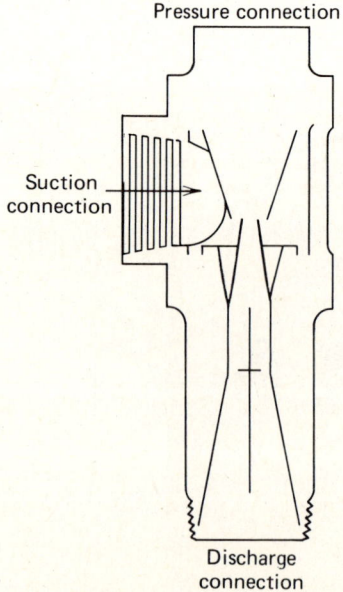

Figure 4.23. Typical water jet eductor. *Source:* Ametek-Schutte and Koerting Div. 1976.

Table 4.7. Suction Capacities of Water Jet Eductors, 1 in. Size Only.[a]

Suction Cap. of Standard 1 in. Water Jet Eductor (gph) Water Temp. 80°F.

Suction Lift in Ft of Water	Discharge Pressure (psi) Gauge	Water Consumption (gph)	Operating Water Pressure psi Gauge							
			10	20	30	40	50	60	80	100
0	0	Suction	351	484	566	627	697	723	733	735
		Operating	213	302	368	423	476	519	600	670
	5	Suction		81	246	360	478	598	675	695
		Operating		297	365	420	471	517	595	665
	10	Suction			17	135	285	387	530	660
		Operating			357	414	463	509	590	663
	15	Suction					74	203	356	516
		Operating					460	505	586	657
	20	Suction						18	207	355
		Operating						495	580	655
	25	Suction							50	232
		Operating							575	650
	30	Suction								101
		Operating								646
5	0	Suction	264	408	516	576	652	681	708	720
		Operating	234	316	381	436	485	529	607	676
	5	Suction		90	192	300	420	540	650	675
		Operating		311	376	432	478	524	603	673
5	10	Suction				114	216	336	515	600
		Operating				425	474	518	597	667
	15	Suction					66	153	348	498
		Operating					470	514	594	662

Table 4.7. *(Continued)*

Suction Lift in Ft of Water	Discharge Pressure (psi) Gauge	Water Consumption (gph)	Suction Cap. of Standard 1 in. Water Jet Educator (gph) Water Temp. 80°F. Operating Water Pressure psi Gauge							
			10	20	30	40	50	60	80	100
	20	Suction							200	337
		Operating							590	660
	25	Suction							28	216
		Operating							585	652
	30	Suction								90
		Operating								648
10	0	Suction	120	276	402	504	540	606	606	612
		Operating	254	331	394	446	494	538	615	685
	5	Suction			120	258	354	462	595	606
		Operating			389	442	490	533	613	679
	10	Suction				66	180	270	487	576
		Operating				435	486	526	601	670
	15	Suction					66	126	336	438
		Operating					480	522	601	670
	20	Suction							168	318
		Operating							595	668
	25	Suction								166
		Operating								663
	30	Suction								66
		Operating								660

Source: Ametek-Schutte and Koerting Div., 1976.

[a]To determine capacities for sizes other than 1 in., multiply these capacities by the proper capacity ratio factor noted in Table 4.8.

Table 4.8.   Relative Capacities of Water Jet Eductors

Size Educator (inches)	½	¾	1	1½	2	2½	3	4	6
Capacity ratio	0.36	0.64	1.00	2.89	4.00	6.25	9.00	16.00	36.00

Source: Ametek-Schutte and Koertig Div., 1976.

The approximate cost in 1982 U.S. dollars of a typical 2-in. eductor ranges from $170 for plastic components to $450 for stainless steel components (Ametek-Schutte and Koerting Div., 1976).

Eductors have been used most often in water treatment plants for conveying gaseous chlorine under vacuum to the point of application, where it is injected as a chlorine-water solution (see section on "Solution-feed Chlorinators"). However, they may also be used to transport and inject other treatment chemicals (such as alum solutions or lime slurries) over relatively long distances, and hence make it possible for all the treatment chemicals to be housed in the same building. They are useful in situations where gravity flow is not practicable, because the chemical solutions can be lifted as they are conveyed to the point of application. If the motive liquid is taken from the raw or finished water lines, the chemicals can be fed in proportion to the flow. Furthermore, eductors can provide sufficient turbulence for rapid mixing of coagulants.

In general, eductors are not recommended for handling dry solids unless provisions are made to prevent material from adhering to and clogging orifices. However, if solids are handled intermittently so as to allow for inspection and cleaning of eductor equipment, for example in the conveyance and cleaning of dirty sand from slow sand filters (see Chapter 8, under "Design of Slow Sand Filters"), then eductors are suitable.

chapter five

HYDRAULIC RAPID MIXING

The function of a rapid mix system is to disperse the coagulant uniformly throughout the entire mass of water with maximum possible rapidity in order to ensure that the coagulation process is as effective as possible. This process is generally followed by a period of flocculation, during which the water is gently mixed to promote agglomeration of the coagulated particles. Rapid mix units are located at the head end of the plant and are designed to generate intense turbulence in the raw water by either mechanical or hydraulic means. Rapid diffusion is necessary because the coagulation process, which comprises the hydrolysis of the coagulant and destabilization of the colloidal material, is completed almost instantaneously (less than one second). Inasmuch as hydraulic rapid mixers are capable of achieving high velocity gradients for rapid diffusion of coagulants without using mechanical equipment, they are preferred in developing countries. Moreover, they require no imported equipment and are easily constructed, operated, and maintained with local materials and personnel.

This chapter examines several types of hydraulic rapid mixers that are designed for use in either open channels or pressure conduits. More space is devoted to open channel mixers because these are generally simpler and less costly, and above-water coagulant diffusers are readily accessible for cleaning.

DESIGN CRITERIA

Two major criteria control the processes of rapid mixing and flocculation: intensity of agitation and the duration of agitation. They are defined for the design of such mixing processes by the velocity gradient (G) and detention time ($t = V/Q$). However, in the design of rapid mixers, as contrasted with flocculators or slow mixers, the velocity gradient is less useful (Vrale and Jordan, 1971). The chemicals must be thoroughly dispersed in the raw water in a fraction of a second, and the stagnant zones of conventional mixers should be avoided. Plug-flow and in-line mixing is preferred. A useful, if unscientific, guideline is that the mixing should appear violent and that the chemicals be added at the point of maximum turbulence.

Although the point of maximum turbulence is often beneath the water surface, coagulant distributors are best located above the water level to permit easier maintenance and to enable the operator to become more quickly aware of any clogging of the diffuser.

RAPID MIXING DEVICES

The primary difference between mechanical and hydraulic rapid mixing devices is the manner by which they impart turbulence in the incoming raw water. For mechanical rapid mixers the degree of turbulence is a function of the equipment's horsepower and is largely independent of flow, whereas the degree of turbulence for hydraulic mixers is measured by the loss in head and is dependent on flow. Mechanical mixers are generally proprietary devices whose major technical advantages are that mixing is not a function of flow and they are flexible in adjusting the degree of turbulence to suit particular treatment needs. However, this advantage is of little consequence in places where the equipment cannot be kept in repair and where skilled operators are unavailable to make necessary adjustments.

Hydraulic rapid mixers are designed for either of two types of flow conditions; namely, open channel flow or pressure flow in pipes. When feasible, open channel flow in gravity channels is preferred, as such designs eliminate costly pipes and fittings, and can reduce the total capital cost of the plant. Moreover, rapid mixers in open channels are relatively simple and have their component parts exposed and accessible for easy operation and maintenance. The general types of open channel hydraulic mixers described in this chapter are (1) hydraulic jump mixers, (2) flumes, and (3) weirs. Rapid mixers that utilize turbulent flow in pressure pipes and that are practical for developing countries are (1) hydraulic energy dissipators and (2) turbulent pipe flow mixers.

Hydraulic Jump Mixers

This type of mixer includes a chute followed by a channel, with or without a drop in the elevation of the channel floor. The chute creates supercritical flow, the gently sloping channel provides a transition from supercritical to tranquil flow, which induces the jump, and the drop in the floor elevation defines the location of the jump. A diagram of a simple hydraulic jump mixer is shown in Figure 5.1.

The relative depths of the upstream and downstream water profile describe the basic conditions required for the formation of a hydraulic jump and can be calculated from the following equation (Fair, Geyer and Okun, 1968):

$$d_2/d_1 = \frac{1}{2} \left[1 + 8F^2 \right)^{\frac{1}{2}} - 1 \right] \tag{5.3}$$

where

$$d_1 = \text{depth of water upstream of the jump } (m)$$
$$d_2 = \text{depth of water downstream of the jump } (m)$$
$$F \text{ (Froude number)} = [v_1/(gd_1)^{1/2}]$$

where

$$v_1 = \text{velocity of flow upstream of jump (m/s)};$$
$$g = \text{gravitational constant } (9.81 \text{ m/s}^2)$$

A hydraulic jump is formed when the depth ratio (d_2/d_1) is greater than 2.4, the Froude number (F) then being greater than 2. When the Froude number is between 4 and 9, the energy consumed in turbulence can be between 45 and 70%, which is quite adequate for rapid mixing (Arboleda, 1973). Typical head losses are 0.3 m or greater.

The elevation of the channel floor must be dropped to assure the location of the hydraulic jump. The drop is generally placed at the end of the expansion of a supercritical flow (see Figure 5.1). The curves shown in Figure 5.2 may be used for design purposes to determine the relative height of the drop required to stabilize a jump for any given combination of discharge, upstream depth, and downstream depth.

Parshall Flume

The Parshall flume, employed conventionally in water and wastewater treatment plants as a flow measurement device, is also effective as a rapid mixer when a hydraulic jump is incorporated immediately downstream of the flume. Advantages of Parshall flumes over other types of rapid mixers are: (1) the hydraulic jump

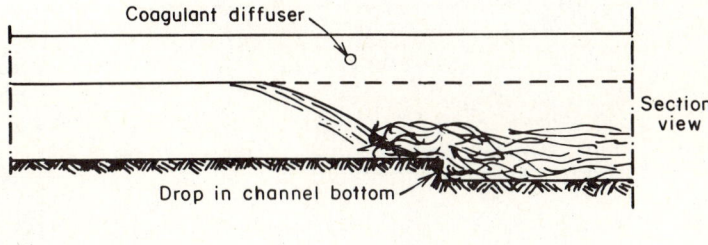

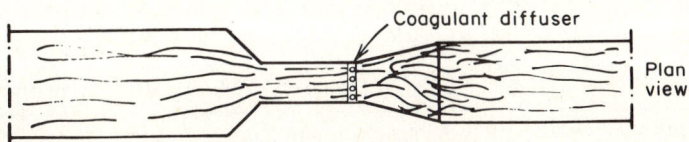

Figure 5.1. Hydraulic jump mixer. *Source:* Adapted from IRC, 1981b.

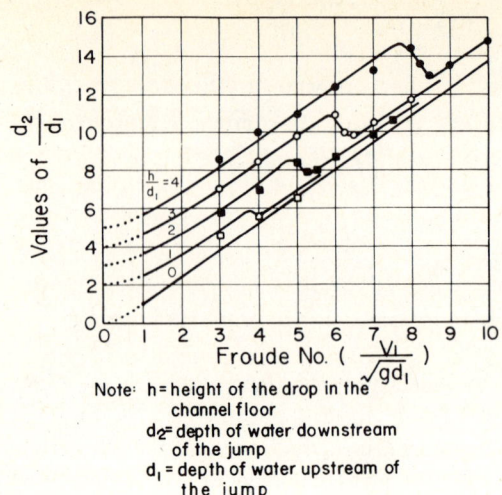

Note: h = height of the drop in the channel floor
d_2 = depth of water downstream of the jump
d_1 = depth of water upstream of the jump

Figure 5.2. Experimental relations among froude number (F), d_2/d_1 and h/d_1 for hydraulic jumps with an abrupt drop. *Source:* Hsu, 1950.

obviates the need for mechanical agitation and minimizes clogging from suspended material in the water that would otherwise accumulate on the floor of the flume, (2) the Parshall flume can be used to measure flows, (3) the Parshall flume operates as a single head device with a minimum loss of head (about one-fourth of that required by a weir under similar flow conditions), and (4) the Parshall flume can be made entirely of materials available locally (e.g., concrete, wood).

The Parshall flume consists of three principal sections: (1) a converging or contracting section at the upstream end leading to (2) a constricted section or throat,

Table 5.1   Dimensions and Capacities of the Parshall Flume for Various Throat Widths[a]

W (mm)	A (mm)	B (mm)	C (mm)	D (mm)	E (mm)	F (mm)	G (mm)	Free-Flow Capacity (Q) (m^3/day) min.	max.
150	620	600	390	400	600	300	600	122	9,550
300	1370	1340	600	850	900	600	900	274	39,500
460	1450	1420	750	1030	900	600	900	367	60,200
610	1530	1500	900	1210	900	600	900	1030	81,100
910	1680	1650	1200	1570	900	600	900	1500	123,000
1220	1830	1790	1520	1940	900	600	910	3190	166,000
1520	1860	1940	1830	2150	900	600	910	3920	210,000

[a]For the significance of the various letters, see Figure 5.3.
Source: Adapted from Okun and Ponghis, 1975.

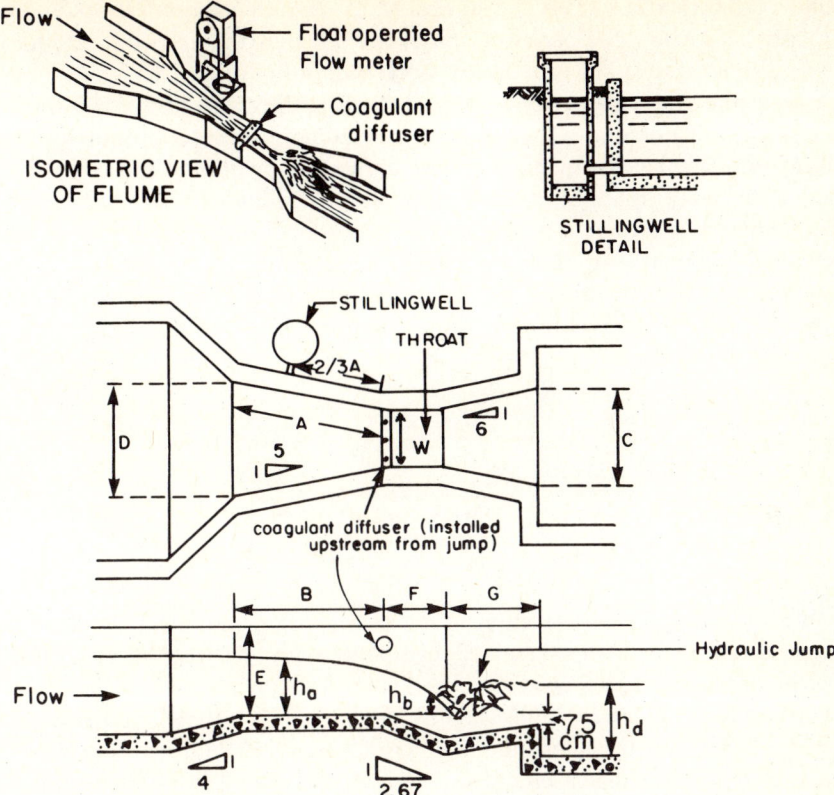

Figure 5.3. Parshall flume rapid mixer. *Source:* Adapted from Okun and Ponghis, 1975.

and (3) a diverging or expanding section downstream. The floor of the converging section is horizontal, the floor of the throat inclines downward, and the floor of the diverging section slopes upward (Figure 5.3).

The Parshall flume can be constructed in a wide range of sizes to handle virtually any flow range that is likely to be encountered in a water treatment plant. The width of the throat (W) is used to designate the size of a flume. Table 5.1 lists standard flume dimensions for various throat widths, designated by letters that appear in Figure 5.3, as well as the range of discharges corresponding to each flume size.

It is important to maintain free-flow conditions in the Parshall flume if it is also to be used for flow measurement. This is defined as the condition under which the rate of discharge for any flume is dependent solely on the depth of water at the gage point h_a in the converging section. The antithesis of free flow is submerged flow, where the elevation of the water surface downstream from the flume is high enough to retard the rate of discharge—a condition that wastes energy and which should be avoided. Nevertheless, it is possible for Parshall flumes to withstand a high degree of submergence without significantly reducing the indicated rate of free flow. It is such partially submerged flow that permits Parshall flumes to serve as effective rapid mix units in water treatment. A partially submerged flow is shown in Figure 5.3, where the backwater raises the downstream water surface, forming a hydraulic jump just downstream from the end of the throat.

The degree of submergence is often defined by the ratio of the two measured heads, h_b/h_a, obtained from the water levels in the throat (h_b) and upstream stilling well (h_a). In practice, however, it is very difficult to determine the value of h_b in advance, for the purpose of calculating the submergence ratio. It has been shown experimentally, however, that when submergence commences, the water surface levels in the downstream channel (h_d) and at the throat (h_b) are about the same. Consequently, the downstream level may be used for design purposes. The submergence ratio must be within the following limits in order to maintain free-flow conditions in the flume:

	Maximum Submergence (h_b/h_a or h_d/h_a)
Width of Throat (W)	Ratio
<0.3 m	0.60
0.3 m < W < 2.5 m	0.70

An abrupt drop in the elevation of the channel floor immediately downstream of the flume is necessary to stabilize a hydraulic jump. The magnitude of the drop can be determined from Figure 5.2.

Parshall flumes may be built of concrete, wood, sheet metal, or plastic. Large flumes are built on the site, but small flumes may be obtained as prefabricated structures to be installed in one piece. A series of Parshall flume rapid mixers for the Guandu plant in Rio de Janeiro, Brazil, is shown in Figure 5.12.

Palmer-Bowlus Flume

A simple modification of the Parshall flume is the Palmer-Bowlus flume, which is similarly formed by constricting the flow in an open channel or pipe. A principal advantage of such a flume is the comparative ease with which it can be installed in existing conduits; it does not require a drop in the conduit invert, as would be required with a Parshall flume. Figure 5.4 shows two cross-sectional shapes of Palmer-Bowlus flumes installed in open channels; the length of the throat for each type is about equal to the average depth. Figure 5.5 shows the location of the hydraulic jump and preferred head measuring point. When installing a Palmer-Bowlus flume, an adequate channel slope (which applies only to the downstream section) is necessary to maintain critical flow through the flume and to prevent submergence. Such conditions are assured as long as the downstream depth of flow is not greater than 85% of the upstream depth. For new installations, a slight drop in the channel floor at the downstream side of the flume will assure free flow and stabilize the jump. In practice, Palmer-Bowlus flumes are not used as widely as Parshall flumes; hence information on their effectiveness in the field as rapid mixers could not be ascertained.

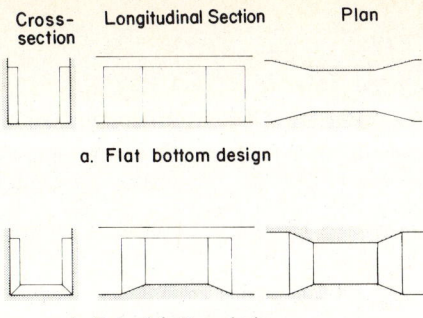

Figure 5.4. Cross-sectional shapes of Palmer-Bowlus flumes. *Source:* Grant, 1979.

Weirs

Flow-measuring weirs are simple but effective methods of rapid mixing for plants having relatively small capacities. A weir is low in cost, relatively easy to install, and can also be used as a flow measuring device. However, a weir normally operates with a rather significant loss of head (about 0.3 to 0.6 m) and must be cleaned periodically to prevent deposition of sediments on the upstream side of the weir. Weirs are generally less expensive to fabricate and install than flumes (particularly Parshall flumes) because of their simpler design and the types of materials required. Weirs are constructed by placing a thin metal plate (3 to 7 mm thick) or concrete wall across the flow and forcing the flow through a specified opening. This opening may be of several configurations, as shown in Figure 5.6.

Triangular weirs are normally used for low flows, whereas rectangular weirs are used for larger flows. Because they distribute the flow more uniformly across the channel width, rectangular weirs are preferred when using above-water coagulant diffusers. Minimum and maximum flow rates for both types of weirs are listed in Tables 5.2 and 5.3. Weirs can be made of wood with steel edges (Figure 5.6), or rigid plastic.

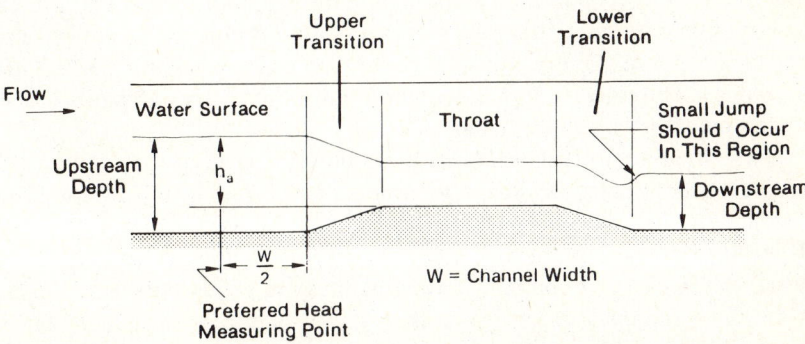

Figure 5.5. Free-flowing Palmer-Bowlus flume. *Source:* Grant, 1979.

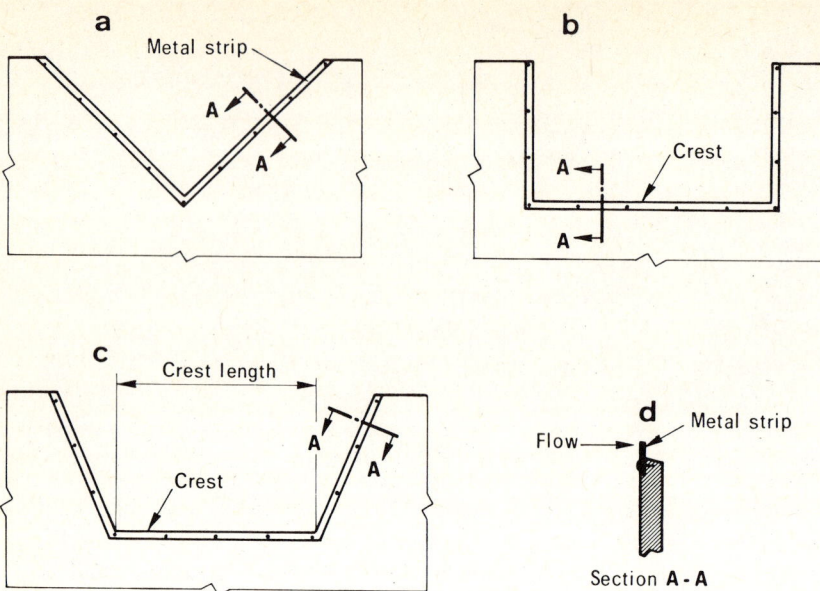

Figure 5.6. Measuring weirs: (a) V-notch weir; (b) rectangular weir; (c) trapezoidal weir, (d) sections A-A. *Source:* Okun and Ponghis, 1975.

The sudden drop in the hydraulic level over the weir induces the turbulence in the water for rapid mixing, and chemicals are added at this "plunge" point with the help of a diffuser. The vertical fall of the raw water over the weir should be at least 0.1 m to ensure sufficient turbulence. The height of the coagulant diffuser over the weir should be at least 0.3 m so that the speed of the falling coagulant solution is high enough to penetrate the nappe thickness. To utilize the energy from the weir effectively, a small receiving chamber should be constructed below the weir, where rapid mixing agitation can take place. A simple chamber consisting of a submerged weir 3 m downstream from the V-notch weir and converging side walls, as depicted in Figure 5.7, is a suitable design. The submerged weir induces a hydraulic jump within the chamber for additional mixing. Another design that is employed in several plants in India incorporates a baffled channel, which immediately follows the measuring weir (Figure 5.8). Turbulence is induced initially by the fall of water over the V-notch weir and then continues in the baffled channel as the water is conveyed to the flocculation basin. The mixing channel is sloped (1:50) and contains baffles turned at 45°.

For large treatment plants, the incoming flow at the head of the plant may be split equally among a set of weirs at the same elevation so as to limit the head loss over each weir. An interesting weir system for rapid mixing is used in a 250,000 m³/day water treatment plant that serves the city of Nairobi, Kenya. In this plant raw water enters the weir chamber through a pressure conduit and discharges onto a concrete pedestal enclosed by sharp-crested weirs along its perimeter. The water flows radially outward, over the weir and into a receiving chamber one meter below

Table 5.2. Minimum and Maximum Recommended Flow Rates for V-Notch Weirs

V-notch Angle	Minimum Head		Minimum Flow Rate		Maximum Head		Maximum Flow Rate	
	ft	cm	MGD	m³/day	ft	cm	MGD	m³/day
22½°	0.2	6.0	.006	22.7	2.0	61.0	1.82	6900
30°	0.2	6.0	.008	30.3	2.0	61.0	2.47	9360
45°	0.2	6.0	.012	45.5	2.0	61.0	3.78	14,300
60°	0.2	6.0	.017	64.4	2.0	61.0	5.28	20,000
90°	0.2	6.0	.029	109	2.0	61.0	9.14	34,600

Source: Adapted from Grant, 1979, p. 21.

Table 5.3. Minimum and Maximum Recommended Flow Rates for Rectangular Weirs with End Contractions

Crest Length		Minimum Head		Minimum Flow Rate		Maximum Head		Maximum Flow Rate	
ft	cm	ft	cm	MGD	m³/day	ft	cm	MGD	m³/day
1	30.5	0.2	6.0	.185	700	0.5	15.2	.685	2,590
1.5	45.7	0.2	6.0	.281	1060	0.75	22.9	1.89	7,160
2	61.0	0.2	6.0	.377	1430	1.0	30.5	3.87	14,700
2.5	76.2	0.2	6.0	.474	1800	1.25	38.1	6.77	25,600
3	91.4	0.2	6.0	.570	2160	1.5	45.7	10.7	40,500
4	122	0.2	6.0	.762	2890	2.0	61.0	21.9	83,000
5	152	0.2	6.0	.955	3620	2.5	76.2	38.3	145,000
6	183	0.2	6.0	1.15	4360	3.0	91.4	60.4	229,000
8	244	0.2	6.0	1.53	5800	4.0	122	124	470,000
10	305	0.2	6.0	1.92	7270	5.0	152	217	822,000

Source: Adapted from Grant, 1979, p. 25.

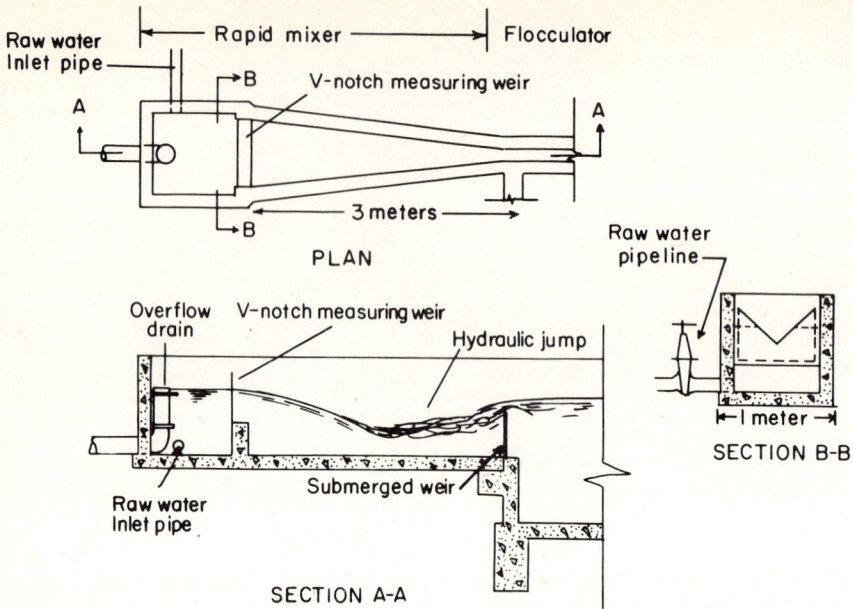

Figure 5.7.   Weir rapid mixer at a treatment plant in Peru. *Source:* Pan American Health Organization.

the pedestal elevation. Turbulence in the form of standing rollers occupies this space with a retention time of about 2 to 5 sec. Figure 5.9 shows the good mixing that is achieved with this type of design.

Baffled Mixing Chambers

In general, baffled mixing chambers are not recommended for rapid mixing because of their plug-flow characteristics, which are not conducive to turbulent mixing in short time periods. Their main contribution in water treatment lies in the flocculation stage, where gentler mixing over a longer period of time is desired (see Chapter

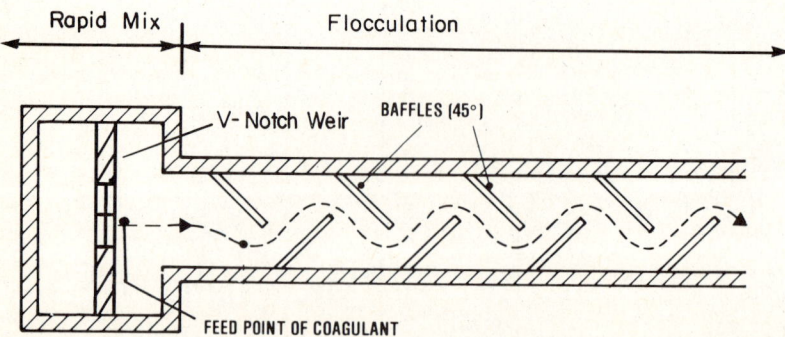

Figure 5.8.   V-notch weir and baffled channel for rapid mixing; plan view. *Source:* Adapted from IRC, 1981b.

Figure 5.9.   Weir rapid mixer at a plant in Nairobi, Kenya. *Source:* Schulz.

Six, under "Baffled Channel Flocculators"). As mentioned above, baffled mixing chambers may be used in conjunction with weirs as a composite rapid mix unit for special configurations in small water treatment plants.

Hydraulic Energy Dissipators

Where the source of water is high above the treatment plant, the available head can be used for rapid mixing. By installing hydraulic energy dissipators, such as stilling basins or jet orifices, sufficient turbulence is created for rapid mixing as the water passes into the mixing chamber. The inlet should be sized to ensure a velocity of at least 1 m/s, with the discharge directed at an end wall so that the flow is changed in direction by 180°.

Turbulent Pipe Flow Mixers

Several recently developed methods for diffusing chemicals in turbulent pipe flow have drawn considerable attention from researchers due to their practicality, simplicity, and relatively low cost (Chao and Stone, 1979). Installations of turbulent pipe flow mixers in a number of water treatment plants have been reported by Kawamura (1976). In the designs listed, adequate mixing is attained in a mixing time of about one second.

Hydraulic mixing in pipes can be achieved in a variety of ways. Modular plants

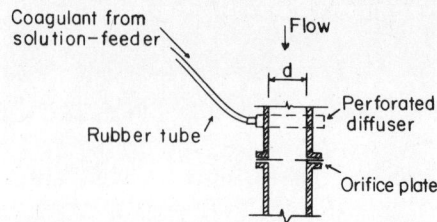

Figure 5.10.   Orifice plate for rapid mixing. *Source:* Adapted from CEPIS, 1982, vol. 2, plan no. 26.

designed by CEPIS (1982) employ an orifice plate placed in the raw-water pipe (see Figure 5.10). Pipes can also be fabricated with fixed, sloping baffles inside them to impart turbulence to the passing water. Eductors can be used effectively as rapid mixers, provided that pressurized water is available to supply the motive force to draw in and entrain the coagulant (see Chapter Four, under "Eductors").

Turbulent pipe flow mixers may present operation and maintenance problems in developing countries because (1) an auxiliary pump is normally required to inject the chemicals into the flow stream unless special arrangements are made for providing enough head for gravity feed, and (2) the small feed openings in the diffuser tend to clog, hence the diffuser should be removable or exposed to allow for easy cleaning. However, they can be effective and economical rapid mixers if pumps are not required and designs permit easy maintenance.

APPLICATION OF COAGULANTS IN OPEN CHANNELS

For open-channel mixers, the coagulant should be applied at a point immediately upstream of the zone of maximum turbulence by means of a perforated trough or

Figure 5.11.   Plastic-pipe diffuser for a weir rapid mixer in Malaysia. *Source:* Ching, 1979.

Figure 5.12. Parshall flume rapid mixer at the Guandu plant in Rio de Janeiro, Brazil. *Source:* Hudson.

perforated pipe diffuser. The pipe diffuser can be easily fabricated from a plastic pipe by drilling ports 0.6 to 1.3 cm diameter and not more than 15 cm apart, so as to distribute the coagulant evenly. In places where low-grade coagulants must be used, a trough fitted with triangular weirs on one side may be preferable to using perforated-pipe diffusers, because in the latter case, impurities in the coagulant are likely to clog the holes frequently. At least two sets of diffusers are desirable so that the coagulant can be fed continuously even when one diffuser is removed from operation for cleaning. Two photographs of perforated pipe diffuser systems for a V-notch weir and Parshall flume rapid mix chamber are shown in Figures 5.11

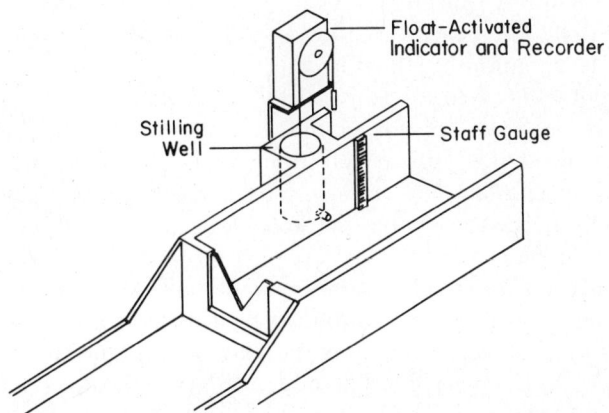

Figure 5.13. Flow measurement system consisting of a stilling well, float activated meter, and staff gauge. *Source:* Grant, 1979.

and 5.12, respectively. The diffusers are located above the plunge point for weir mixers, and directly upstream from the hydraulic jump for Parshall flumes.

FLOW MEASUREMENT SYSTEMS IN OPEN CHANNELS

The flumes and weirs described in this chapter for use as hydraulic mixers can be readily adapted for flow measurement. The constrictions that are formed in the channel by flumes and weirs change the water level upstream from the constriction by a known function. Thus, the flow rate through an open channel can be derived by a measurement of this water level.

The following discharge equations are applicable for the given devices as long as free-flow conditions are maintained in the constriction (Grant, 1979):

90° V-notch weir	$Q = 1.38\, h_w^{5/2}$	(5.4)
rectangular weir	$Q = 1.84\, Bh_w^{3/2}$	(5.5)
Parshall flume	$Q = 2.27\, Wh_a^{3/2}$	(5.8)
Palmer-Bowlus flume	$Q = 1.66 Wh_a^{3/2}$	(5.9)

where

Q = discharge (m^3/s)
h_w = head on weir (m)
h_a = depth at entrance to the flume at specified measuring point (m)
B = depth at entrance of weir (m)
W = width of throat (m)

Measuring of open channel flow rates may be done by using a stilling well and float-actuated indicator and/or recorder or a staff gauge (see Figure 5.13). The stilling well suppresses any surges that are present in the water flowing through the channel due to wind action, waves, and so on. The head connection line between the open channel and the stilling well should have a small cross-sectional area with respect to the stilling well. The float is usually conically shaped to provide stability. The wire or chain leading from the float is draped over a pulley behind the indicator. A counterweight is attached to the free end of the chain. As the float travels up with water level and down with the aid of the counterweight, the flow is read manually and/or recorded. The stilling well should be large enough and provided with a drain to facilitate cleaning. For accurate measurement of rapidly fluctuating flows, smaller wells are necessary, so that the water level in the well adjusts quickly to changing flows in the channel.

A staff gauge can aid in the zero adjustment of the flow meter or, in the simplest applications, can be used directly without a float. A fixed scale is placed securely on one side of the open channel so that the water level in the channel can be read directly in order to calculate the flow or to calibrate a float-actuated meter. For convenience, the staff gauge should be calibrated by the engineer to read in flow units, so the conversion need not be made by the operator.

chapter six

HYDRAULIC FLOCCULATION

Flocculation is the process of gentle and continuous agitation, during which suspended particles in the water coalesce into larger masses so that they may be removed from the water in subsequent treatment processes, particularly by sedimentation. Flocculation follows directly after the rapid mix process and, like rapid mixing, the agitation may be induced either by mechanical or hydraulic means. Flocculation is also generally required in direct filtration (see Chapter Eight, under "Direct Filtration"). It is necessary to form a fine "pinpoint" floc prior to filtration. This may be achieved in-line with floc formation in the filter or conventionally with 5 to 20 min. of mechanical or hydraulic flocculation.

Mechanical flocculators are preferred in the industrialized countries because of their greater versatility; that is, the speed of the mechanically operated paddles can be adjusted to suit variations in flow, temperature, or raw water quality. Also, the flocculation is independent of the flow through the unit. Mechanical flocculators are readily available from proprietors in industrialized countries in a variety of designs to suit any plant layout or mode of operation. The principal elements of mechanical flocculator systems are agitator impellers, drive motors, speed controllers and reducers, transmission systems, shafts, and bearings. The relatively high cost and complexity of mechanical flocculator systems, particularly with regard to operation and maintenance, render them less suitable for developing countries.

A more practical approach is to use hydraulic flocculators that do not require mechanical equipment, nor a continuous power supply if gravity flow is available, and which can be built primarily from concrete, brick, wood, or masonry by local labor at relatively low cost. Moreover, several hydraulic flocculation systems operate under plug-flow conditions (plug-flow, under ideal conditions, is achieved when water flows through a chamber at a uniform rate without intermixing) which minimizes short-circuiting of the flow (i.e., when a portion of the incoming flow of water traverses the flocculation chamber in a much shorter time than the nominal detention period). Short-circuiting, an inherent problem of mechanical flocculators, is alleviated somewhat in practice by installing a series of successive compartments in the flocculation chamber.

The major shortcomings of hydraulic flocculators have been reported widely in the technical literature.

1.   No flexibility to respond to changes in raw water quality.
2.   Hydraulic and consequent flocculation parameters are a function of flow and cannot be adjusted independently.
3.   Head loss is often appreciable.
4.   Cleaning may be difficult.

These shortcomings are the reasons why hydraulic flocculators have not continued to be used extensively in the industrialized countries. Of 42 U.S. plants built between 1908 and 1932, 30 had hydraulic flocculators (American Society of Civil Engineers, 1940). It is possible, however, to mitigate these shortcomings with properly designed systems that will function under a reasonably wide range of operating conditions. In fact, new designs for hydraulic flocculators and improvements in older designs have been implemented and are operating successfully in water treatment plants in Latin America (Azevedo-Netto) and, interestingly enough, in several plants in California, where high technology is readily available (MacDonald and Streicher, 1977).

This chapter examines several types of hydraulic flocculators that are appropriate for water treatment plants in developing countries. Baffled channel flocculators are the most widely-used hydraulic method, particularly in Latin America. Gravel-bed flocculators have been installed during the last 10 years in several small water treatment plants in India (Kardile, 1981), and have been tested experimentally to ascertain their potential for use in Brazil (Richter and Moreira, 1981). Flocculators that use the jet action of the water to impart turbulence, such as the Alabama type and heliocoidal-flow type, have also been used in small treatment plants. Staircase-type flocculators, recently developed in Brazil, and surface-contact flocculators, recently developed in India, are also examined in this chapter.

DESIGN CRITERIA

The velocity gradient (G) for flocculators is determined from the equations developed by Camp and Stein (1943).*

$$G = (Q \rho g \, h_l/\mu V)^{1/2} = (\rho g \, h_l/\mu t)^{1/2} \qquad (6.1)$$

*These equations are applicable for rapid mixing, especially for matching jar test data to plant operation in conjunction with Figure 4.2.

for hydraulic flocculation, and

$$G = [P/(\mu V)]^{1/2} \qquad (6.2)$$

for mechanical flocculation,

where

G = velocity gradient (s^{-1})
ρ = density of water (kg/m^3)
h = head loss (m)
μ = dynamic viscosity $(kg/m \cdot s)$
t = detention time, Q/V, (s)
Q = flow (m^3/s)
P = power, $Q\rho gh$ (watts, $kg \cdot m^2/s^3$)
V = volume of unit (m^3)
g = gravitational constant $(9.81 \ m/s^2)$.

Values of the density (ρ) and dynamic viscosity (μ) for water of various temperatures are listed in Table 6.1. The head loss in hydraulic flocculators results primarily from the turbulence created within the unit. The power required to provide the turbulence is a function of the head loss. The above equations are the basis for Figure 6.1, which allows determination of the velocity gradient (G) for a known head loss (h) and detention time (t) in hydraulic flocculators. The graph is calibrated for 12°C water temperature. Conversion factors for other water temperatures are listed in the table within the figure. For example, a hydraulic flocculator having a detention time of 15 min (900 s) and a head loss of 0.3 m would give a velocity gradient of about 55 s^{-1} at 20°C. G-values for particular types of flocculator designs may also be obtained from formulas presented later in this chapter.

In the design of flocculation systems, the total number of particle collisions, and thus the floc formation action, is indicated as a function of the product of the velocity gradient and the detention time, Gt. The range of velocity gradient (G) and Gt values given in Table 6.2 have been shown in practice to be the most

Table 6.1. Variations of the Specific Gravity (Density) and Viscosity of Water with Temperature

Temperature, °C	0	5	10	15	20	25	30
Density (ρ) (kg/m^3)	999.9	1000	999.7	999.1	998.2	997.1	995.7
Dynamic viscosity, poises (μ) $(kg/m \cdot s)$	0.0179	0.0152	0.0131	0.0114	0.0101	0.0089	0.0080

Source: Adapted from Fair, Geyer, and Okun, 1968.

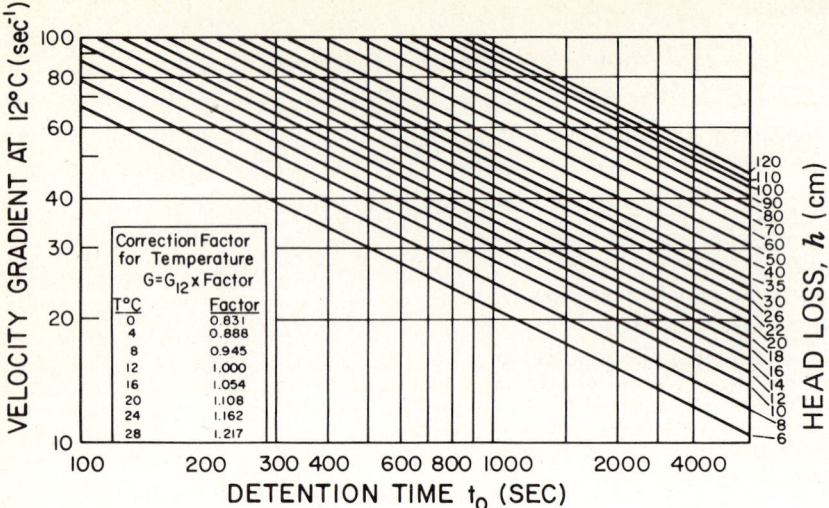

Figure 6.1.   Velocity gradients in hydraulic flocculators for different detention times (t_0) and head losses (h) at a temperature of 12°C. *Source:* Arboleda, 1973.

effective for plants using mechanical flocculators. Nonetheless, in order to obtain appropriate values for particular designs and water characteristics to provide for the optimal formation of flocs, laboratory jar testing or pilot-plant studies should be conducted on the water to be treated.

Velocity gradients in a flocculation basin can be tapered to be high at the inlet end and low at the outlet end to achieve more efficient mixing and agglomeration of the floc particles. Such a design can reduce the magnitude of the shearing forces on the flocs as they agglomerate, and thereby reduce the chance of floc breakup. Designs that yield tapered velocity gradients are discussed below.

Table 6.2.   Recommended G and GT Values for Flocculators

Type	Velocity Gradient G (s^{-1})	GT
Turbidity or color removal (without solids recirculation)	20 to 100	20,000 to 150,000
Turbidity or color removal (with solids recirculation)	75 to 175	125,000 to 200,000
Softeners (solids contact reactors)	130 to 200	200,000 to 250,000

Source: Smethurst, 1979, p. 57.

BAFFLED CHANNEL FLOCCULATORS

In baffled channel flocculation, mixing is accomplished by reversing the flow of water through channels formed by around-the-end or over-and-under baffles (Figures 6.2 and 6.3). Baffled channel flocculators are limited to relatively large treatment plants (greater than 10,000 m³/day capacity) where the flowrates can maintain sufficient head losses in the channels for slow mixing without requiring that baffles be spaced too close together (which would make cleaning difficult). A distinct advantage of such flocculators is that they operate under plug-flow conditions that free them from short-circuiting problems.

Horizontal-flow flocculators with around-the-end baffles are sometimes preferred over vertical-flow flocculators with over-and-under baffles because they are easier to drain and clean; also, the head loss, which governs the degree of mixing, can be changed more easily by installing additional baffles or removing portions of existing ones. However, vertical-flow units have been used successfully in Brazil and in the United States (see Figure 6.4) and are appropriate for specific applications, such as, for example, where a scarcity of land prohibits the use of larger horizontal-flow flocculators.

The water depth in the channels of vertical flow units can be as high as 3 m, and therefore less surface area is required than with horizontal units. The major problem with such flocculators is the accumulation of settled material on the chamber floors and the difficulty in removing it. To mitigate this problem, the Brazilian designs have included small openings (weep holes) in the base of the lower baffles of a size equivalent to 5% of the flow area of each chamber. The purpose is to

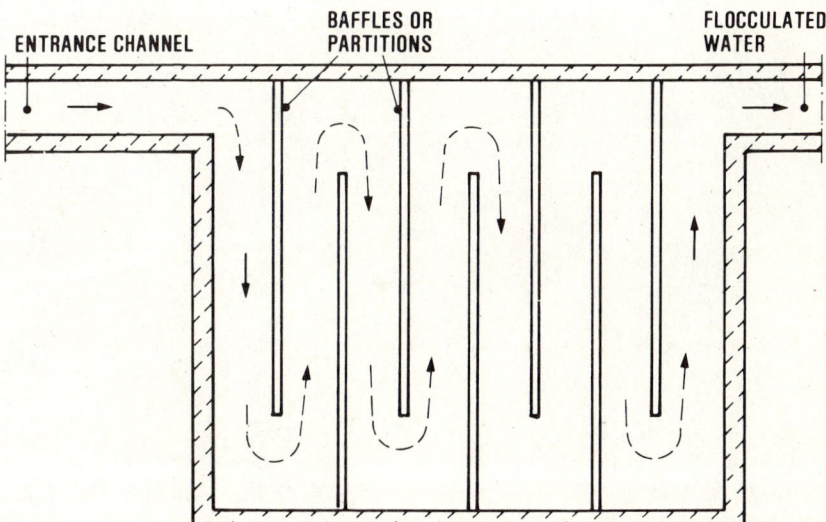

Figure 6.2.   Horizontal-flow baffled channel flocculator (plan). *Source:* IRC, 1981b.

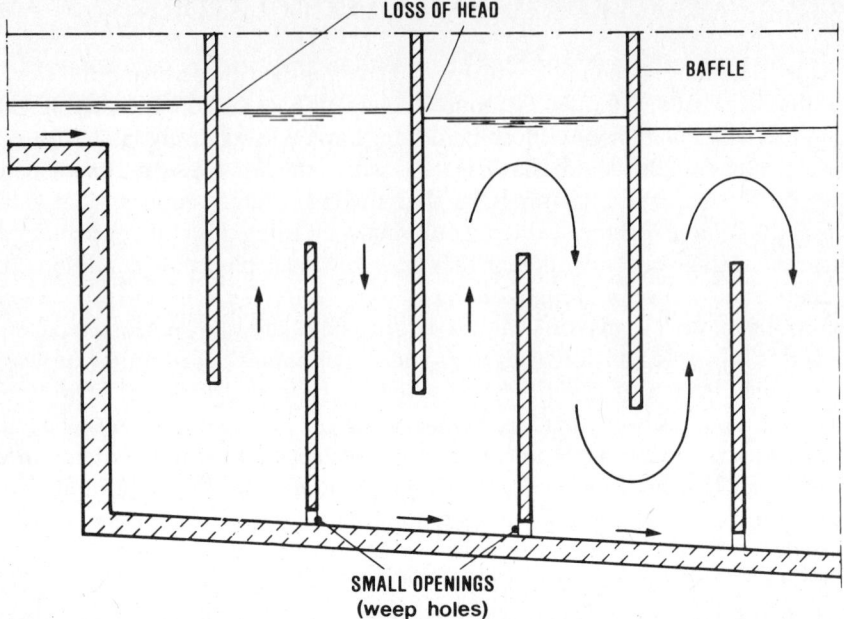

Figure 6.3. Vertical-flow baffled channel flocculator (cross-section). *Source:* IRC, 1981b.

Figure 6.4. Vertical-flow baffled channel flocculator at a plant in Danville, Virginia, USA. *Source:* G. Robinson.

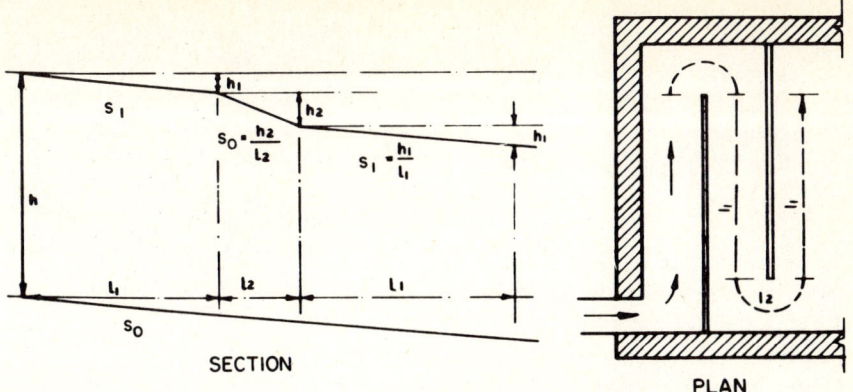

Figure 6.5. Energy gradient for horizontal-flow baffled channel flocculators. *Source:* Arboleda, 1973.

allow the major portion of the flow of water to follow the over-and-under path created by the baffles, whereas a smaller portion flows through the hole, creating some additional turbulence and preventing the accumulation of material (Arboleda, 1973). Weep holes also facilitate manual cleaning of the over-and-under flocculator. An over-and-under flocculator for a plant in Danville, Virginia, which has been operating satisfactorily for over 40 years, is shown in Figure 6.4.

The energy gradient for a horizontal flow unit is shown in Figure 6.5, revealing a relatively large head loss (h_2) across the bend (l_2) as compared with the head loss (h_1) in the channel (l_1). Studies by Arboleda (1973) and MacDonald and Streicher (1977) have suggested a reliance on the velocity gradients produced in the bend for mixing, and reducing the length of the channel (l_1) so as to prevent quiescent flow. For design purposes, the head loss in the bend is approximated by the following formula:

$$h = k(v^2/2g) \qquad (6.3)$$

where

h = head loss (m)
v = the fluid velocity (m/s)
g = the gravitational constant (9.81 m/s^2)
k = empirical constant (varies from 2.5 to 4)

The value of k cannot be determined precisely in advance; therefore it is better to design for a low k value, because boards can always be added to the baffles if additional head loss is needed.

The number of baffles needed to achieve a desired velocity gradient for both horizontal and vertical flow units can be calculated from equations 6.4 and 6.5, which are adapted from formulas derived by Richter (1981).

$$n = \{[(2\mu t)/\rho(1.44 + f)] \, [(HLG)/Q]^2\}^{1/3} \qquad (6.4)$$

for horizontal units, and

$$n = \{[(2\mu t)/\rho(1.44 + f)] \, [(WLG)/Q]^2\}^{1/3} \tag{6.5}$$

for vertical units,

where

n = number of baffles in the basin
H = depth of water in the basin (m)
L = length of the basin (m)
G = velocity gradient (s^{-1})
Q = flow rate (m^3/s)
t = time of flocculation (s)
μ = dynamic viscosity (kg/m·s)
ρ = density of water (kg/m^3)
f = coefficient of friction of the baffles
W = width of the basin (m)

The water velocity in both horizontal-flow and vertical-flow units generally varies from 0.3 to 0.1 m/sec. Detention time varies from 15 to 30 min (IRC, 1981b). In general, velocity gradients for both types of baffled channel flocculators should vary between 100 to 10 s^{-1}. In addition to the foregoing design criteria, the practical

Table 6.3.   Guidelines for the Design and Construction of Baffled Channel Flocculators

---

A. Around-The-End (Horizontal Flow)

---

1. Distance between baffles should not be less than 45 cm to permit cleaning.
2. Clear distance between the end of each baffle and the wall is about 1½ times the distance between baffles; should not be less than 60 cm.
3. Depth of water should not be less than 1.0 m.
4. Decay-resistant timber should be used for baffles; wood construction is preferred over metal parts.
5. Avoid using asbestos-cement baffles because they corrode at the pH of alum coagulation.

B. Over-and-Under (Vertical Flow)

---

1. Distance between baffles should not be less than 45 cm.
2. Depth should be two to three times the distance between baffles.
3. Clear space between the upper edge of a baffle and the water surface, or the lower edge of a baffle and the basin bottom, should be about 1½ times the distance between baffles.
4. Material for baffles is the same as in around-the-end units.
5. Weep holes should be provided for drainage.

---

guidelines enumerated in Table 6.3 should be considered in the design and construction of baffled channel flocculators, although they are somewhat general and should not be interpreted as necessarily binding in all cases.

Tapered energy flocculation in baffled channels generally is achieved by varying the spacing of the baffles, that is, close spacing of baffles for high velocity gradients, and wider spacing for low velocity gradients. The configuration of baffles that will induce a specific tapered velocity gradient is best determined under actual plant operating conditions, by either respacing or changing the number of baffles in the flocculation basin to attain the desired head loss. Arboleda (1973) recommends a tapered velocity gradient from about $75 \ s^{-1}$ at the inlet to 10 to $15 \ s^{-1}$ at the outlet of the flocculators. The Cochabamba water treatment plant in Bolivia has a tapered horizontal-flow flocculator consisting of three chambers, with each chamber containing baffles at different spacing, as shown in Figure 6.6.

Typical hydraulic calculations for the design of an around-the-end (horizontal-flow) flocculator are presented in Appendix B.

An innovative baffled channel flocculator with a tapered design has been installed

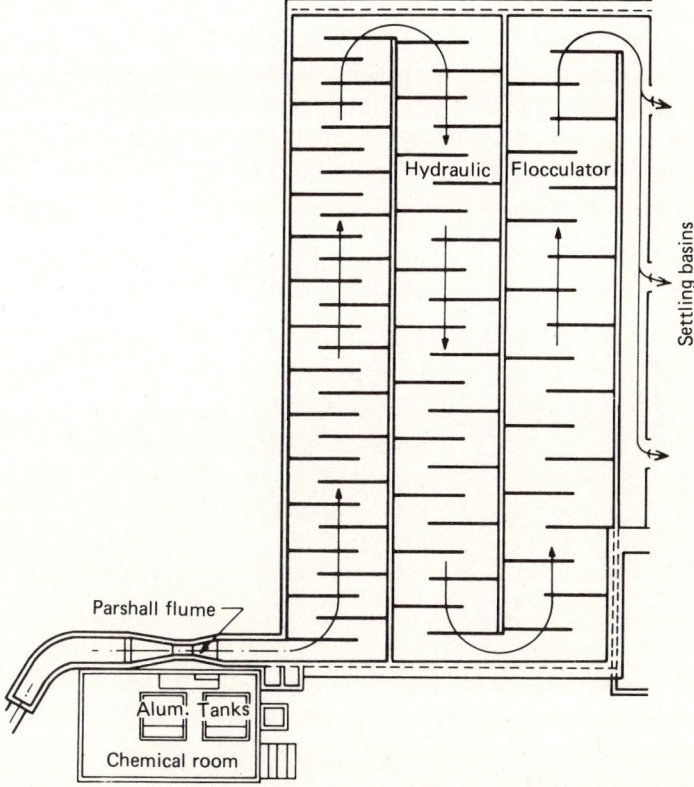

Figure 6.6.   Tapered horizontal-flow flocculator at a plant in Cochabamba, Bolivia. *Source:* Arboleda, 1976.

in a plant in Oceanside, California (MacDonald and Streicher, 1977), and is illustrated in the plant layout shown in Figure 6.7. Two independent flocculation basins encompass the sedimentation and filtration units, sharing a common sidewall. Modified baffles and a sloped basin floor are arranged in such a way that the minimum water depth is at the inlet to the flocculators and the depth gradually increases to a maximum at the outlet. This results in tapered flocculation that promotes relatively high velocity gradients at the entrance, even at reduced flow rates, and decreasing gradients toward the outlet. Moreover, the overall reduced level of energy at lower flows is counterbalanced by the increased detention time. The mean velocity gradient in the two flocculator basins varies from 200 to 20 s^{-1} at a flow rate of 38,000 m^3/day per basin, and from 208 to 8 s^{-1} at a flow rate of 15,000 m^3/day per basin. The value of Gt over a plant flow range of 30,000 m^3/day to 60,000 m^3/day is about 65,000 to 79,000 with both flocculation channels in service. For plant flow rates less than 30,000 m^3/day, only one basin is used. Consequently, effective flocculation can be achieved over a plant flow range of 15,000 to 80,000 m^3/day. Figure 6.8 shows the variation of the velocity gradient along the flocculator basin for high and low flow rates. Figure 6.9 shows the effect of the variation in the flow on Gt values. The flocculator performance data for the plant in Oceanside, California, clearly indicate that properly designed hydraulic flocculators can operate effectively under variable flow rates; this refutes a heretofore general criticism of hydraulic flocculation systems, that maintenance of velocity gradients is not possible with changes in raw water flow rates. Where the required G-value is a function of water quality, however, hydraulic flocculation systems are less flexible than mechanical systems.

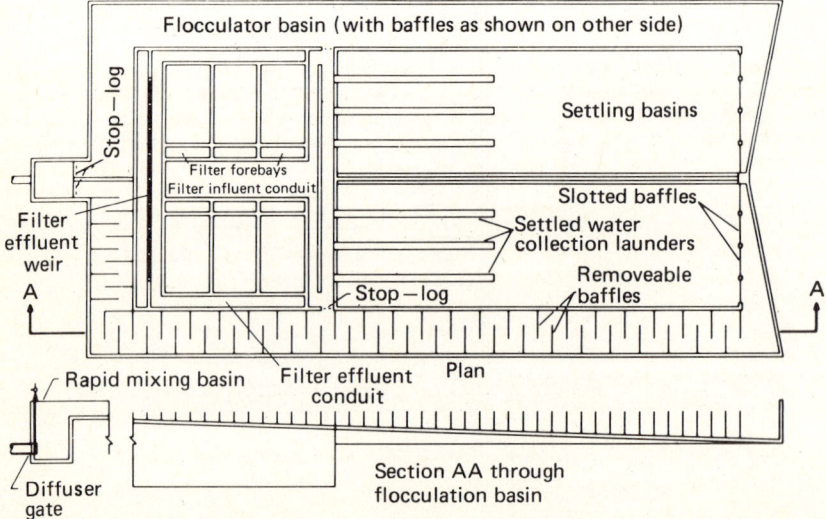

Figure 6.7. Tapered-energy flocculator at the Oceanside plant in Arcadia, California. *Source:* MacDonald and Streicher, 1977.

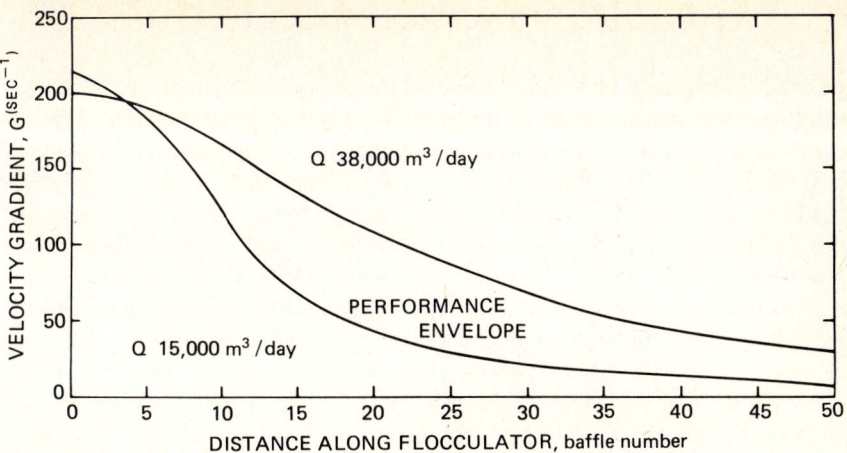

Figure 6.8. Mean velocity gradient variations with flow for the tapered-energy flocculator. *Source:* MacDonald and Streicher, 1977.

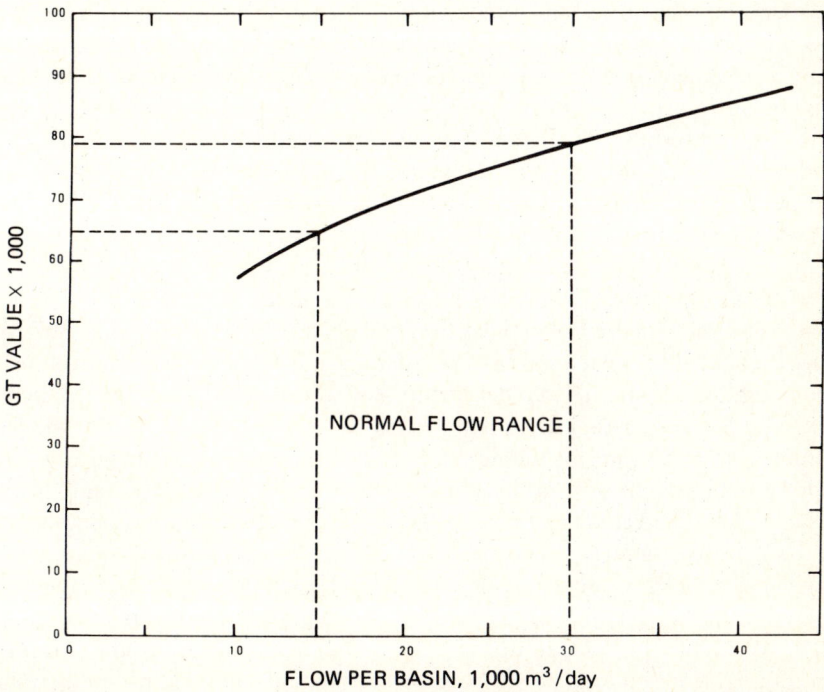

Figure 6.9. GT variations with flow for the tapered-energy flocculator. *Source:* MacDonald and Streicher, 1977.

HYDRAULIC JET-ACTION FLOCCULATORS

A less well-known type of hydraulic flocculator suitable for small treatment plants is one that uses the jet action of the influent water to cause agitated flow. Two types of jet-action flocculators are considered here: (1) the heliocoidal-flow type, and (2) the Alabama type. Both are presently used in small plants in Latin America, particularly in Brazil where Alabama-type flocculators are used widely (Azevedo-Netto). The heliocoidal-flow units were extensively used in the United States in the early years of this century, and are now widely used in China.

Heliocoidal-flow flocculators (also called tangential-flow or spiral-flow) impart a rotational movement to the water, which creates turbulence for mixing and ag-glomeration of the flocs. This is accomplished in a series of rectangular or cylindrical chambers by allowing a stream of water to enter tangentially into each chamber so as to cause heliocoidal flow toward the outlet. The turbulence that is created by this jet action is governed by the inlet velocities and the size and shape of the chamber. For example, a series of square chambers provide some resistance to the spiral flow of water in each of the chambers, thereby causing additional agitation. An effective design for this type of flocculator employs a series of small chambers interconnected by pipes or box conduits, carefully sized and arranged to produce the desired entry velocity for jet action. The direction of flow (upward or downward) is alternated in successive chambers. Tapered velocity gradients are easily provided for by increasing the area of the inlet opening for each successive chamber, thereby decreasing the jet action of the water and the intensity of mixing. The size of the opening for each chamber can be adjusted manually by using sluice gates, removable boards, or orifices.

The size and number of chambers are a function of the plant flow rate and the desired time for flocculation. Five to seven basins are usually required to provide for an adequate detention time and to mitigate short-circuiting effects. For larger plants the number may be increased by installing two or more groups of chambers operating in parallel. Recommended inlet velocities range from 0.5 m/s to 0.7 m/s for the first chamber to 0.1 to 0.2 m/s for the latter chambers.

Cox (1960) designed a heliocoidal-flow flocculator for a small plant (3020 m³/day) in Brazil, with tapered-energy flocculation and flexibility in controlling the degree of agitation. The heliocoidal flocculator shown in Figure 6.10 is based on this design. Six rectangular chambers having a total volume of 60 m³ yielded a detention time of 28 min for the design flow rate. Locally-made sluice gates were provided to control the size of the opening and hence the water velocity at each chamber inlet. The velocities ranged from 0.5 to 0.2 m/s for each chamber and could be adjusted manually by turning handwheel operated sluice gates. A drain was provided for each chamber for dewatering and cleaning purposes. Arrangements were made to allow for the construction of six additional chambers to operate in parallel with the first group, so that the plant could be enlarged in the future.

An inherent shortcoming of heliocoidal-flow type flocculators, which they share with mechanical flocculators, is short-circuiting of the flow within each chamber. For mechanical flocculation, this problem has been commonly solved by installing

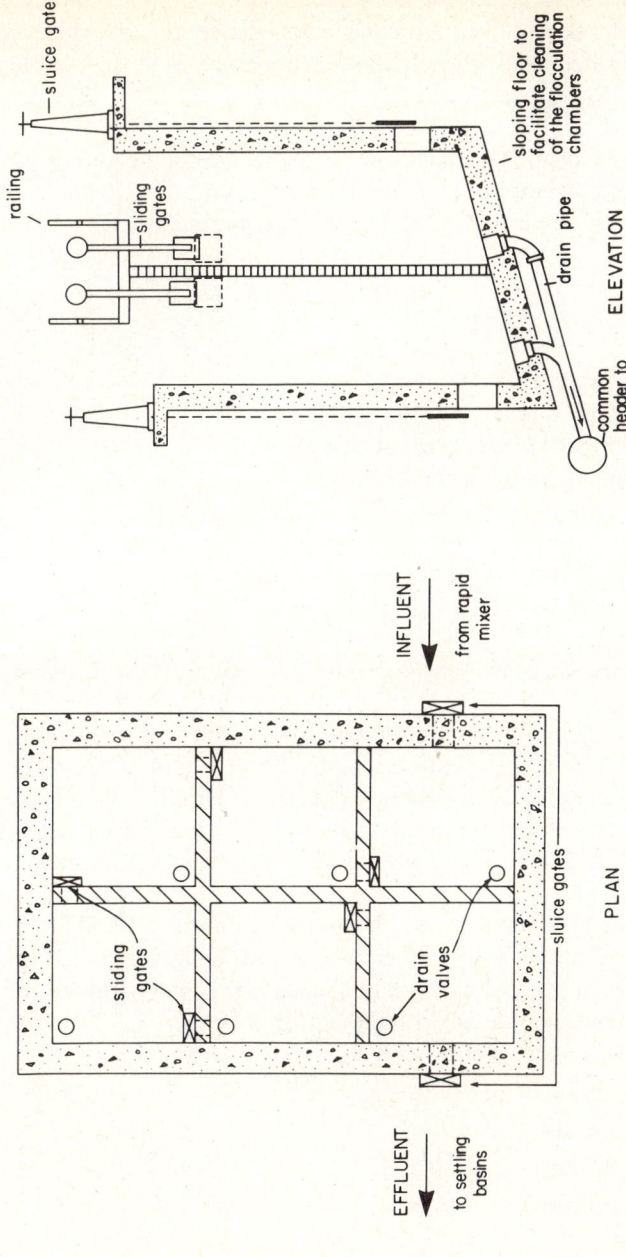

sluice gate

railing

sliding gates

sloping floor to facilitate cleaning of the flocculation chambers

drain pipe

ELEVATION

common header to drain

INFLUENT

from rapid mixer

sliding gates

sluice gates

drain valves

PLAN

EFFLUENT

to settling basins

Figure 6.10. Heliocoidal-flow flocculator. *Source:* Adapted from, IRC, 1980.

a series of compartments within the flocculation basin. Similarly, a "staircase"-type design, developed in Brazil (see Figure 6.11) has been found effective in controlling short-circuiting in heliocoidal flocculation chambers as well as providing for more controlled hydraulic agitation within each chamber (Pinheiro). This device causes a heliocoidal movement of the liquid around an axis with constant G-values at the center and periphery of any horizontal cross-sectional area of the flocculator chamber. Pinheiro developed an empirical equation for calculating G-values in staircase-type flocculators. This formula is adapted for square chambers:

$$G^2 = (2\rho k Q^3)/(\mu L^4 h^3) \tag{6.6}$$

where

G = velocity gradient (s^{-1})
ρ = density (kg/m^3)
k = friction loss coefficient (about 7.5)
g = gravitational constant (9.81 m/s^2)
μ = dynamic viscosity $(kg/m \cdot s)$
Q = flow rate (m^3/s)
L = length of the side of square chamber (m)
h = pitch (m)

The staircase-type flocculator can be made from marine plywood and is assembled like a spiral staircase with the treads around a central column. The flight of four or eight steps corresponds to a spire, and all treads are equal trapeziums; therefore each rise is one-fourth or one-eighth of the pitch (see Figure 6.11b). The inlet and the outlet of the chamber are positioned opposite each other. With some modifications, staircase-type flocculators may be retrofitted in conventional heliocoidal-flow flocculation chambers. Hydraulic calculations for staircase flocculators are presented in Appendix B.

The Alabama-type flocculator is illustrated in Figure 6.12. The jet action is provided in each chamber via a cast iron pipe with its outlet turned upward. For effective flocculation, the outlet should be placed at a depth of about 2.5 m below the water level. Common design criteria are listed below:

Rated capacity per unit chamber	25 to 50 l/s per m^2
Velocity at turns	0.40 to 0.60 m/s
Length of unit chamber (L)	0.75 to 1.50 m
Width (W)	0.50 to 1.25 m
Depth (H)	1.50 to 3.0 m
Detention time (t)	15 to 25 min

The head loss with this type of flocculator is estimated at two velocity heads per chamber, generally about 0.35 to 0.50 m of head loss for the entire unit. Velocity gradients range from 40 to 50 s^{-1}. Arrangements should be made for draining each

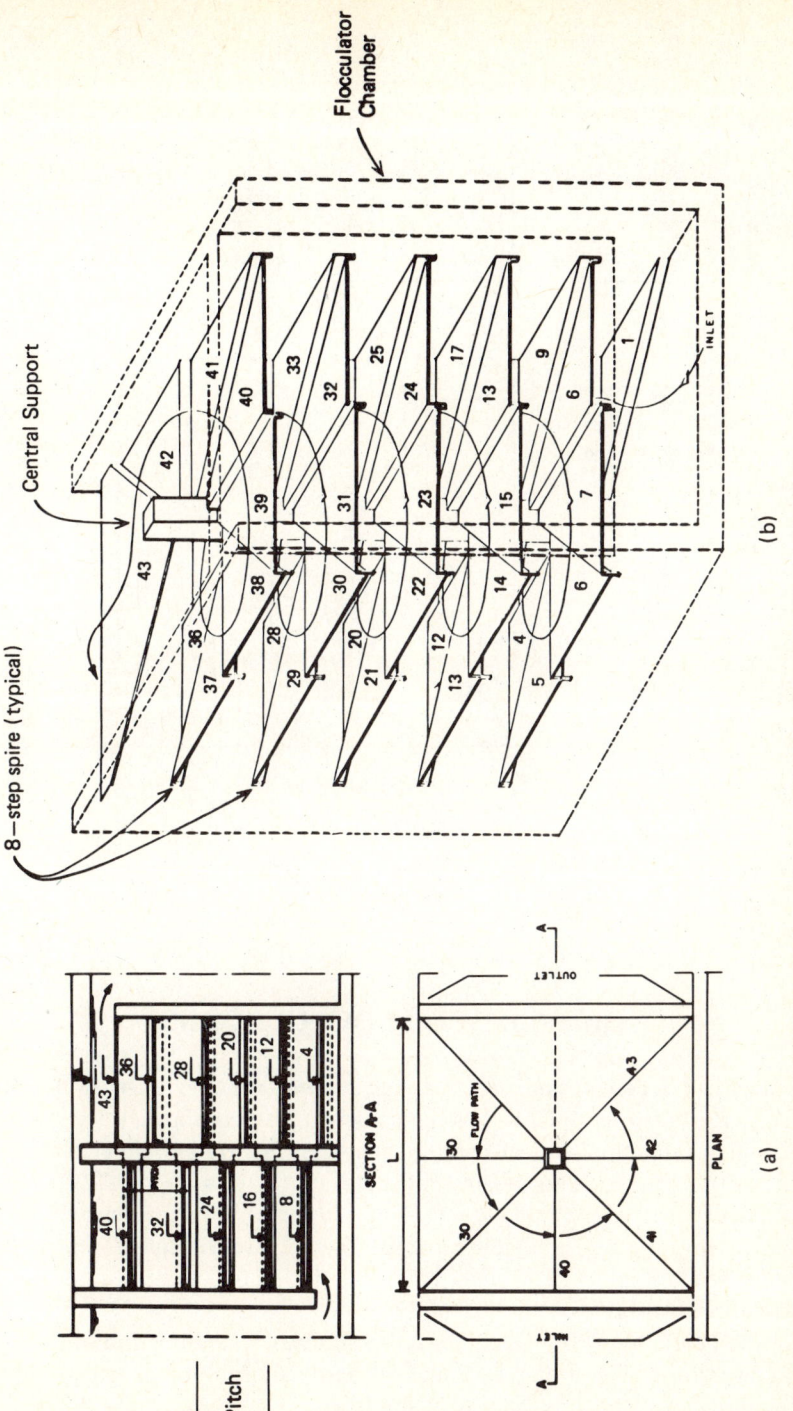

Figure 6.11. Staircase-type heliocoidal-flow flocculator. *Source:* Pinheiro. (Note: stairs are numbered sequentially)

119

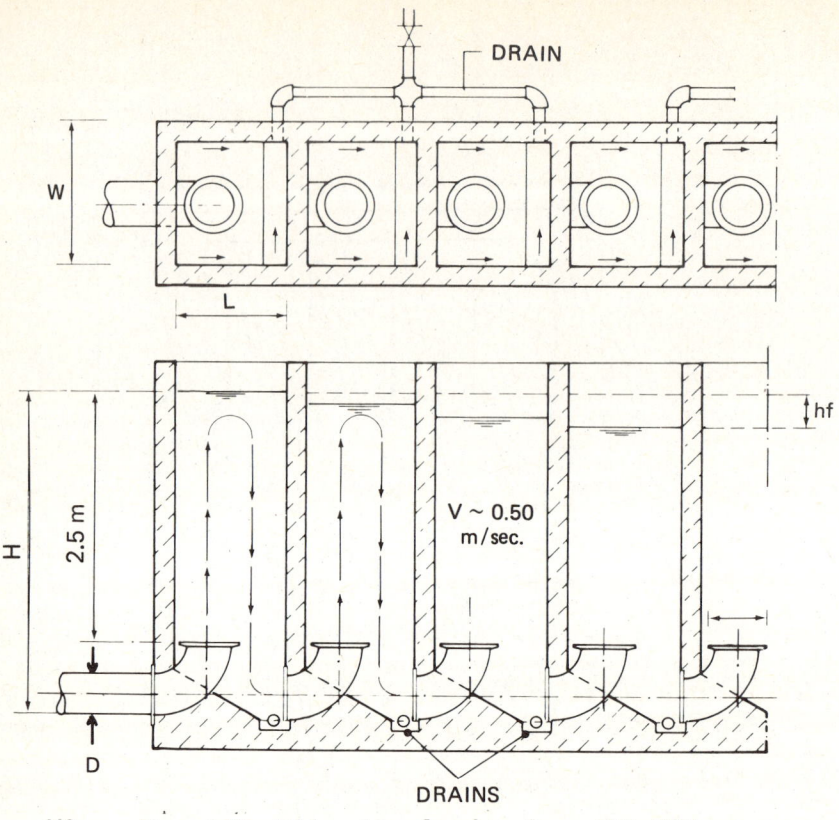

103   **Figure 6.12.**   "Alabama"-type flocculator. *Source*: IRC, 1981b.

chamber, because accumulated material tends to collect at the bottom and must be removed occasionally.

GRAVEL-BED FLOCCULATORS

The gravel-bed flocculator provides a simple and inexpensive design for flocculation in small water treatment plants (less than 5000 m³/day capacity). It has been tested experimentally and employed successfully in several upflow-downflow plants in India (Kardile, 1981) and in package plants in Parana, Brazil (Wagner, 1982b). The packed bed of gravel provides ideal conditions for the formation of compact settleable flocs because of continuous recontacts provided by the sinuous flow of water through the interstices formed by the gravel. The velocity gradients that are introduced into the bed are a function of (1) the size of the gravel, (2) rate of flow, (3) cross-sectional area of the bed, and (4) the head loss across the bed. The direction of flow can be either upward or downward, and is usually determined from the design and hydraulic requirements of other process units in the plant.

A unique characteristic of this type of hydraulic flocculator is its ability to store agglomerated flocs within the interstices or to settle flocs on top of or below the gravel bed (depending on the direction of flow) due to the sudden drop in velocity as the flow of water emerges from the bed. Moreover, the sludge storage capabilities of gravel-bed flocculators make them ideal pretreatment units prior to filtration in small plants, often eliminating the need for a separate sedimentation step (see Chapter Eight, under "Upflow-Downflow Filtration").

Assuming laminar flow across the gravel bed, velocity gradients, and head losses in gravel-bed flocculators can be estimated from the following equations (adapted from Fair, Geyer, and Okun, 1971):

$$G = [(h\rho g Q)/(\mu \alpha V)]^{1/2} \tag{6.7}$$

$$h = \frac{f}{\theta} \left(\frac{1 - \alpha}{\alpha^3} \right) \frac{L}{d} \frac{v^2}{g} \text{ (Carmen-Kozeny Equation)} \tag{6.8}$$

$$f = 150 \left(\frac{1 - \alpha}{R_N} \right) + 1.75 \tag{6.9}$$

$$R_N = \frac{d v \rho}{\mu} \tag{6.10}$$

where

G = velocity gradient $(\sec^{-1})$
h = head loss (m)
ρ = specific gravity of water (kg/m^3)
g = gravity constant $(9.8 \ m/\sec^2)$
Q = flow rate $(m^3/\sec)$
μ = dynamic viscosity $(kg/m{\cdot}s)$
α = porosity (~ 0.4)
V = volume of gravel bed (m^3)
f = friction factor
L = depth of gravel bed (m)
θ = shape factor (~ 0.8)
R_N = Reynolds number
d = average size of gravel (m)
v = face velocity (m/sec)

A design for a gravel-bed flocculator using the above equations is presented in Appendix B.

When greater accuracy is desired, G-values may be determined from bench-scale experiments. Plastic cylinders are filled with the desired gravel medium at the same depth as the full-scale gravel-bed flocculator, and arrangements are made for measuring head loss at several points along the length of the cylinder. After sufficient head loss data are collected for a range of flows, the corresponding velocity gradients can be calculated from equation 6.7.

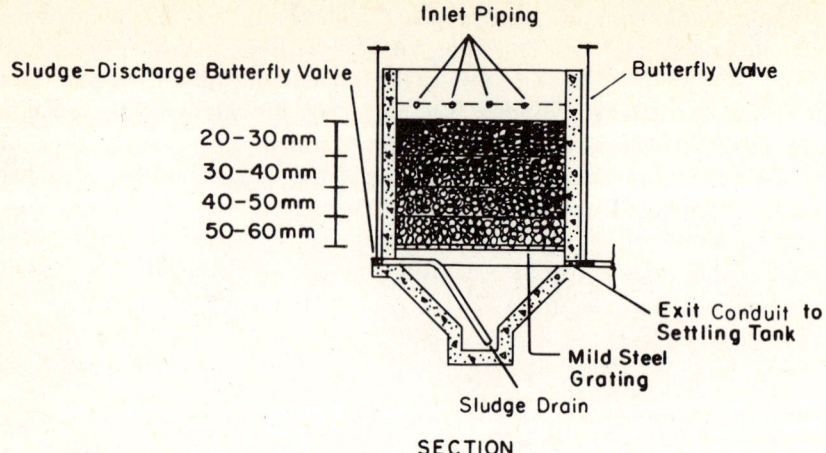

SECTION

Figure 6.13. Downward-flow gravel bed flocculator. *Source:* Adapted from Kardile, 1981.

Tapered velocity gradients are achieved in gravel-bed flocculators by changing the cross-sectional area of the bed and/or by grading the bed with different-sized layers of gravel. The downward flow unit in Figure 6.13 is comprised of a graded gravel bed ranging in size from 20 to 60 mm from top to bottom inside a concrete masonry chamber, and supported on mild steel grating. The hopper bottom in the chamber has 45° slopes, and is used to drain sludge under hydrostatic pressure.

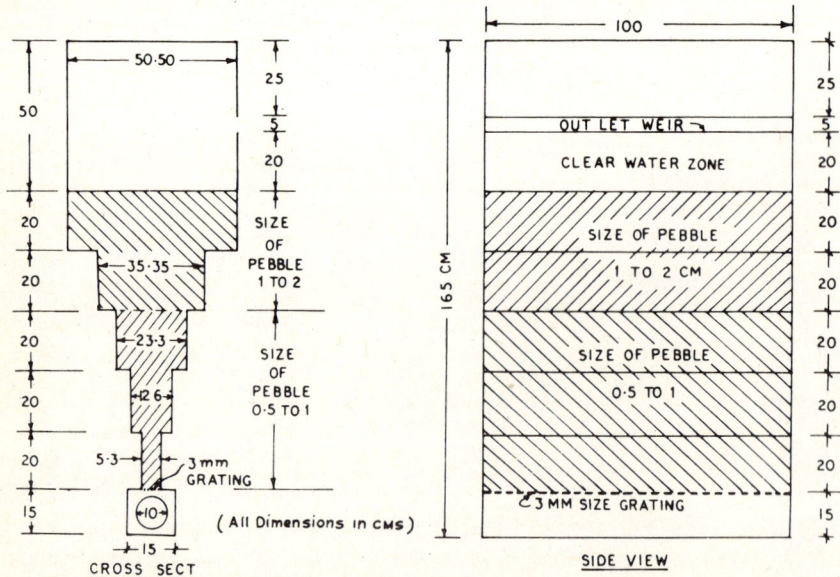

Figure 6.14. Upward-flow gravel bed flocculator. *Source:* Bhole, 1981.

The upward flow unit, shown schematically in Figure 6.14, combines two sizes of layered gravel (5 to 10 mm and 10 to 20 mm) with sections of increasing cross-sectional area to produce the desired tapering. The velocity gradients range from 1230 s^{-1} at the inlet (where rapid mixing occurs), to 35 s^{-1} in the uppermost and largest section for a flow rate of 270 m^3/day (see Appendix B for corresponding hydraulic calculations). The terraced shape of the flocculator is formed from mild steel and is protected by corrosion-proof paint and supported by horizontal rods attached to an outer concrete chamber. This design has been used in package plants in India (see Chapter 9, under "Package Water Treatment Plants").

Flocculation time can be reduced considerably by using gravel beds, because the entire bed is effective in the formation of sizable flocs and there is very little short-circuiting. Three to 5 min flocculation in the gravel bed is equivalent to 15

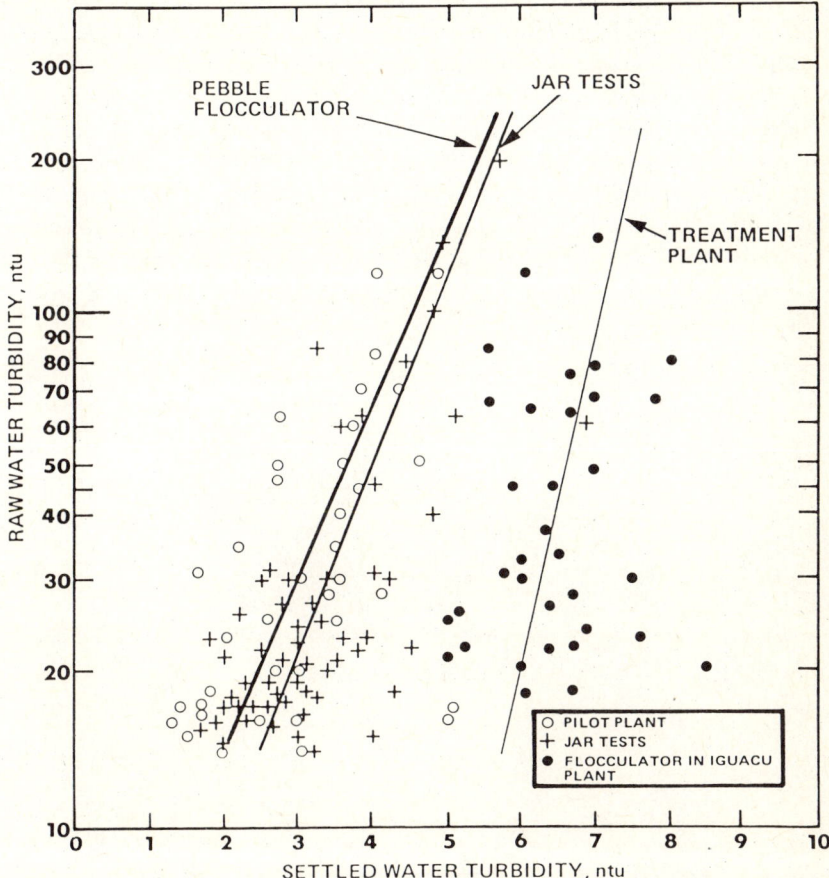

Figure 6.15. Comparison of results of gravel bed (pebble) flocculation in the pilot plant with results of jar tests with the full-scale plant flocculator at the Iguacu plant in Curitiba, Brazil. *Source:* Adapted from Richter, 1981.

min in jars under laboratory conditions, and to 25 min in noncompartmented plant flocculation basins, as revealed in Figure 6.15 (Wagner, 1982a; Richter and deBarro Moreira, 1981). Depth of the gravel bed generally varies from 1.5 to 3 m. Flocculated water may be conveyed from the flocculation chamber to the settling tanks via submerged perforated pipes or channels. The sedimentation step is often omitted in small plants, and the flocculated water is applied directly to the filter, as in direct filtration.

The main problem with gravel-bed flocculators is likely to be one of fouling, either by intercepted flocs or biological growth in the gravel. Therefore, sludge collection and removal is an important consideration in the design of such units. For downward flow units, hopper bottoms, such as that depicted in Figure 6.13, drain the sludge by hydrostatic pressure. Upward flow units often rely on a perforated drainage pipe grid, located just above the top of the bed, for removing the sludge that is deposited on the surface of the gravel. Both types of flocculator units should include arrangements for draining the water from the flocculator chamber to waste, and backwashing capabilities to remove sludge settled within the bed.

Gravel-bed flocculators have proven to be simple, low-cost, and an effective method of flocculation for several small water treatment plants in India (Kardile, 1981), and they have been used recently in modular plants in Latin America (CEPIS, 1982). They have also been installed in low-cost package water treatment plants designed and manufactured in India (Bhole, 1981). Plant designs that employ gravel-bed flocculators are described in Chapter 8 (under "Upflow-Downflow Filtration") and Chapter 9 (under "Package Water Treatment Plants").

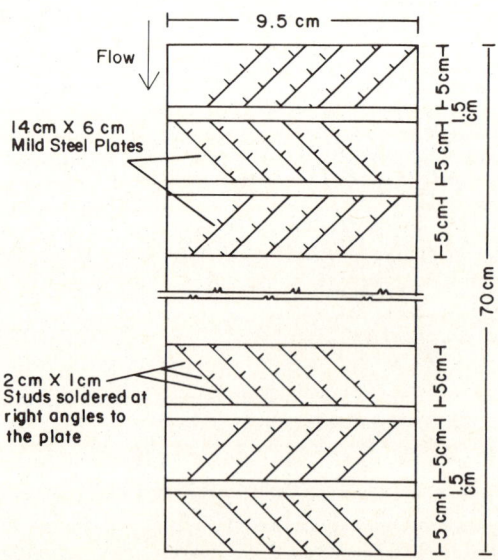

Figure 6.16.   Surface-contact flocculator—India. *Source:* Bhole, and Ughade, 1981b.

SURFACE-CONTACT FLOCCULATORS

Surface-contact flocculators have been studied experimentally in India (Bhole and Ughade, 1981) as a way to overcome the inherent problem of sludge choking in gravel-bed flocculators, which increases the head loss over time in such systems, so that periodic cleaning of the gravel bed is necessary.

Surface-contact flocculators consist mainly of studded plates placed in a zigzag form along the direction of flow, as shown in Figure 6.16. The experimental flocculator used in the Indian study was comprised of 55 mild steel plates, 14 X 6 cm in size, arranged in 11 rows of five plates each. These plates were fixed at 45° to a base plate in a zigzag fashion. Each plate was studded with 14 strips, each 2 X 1 cm in size.

The flocculator was tested in a continuous down-flow system, with velocities of flow ranging from 5 m/hr to 25 m/hr and turbidities ranging from 50 to 1600 NTU. The results showed surface-contact flocculators to be most effective for low turbidity waters and low rates of flow. The buildup of head loss was negligible, indicating that sludge choking was not a problem. It was concluded that this type of flocculator could be used effectively for surface waters containing about 100 NTU turbidity, with flow rates as high as 25 m/hr. Presently, no information is available on the effectiveness of these units under plant operating conditions.

chapter seven

SEDIMENTATION

The sedimentation process in water treatment provides for the settling and removal of heavier and larger suspended particles from water. Most commonly, it is used for removal of flocculated particles prior to filtration. The removal efficiency in the sedimentation basin determines the subsequent loadings on the filters and, accordingly, has a marked influence on their capacity, the length of filter runs, and the quality of the filtered water. The two major classifications for the design of sedimentation basins are (1) horizontal-flow units and (2) upflow units. The design of both types of units involves such factors as shape, number of basins, dimensions, velocity and direction of flow, detention time, volume of sludge storage, method of sludge removal, inlet and outlet arrangements, and the characteristics of the incoming flocculated water.

The horizontal-flow sedimentation basin has performed admirably in numerous water treatment plants in the United States and other parts of the world for decades, and is still advocated by water treatment experts because of its efficiency and inherent simplicity (Sanks, 1978; Hudson, 1981; Smethurst, 1979). In the United States, however, horizontal-flow units are almost matched in popularity by proprietary upflow clarifiers, such as solids-contact reactors and slurry-recirculation units, which combine the processes of mixing, flocculation, and sedimentation into a single unit. The advantages of upflow units are largely in cost: by combining the pretreatment processes that precede filtration, substantial savings can be realized in construction costs and manpower requirements. Upflow clarifiers perform quite well under suitable conditions and skilled supervision, provided their hydraulic capacity is not exceeded. When upflow clarifiers are overloaded, sludge escapes from the blanket in large volumes and clogs the filters, interfering with the entire treatment process. For developing countries, horizontal-flow tanks without mechanical sludge removal are much to be preferred, because they require no importation of equipment, and labor for cleaning the tanks is readily available. Equally important, when horizontal-flow tanks are overloaded, most of the settleable solids will still be removed, so that the filters can continue to operate normally. Overloading of plants is a chronic condition in developing countries because the financing of plant expansion rarely keeps up with demand.

The principles governing the design and construction of horizontal-flow sedimentation basins are well documented in standard texts (American Water Works Association, 1971; Cox, 1964; Fair, Geyer, and Okun, 1968; Hudson, 1981; IRC, 1981b; Sanks, 1978; Smethurst, 1979). The topics covered in this chapter on horizontal-flow sedimentation include design criteria, inlet and outlet arrangements, methods for sludge removal, and the application of tube and inclined plate settling. In addition, upflow-type clarifiers are presented briefly, because such designs may be appropriate in places where large horizontal-flow tanks are impractical.

HORIZONTAL-FLOW SEDIMENTATION

Horizontal-flow sedimentation is a gravity separation process in which a settling basin provides a quiescent environment that enables particles of specific gravity greater than water to settle to the bottom of the tank. A well-designed horizontal-flow sedimentation basin can remove up to 95% of raw water turbidity following effective coagulation and flocculation; the remaining turbidity is removed in the filters. Rectangular horizontal-flow clarifiers without mechanical sludge removal are advantageous for communities in developing countries because of their simplicity and ability to adapt to various raw water conditions, such as sudden changes in turbidity, flow increases, or too-high flow rates. Circular-shaped basins are not appropriate, and their main advantage over rectangular basins is where circular mechanical sludge removal equipment is to be used. In developing countries, manual sludge removal is preferred over the importation of mechanical sludge removal equipment, because the latter is difficult to maintain under the technical and climatic conditions prevalent in those countries, and it is more costly than employing laborers to clean the tanks manually. Many plants in the United States still use manually cleaned horizontal-flow rectangular basins.

A rectangular horizontal-flow sedimentation basin is shown in Figure 7.1. Flocculated water is distributed uniformly across the inlet zone through a diffusing system, such as a perforated inlet baffle. The water slowly traverses the length of the basin, depositing settled floc on the tank bottom, forming a sludge layer in a fashion outlined by the sludge profile in Figure 7.1. The clarified supernatant is collected by outlet weirs or submerged launders. The sloping floor of the bottom facilitates manual cleaning and drainage of sludge, usually by means of high pressure hoses or fixed nozzles on the basin floor.

There are several advantages of horizontal-flow units over upflow units.

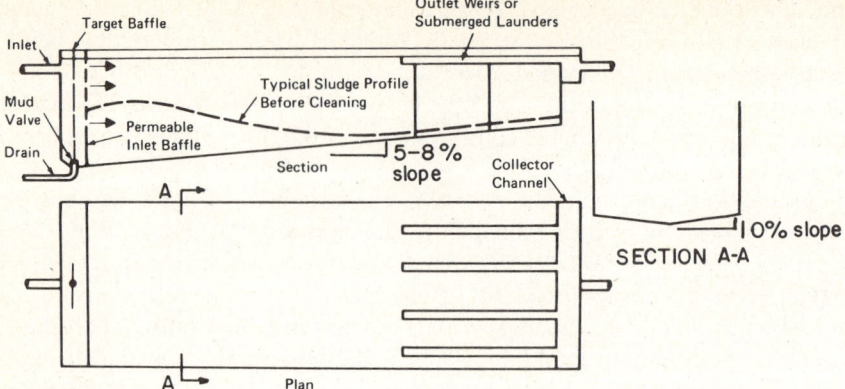

Figure 7.1.   Conventional horizontal-flow settling basin. *Source:* Adapted from Hudson, 1981.

1. The process is more tolerant of hydraulic and quality variations.
2. The process gives predictable performance under most operational and climatic conditions.
3. The process "scales-up" well.
4. The process works exceptionally well when silt loads are very high.
5. Construction costs are low, permitting oversizing.
6. Operation and maintenance are simple.

Horizontal-flow units may be more expensive to construct than upflow clarifiers in industrialized countries (because of lower surface loadings that require larger-sized tanks). Also, space may be limited. This is seldom the case in developing countries, where these units can be built quite cheaply using local materials such as concrete or masonry; also, lower-cost labor is available for construction and cleaning, and space is seldom a problem. Equipment for upflow units would need to be imported from foreign manufacturers.

Design Criteria

The design of sedimentation basins is governed by three basic criteria: (1) the quantity of water to be treated, (2) the selected detention period, and (3) the selected surface loading rate (or overflow rate). The surface loading rate is defined as the ratio between the influent flow rate and the surface area of the tank and can be expressed in units of flow rate per unit of basin surface area (e.g., $m^3/day/m^2$). This is equivalent to a velocity; hence some design books prefer to use settling velocity as a loading parameter (Smethurst, 1979; Hudson, 1981). The basic formulas that pertain to sedimentation basin design are:

$$t = 24 \ V/Q \tag{7.1}$$

$$v_s = 24 \ H/t \tag{7.2}$$

$$v_s = Q/WL \tag{7.3}$$

where

v_s = surface loading rate or settling velocity ($m^3/day/m^2$ = m/day)
t = detention time (hr)
Q = flow rate (m^3/day)
V = basin volume (m^3)
H, W, L = depth, width and length of basin (m)

These formulas can be used in conjunction with certain graphical methods (e.g., cumulative frequency distributions) to determine settling velocities for the settleable particles in the raw water. Settling data may be obtained by running preliminary bench scale experiments utilizing plastic cylinders equal in depth to the proposed basin, with draw-off points at different levels, and filled with test samples of the raw water. Samples are taken at regular time intervals to measure the turbidity at various depths in the cylinder, which is an indication of the rate of settling. If the period of settling is short, and a distinct separation forms between the upper clarified zone and the lower zone of settled solids, then flocculation may not be necessary. If, however, the period is relatively long, and the two zones are not well-defined, then it is likely that colloidal material is present, and flocculation is essential. The settling test should be repeated, using coagulated water, after jar tests have been run to determine the optimum dose of coagulants and, if necessary, coagulant aids. The settling test is covered in more detail by Camp (1946) and the IRC (1981b). Jar testing may also be used for determining settling data, as described in Appendix D.

An inherent assumption in the settling test is that the settling process is not hindered by density currents, eddies, temperature changes, or other conditions found in actual practice. Practical experience has shown that a discrepancy exists between the design values predicted by theoretical formulas and bench scale testing and the design values found most effective in practice. For example, the theoretical detention period is computed by dividing the volume of the basin by the rate of flow through it. However, short-circuiting caused by density currents and eddies at inlet and outlet structures, baffles, and the shape and dimensions of the basin, makes the observed "flowing-through period" for some of the particles of water in the basin shorter (and some longer) than the theoretical detention time. Because removals increase with detention time, the removals will be considerably less with short-circuiting than would occur if all the water particles were held for the theoretical detention period. Although most designs try to minimize short-circuiting in sedi-

Table 7.1.   Design Guidelines for Horizontal-Flow Settling Basins

Type	Description	Surface Loading Rate (Settling velocity) (m/day)	Detention Period (hr)
A	Small installations with precarious operation	20 to 30	3 to 4
B	Installations planned with new technologies[a] and reasonable operation	30 to 40	2½ to 3½
C	Installations planned with new technologies[a] and good operation	35 to 45	2 to 3
D	Large installations with new technologies[a] and excellent operation, with provisions for adding coagulant aids whenever necessary	40 to 60	1½ to 2½

Source: Adapted from Azevedo-Netto, 1977.
[a]Properly designed hydraulic rapid-mix and flocculation units.

mentation basins, it cannot be eliminated completely. Hence, the design of sedimentation basins rests largely on experience and should integrate the results from experimental settling tests with established guidelines that have proven successful in practice.

One such rule-of-thumb guideline has been suggested by Smethurst (1979). In bench scale settling tests, the time it takes for the average suspended solids of the water at all draw-off points above the sludge zone to fall to a concentration of 2 mg/l should be multiplied by factor of safety of 3 to obtain the detention time of the proposed settling basin.

Table 7.1 lists recommended surface loading rates (settling velocities) and detention times for various conditions likely to be encountered in practice. The values

Table 7.2.   Design Parameters for Horizontal-Flow Settling Basins in Brazil

Location of Water Treatment Plant	Capacity (m³/s)	Settling Velocity (m/day)	Detention Period (hr)
Guarau, Sao Paulo	33.0	40	3.0
Rio das Velhas, Belo Horizonte	9.0	49	2.1
Rio Descoberto, Brasilia	6.0	41	2.2
Campinas	2.1	34	3.0
T. Ramos Sao Paulo[a]	2.0	59	2.0

Source: Azevedo-Netto, 1977.
[a]Experimental operation.

Table 7.3. Performance of Horizontal-Flow Settling Basins in Colombia (1959)

Location of Water Treatment Plant	Detention Period (hr)		Settling Velocity (m/day)		Temperature (°C)		Influent Turbidity (NTU)		Effluent Turbidity (NTU)	
	Max.	Min.	Max.	Min.	Max.	Min.	Max.	Min.	Max.	Min.
Cali	5.05	3.92	22.2	17.3	18	2	600	5	8	2
Pasto	3.96	3.1	27.2	21.3	14	14	120	5	4	0.9
Pereira	2.66	2.13	35.4	28.2	18.5	18.5	130	7	9	4
Santa Marta	3.31	1.83	50.2	27.6	28	28	4690	59	6	2

Source: Adapted from Arboleda, 1973.

vary from 20 to 60 m/day, depending on the type of unit and the pretreatment afforded. Table 7.1 also reveals that effective mixing and flocculation prior to sedimentation (condition D) can substantially reduce required detention times and increase surface loadings, thereby enabling the design of smaller, less costly settling basins. Smethurst (1983) recommends somewhat more conservative overflow rates of 9 to 24 m/day without coagulant aids and 18 to 36 m/day with coagulant aids.

Design parameters for several water treatment plants in Latin America are tabulated in Tables 7.2 and 7.3, and include specifically plant capacity, surface loadings, and detention times for settling basins. In addition, Table 7.3 lists plant operating temperature ranges and settling efficiency—the latter obtainable by comparing raw water versus settled water turbidity.

The remaining design criteria are concerned primarily with mitigating the problems of turbulence, short-circuiting, and bottom scour (i.e., disturbance of the sludge layer). A basin depth of 3 m is recommended to allow for sludge deposits and storage in the basin. The relationship among basin depth, detention time, and surface loadings are revealed in the graph of Figure 7.2. For example, a detention time of 3 hr and an overflow rate of 30 m/day would fix the clarifier depth at 3.75 m. To reduce the likelihood of short-circuiting, a length to width ratio (L/W) of 3 or more is recommended. Horizontal-flow velocities are fixed by these constraints and should range from 4 to 36 m/hr. It should be noted that these velocities are not uniform across the basin cross-section, due to the influence of drag from the floor and walls of the basin. Basin drag is also a contributing factor in the formation of density currents (Arboleda, 1973; Hudson, 1981; Sanks, 1978). The number of basins that should be selected for a particular plant is influenced by (1) the effect upon the production of water if one basin is removed from service, and (2) the largest size that can be expected to produce satisfactory results. A treatment plant

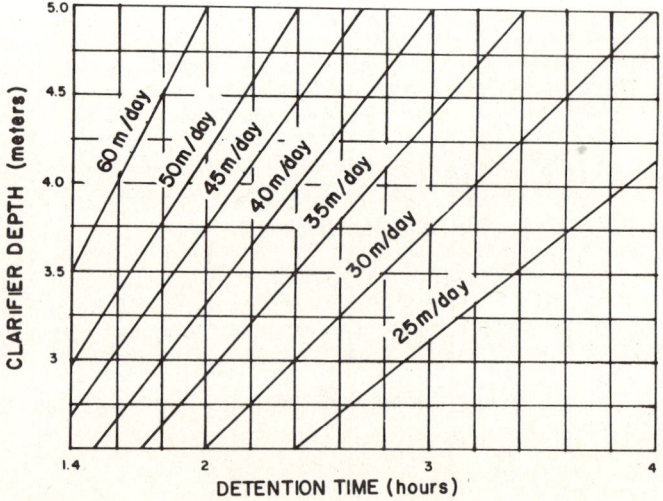

Figure 7.2. Detention times for different clarifier depths and overflow rates. *Source:* Arboleda, 1973.

should contain a minimum of two basins, and more where feasible, to reduce the effects of higher velocity and shorter detention period when one basin is removed from service for cleaning.

Inlet Arrangement

Inlet arrangements should be so designed that the flocculated water entering the basin is distributed uniformly across the full cross-sectional area of the inlet zone without causing excessive turbulence, which would break up the floc. It is important to attain good distribution of flows among parallel basins. The proper sizing of ports and manifolds is much to be preferred over regulating devices.

An efficient type of inlet arrangement employs a perforated baffle in a diffusion wall following target baffles, as shown in Figure 7.3. Hudson (1981) suggests velocities through the perforated baffle of about 0.2 to 0.3 m/s. The head loss through the ports is estimated to be 1.7 times the velocity head ($v^2/2g$). The perforated baffle wall is usually constructed with concrete, but timber baffles or brick masonry may also be used. Four basic requirements should be strived for in the design of perforated baffles for settling basins (Hudson, 1981).

1. The velocity through the ports should be about four times higher than any approaching velocities in order to equalize flow distribution both horizontally and vertically.

2. To avoid breaking up floc, the velocity gradient through inlet conduits and ports should be held down to a value close to or a little higher than that in the last portion of the flocculator.

3. The maximum feasible number of ports should be provided in order to minimize the length of the turbulent entry zone produced by the diffusion of the submerged jets from the ports in the perforated baffle inlet.

4. The port configuration should be such as to assure that the discharge jets will direct the flow toward the basin outlet.

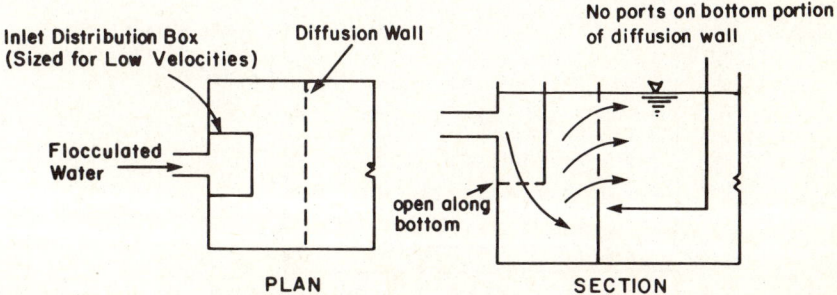

Figure 7.3. Inlet arrangement consisting of a flow-distribution box, followed by a diffusion wall. *Source:* Hudson.

Figure 7.4. Timber diffusion wall at the Guandu plant in Rio de Janeiro, Brazil. *Source:* Hudson.

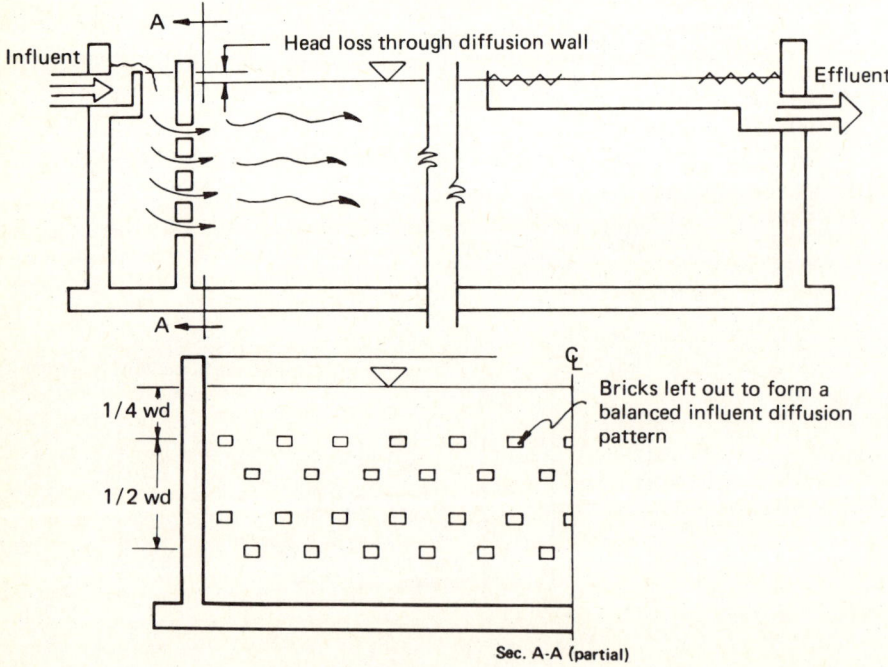

Figure 7.5. Checkerwork influent diffusion wall. *Source:* Sanks, 1979.

In practice, one can rarely meet all four basic requirements because they conflict with one another; thus, a reasonable compromise must be attained.

To ensure proper dimensioning of the ports for timber baffles, tubular inserts made of plastic or of wood construction can be fastened to the openings of the timber wall. The latter type of insert is shown in Figure 7.4, which is a photograph of a timber diffusion wall for the Guandu plant in Rio de Janeiro, Brazil. For masonry walls, a checkered configuration may be constructed by intentionally leaving bricks out of the wall at certain spacings (Figure 7.5). After plant start-up, if necessary, loose half bricks can be added to improve the distribution of the incoming water.

Outlet Arrangement

Weirs or perforated launders are the most common structures for withdrawing the effluent water from the basin. Weir lengths should be selected to prevent high velocities of approach and disturbance of the sludge layer. The following formula is useful in determining an acceptable weir length (adapted from IRC, 1981b):

$$L = 0.2 \, Q/H \, v_s \tag{7.4}$$

where

L = combined weir length (m)
Q = flow rate (m³/day)
H = depth of tank (m)
v_s = settling velocity (m/day)

Burlingame, however, points out that the importance of outlet weir length has been overemphasized. Simple end or circumferential weirs or submerged orifices (see Figure 3.4) are apt to serve as well as longer and more elaborate configurations. The outlet weirs or launders may be arranged either parallel to or transverse to the direction of flow in the basin. A center-to-center distance of one to two times the depth of the tank is reasonable for outlet conduit channel spacing.

Adjustable V-notched weirs are convenient for ensuring uniform flow throughout the collecting trough, especially when low overflow rates are used. They are constructed from metal strips containing V-notches about 5.0 cm deep and 15 to 30 cm apart, which are fastened with bolts to the concrete wall of the collecting trough (Figure 7.6). However, overflow weirs must be leveled accurately, which may be difficult in places where skilled plant personnel are not available.

Perforated launders, on the other hand, have ports submerged 30 to 90 cm below the surface, and hence do not require precise leveling. Submerged launders are useful in preventing floating debris from entering the outlet conduits and can readily handle changes in water levels in the basin. Storage in the settling basin is often used to permit some temporary differences between inflow to the plant and the discharge from the plant. This cannot be done when overflow weirs are used in

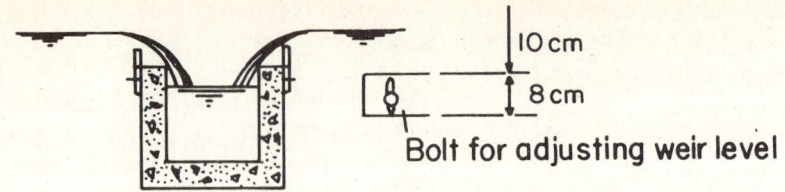

10 cm

8 cm

Bolt for adjusting weir level

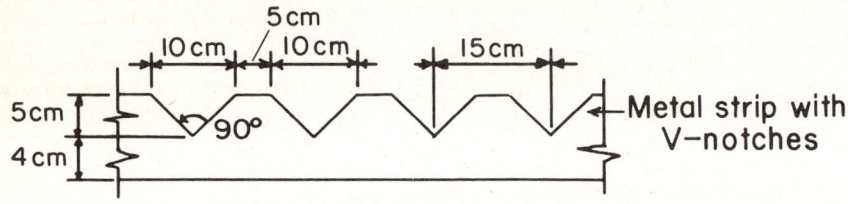

5 cm

10 cm

10 cm

15 cm

5 cm

4 cm

90°

Metal strip with V-notches

Figure 7.6. Adjustable V-notch weirs attached to effluent launders. *Source:* Azevedo-Netto, 1977.

Figure 7.7. Perforated effluent launders at a treatment plant in Latin America. *Source:* Okun.

the basin outlet. Perforated launders may be tapered to prevent velocities from increasing too much along them. A perforated launder for a settling basin at a plant in Latin America is shown in Figure 7.7.

Manual Sludge Removal

The sludge collection and removal mechanisms that are commonly employed in horizontal-flow sedimentation basins in industrialized countries, such as chain and sprocket scrapers or vacuum-type systems, are not practical in most developing countries. Manual rather than mechanical sludge removal is preferred because it does not require imported equipment nor spare parts, and the labor required is low-cost and abundant. Although manual sludge removal requires the periodic shutdown of a basin, this should not pose a problem when two or more basins are available.

For sedimentation basins that are manually cleaned, a major portion of the volume is reserved for sludge accumulation between cleanings. For plants having good mixing and flocculation, the sludge layer will tend to be deeper at the inlet end than the outlet end (see Figure 7.1). Hence, a sludge storage depth tapering from 2 m at the inlet end to about 0.3 m at the outlet end is desirable (Hudson, 1981). When the sludge layer exceeds the basin's storage limitations, the settling basin should be removed from service and cleaned. To facilitate efficient drainage of the basin, the floor should slope about 10% from the side walls to the centerline, and 5 to 8% from the outlet end to the inlet end.

The removal of sludge is accomplished expeditiously by first draining the basin by opening one or more mud valves located at the inlet end. Afterward, the sludge

Figure 7.8. Manually cleaned settling basin with fixed nozzles on the floor bottom—Latin America. *Source:* Okun.

remaining in the basin can be flushed to drainage with the help of either high-pressure hoses or fixed nozzles attached to the basin floor; the latter type is shown in Figure 7.8. A fixed nozzle arrangement for a sedimentation basin at Grand Rapids, Michigan, called for pressure flushing for about 1/2 hour, through 1/8 in. holes in 2 1/2 in. pipes laid on the floor of the basin, prior to basin draining (Flinn, Weston, and Bogert, 1927). If water pressure is inadequate, the flow from the adjacent basins may be used to flush the basin being cleaned by partly opening the effluent valves or gates. The entire cleaning operation, if done by plant personnel or laborers familiar with the procedure, should take no longer than 12 to 18 hr. The frequency of sludge removal varies considerably depending on basin capacity, the turbidity of the raw water, and tendencies toward septicity with resulting tastes and odors or sludge flotation problems. A reasonable estimate for most plants, though, is about once every three to four months.

The installation of inclined-plate or tube settlers in sedimentation basins permits much higher surface loading rates, and hence may result in higher amounts of sludge generated in a small space when treating water of similar turbidity. Consequently, units that employ inclined-plate or tube settlers require frequent sludge removal beneath the settling modules. In such situations, hydraulically drained hoppers can be used to avoid having to drain the tank at relatively short intervals. The floors of the hoppers must be sloped no less that 55° above the horizontal, because sludge usually will not move down flatter slopes under water.

A hopper-bottomed horizontal flow settling tank that has been used in several treatment plants in North Carolina (Piatt and Associates) is shown in Figure 7.9. The design employs conventional horizontal-flow sedimentation, manual sludge removal by hydraulic discharge from hopper bottoms, and over-and-under baffles that serve to create a sludge-blanket effect above the first hopper and promote better settling in the remaining portion of the basin. The three hopper bottoms and two baffles are located at the inlet side of the basin where most of the floc is likely to settle out. With this type of design, the horizontal-flow portion of the basin needs to be manually cleaned only about once every six months.

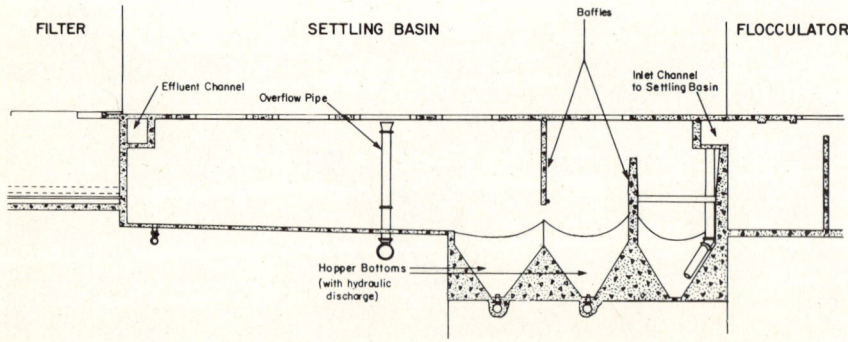

Figure 7.9.   Hopper-bottom settling basin with over-and-under baffles—North Carolina, USA. *Source:* Adapted from construction drawings by Piatt & Assoc. Consulting Engineers, Raleigh, NC.

INCLINED-PLATE AND TUBE SETTLING

Inclined-plate and tube settling have become important components in water treatment in recent years. When installed in either upflow sludge blanket-type clarifiers or horizontal-flow basins, these units can improve clarifier performance and increase the capacity of conventional clarifiers by 50 to 150%. Furthermore, they may also be incorporated into the design of new sedimentation basins, reducing the settling area to one-fourth to one-sixth of that required by conventional basins. A number of such installations in the United States, Europe, and Latin America utilize commercially fabricated modules, and the technology for their design and construction is fairly well-developed (Culp and Culp, 1974).

Yao (1973) has concluded that the efficiency of inclined-tube and plate settlers exceeds that of conventional clarifiers under similar loadings. The design of an inclined-plate settler unit for a small package water treatment plant using Yao's approach was presented by Bhole (1981). Rao and Paramasivam (1980) summarized the existing knowledge available on tube and plate settlers to determine their applicability for developing countries. Other authors have also addressed the subject (Arboleda, 1973; Hudson, 1981; Sanks, 1978; Smethurst, 1979).

For developing countries, the use of inclined-plate or tube settlers should be limited, in most cases, to expanding settling basin capacity and/or improving plant effluent quality. Also, their use may be limited in hot, sunny climates where algae growth on tubes and plates can be a troublesome maintenance problem. The incorporation of settlers in the design of new plants in developing countries to reduce basin size and cost is usually not justified, because land is generally not restricted and low-cost labor and materials are available for construction of the settling tanks. Moreover, when conventional settling tanks are installed during initial plant construction, the option remains for installing inclined-plate or tube settlers in the future, when the plant undergoes a capacity expansion; at little additional cost. If this is not done (if tube or plate settlers are installed initially), then the next plant expansion is likely to require the construction of a new settling tank. There may be certain situations, of course, where land and/or cost are the overriding constraints in design; under these circumstances they should be considered.

The following equations, which are derived from geometrical considerations and laboratory performance studies, can be used for designing inclined-plate and tube settlers (Arboleda, 1983). A nomenclature diagram is presented in Figure 7.10.

$$v_{sc} = \frac{k\,v_o}{\sin\theta + L_u \cos\theta} \tag{7.5}$$

$$L_R = L/e \tag{7.6}$$

$$L_u = L_R - 0.013\,R_N \tag{7.7}$$

where

v_{sc} = critical surface loading rate, or settling velocity (m/day)

k = efficiency factor (1.0 for inclined plates; 1.33 for circular tubes; hence inclined-plate settlers are inherently more efficient than tube modules)

v_o = surface loading rate for area of high-rate settling (m/day)
θ = angle of inclination of settler
L_R = relative depth
L = length of plates, or tubes (m)
e = perpendicular distance between plates, or diameter of tube (m)
L_u = effective relative depth*
R_N = Reynolds No. (~280)

Equation 7.5 can be simplified as follows (ignoring the settler efficiency factor k):

$$v_{sc} = \frac{Q}{A_o f} = \frac{Q}{A} \tag{7.8}$$

where

f = $\sin \theta - L_u \cos \theta$
A = surface area of conventional horizontal settling tank (m^2)
A_o = surface area of high rate settling (m^2)

Hence, for a unit flow Q:

$$A_o = \frac{A}{f}$$

The f factor then becomes the number of times the area of a horizontal settling tank must be divided to obtain the area of an inclined-plate or tube settler. For example, for an effective relative depth L_u = 20, and angle θ = 60; the area factor f = 10.9.

Typical hydraulic calculations for the design of inclined-plate settlers and tube modules in horizontal-flow basins are presented in Appendix B.

Recommended surface loading rates for horizontal-flow sedimentation basins equipped with inclined-plate or tube settlers are listed in Table 7.4 for two categories of raw water turbidity; 0 to 100 NTU and 100 to 1000 NTU. These loadings apply specifically to warm water areas (temperature nearly always above 10°C) and apply to most developing countries. For efficient self-cleaning, tubes or inclined plates are usually arranged at an angle of 40 to 60° to the horizontal. The most suitable angle for a particular design depends on the sludge characteristics of the water being treated, usually 55° above the horizontal. The distance between parallel-inclined plates or, similarly, the diameter of settling tubes, is about 5 cm. The passageways

---

*This expression takes into account the transitions from turbulent to laminar flow in the settler unit. Only after a distance L_u (measured from the inlet edge of the plate or tube) can full laminar flow be considered. The transitional zone must be discarded in design calculations (according to Yao, 1973), due to possible turbulence which may hinder the sedimentation of particles. For safe design, a value higher than R_N = 280 should not be used for plate settlers; this value is easily attainable for normal flow rates (Arboleda, 1983).

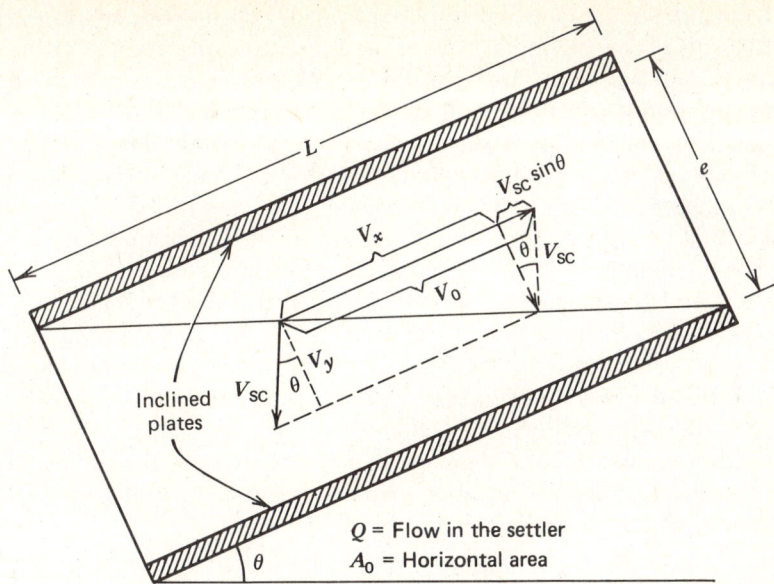

Figure 7.10. Geometrical relationships of an inclined-plate settler. *Source:* Adapted from Arboleda, 1983.

formed by the plates, or inside the tubes, are commonly about one meter long. Care needs to be exercised in the design of outlet collection systems to assure even distribution of flow through plate or tube modules. This is more easily done with overflow weir outlets than with submerged launders.

Local materials and labor may be used for the construction of inclined-plate or tube settlers. For inclined-plate settlers, the individual trays can be fabricated from

Table 7.4. Loading for Horizontal-Flow Settling Basins Equipped with Inclined-Plate or Tube Settlers in Warm-Water Areas (above 10°C)

Settling Velocity Based on Total Clarifier Area (m/day)	Settling Velocity Based on Portion Covered by Plates or Tubes (m/day)	Probable Effluent Turbidity (NTU)
(A) Raw Water Turbidity 0 to 100 NTU		
120	140	1 to 3
120	170	1 to 5
120	230	3 to 7
170	200	1 to 5
170	230	3 to 7
(B) Raw Water Turbidity 100 to 1000 NTU		
120	140	1 to 5
120	170	3 to 7

Source: Culp and Culp, 1974.

polyethylene (or similar type of plastic) or of wood. Asbestos-cement plates should be coated with plastic or similar type of protective covering because of their susceptibility to corrosion from alum-treated water. Where wood is used on low slopes, trays are commonly 30 cm apart. It may also be necessary to drain the tank for cleaning occasionally, because sludge does not readily slide down wooden trays while the basin is in service. Tube settlers, on the other hand, are easily fabricated from PVC pipes (3 to 5 cm internal diameter), which are packed closely together to form a module. In countries with indigenous plastics industries, commercially available tube modules that are prefabricated at the factory are suitable for larger installations. A Brazilian-built tube module is shown in Figure 7.11.

When installing inclined-plate or tube settlers in horizontal-flow sedimentation basins, it is advisable not to locate them near the inlet zone where turbulence could inhibit the effectiveness of the settlers. Furthermore, it is sometimes necessary to supplement the existing effluent collection system with additional weirs or launders so it can carry the increased loading and to allow for additional head loss through the influent flume. Figure 7.12 shows a typical tube module installation in a conventional sedimentation basin.

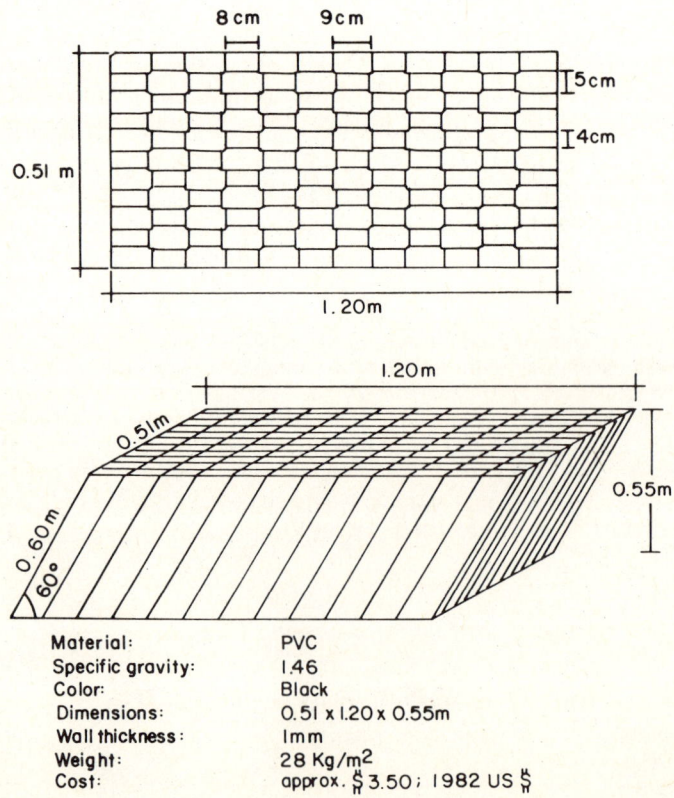

Material:	PVC
Specific gravity:	1.46
Color:	Black
Dimensions:	0.51 x 1.20 x 0.55m
Wall thickness:	1mm
Weight:	28 Kg/m^2
Cost:	approx. $3.50; 1982 US$

Figure 7.11.   Plastic-tube module fabricated in Brazil. *Source: Azevedo-Netto, 1977.*

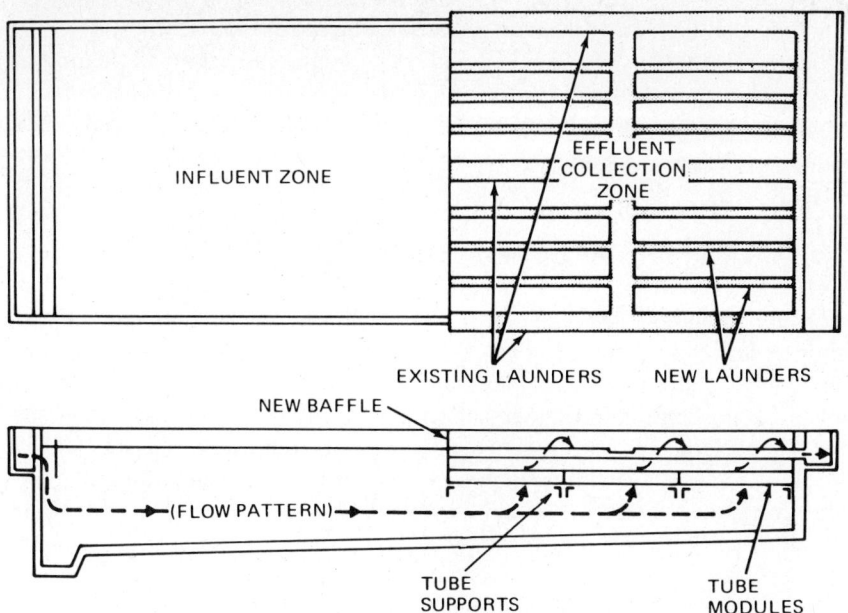

Figure 7.12. Typical tube-settler installation in a rectangular basin. *Source:* Culp and Culp, 1970.

Figure 7.13. Inclined-plate settlers with perforated plastic pipe outlet system at a plant in Zarzal, Colombia. *Source:* H. Lora.

Inclined-plate settlers are now in general use in Latin America; they are designed with settling velocities between 120 and 185 m/day, and reportedly can achieve removal efficiencies greater than 90% (Arboleda, 1983). Figure 7.13 shows inclined-plate settlers made from asbestos-cement sheets installed at a treatment plant in Zarzal, Colombia.

UPFLOW SEDIMENTATION

Combining sludge-blanket type clarifiers with flocculation may be an appropriate technology for larger plants in urban centers in more advanced developing countries. These clarifiers have no moving parts and, except for a few valves, require no mechanical equipment. They may be appropriate under the following conditions: (1) relatively constant raw water quality with turbidity not exceeding 900 NTU, (2) plants that are designed with enough excess capacity so that the unit processes will not be overloaded, and (3) availability of skilled supervision. Also, the compact nature of upflow clarifiers may make them attractive for package plants or modular-type designs (see Chapter 9), or where land is not available to build larger horizontal-flow basins. Smethurst (1979) recommends overflow rates ranging from 36 m/day for upflow basins with normal coagulation up to about 100 m/day, when coagulant aids are used.

A typical upflow sludge-blanket clarifier is shown in Figure 7.14. Flocculation takes place at the inlet at the bottom of the unit, where the upflow velocity is greatest. The upflow velocity decreases with increased cross-sectional area of the tank. The units are square, with effluent weirs along the outside. Sludge can be

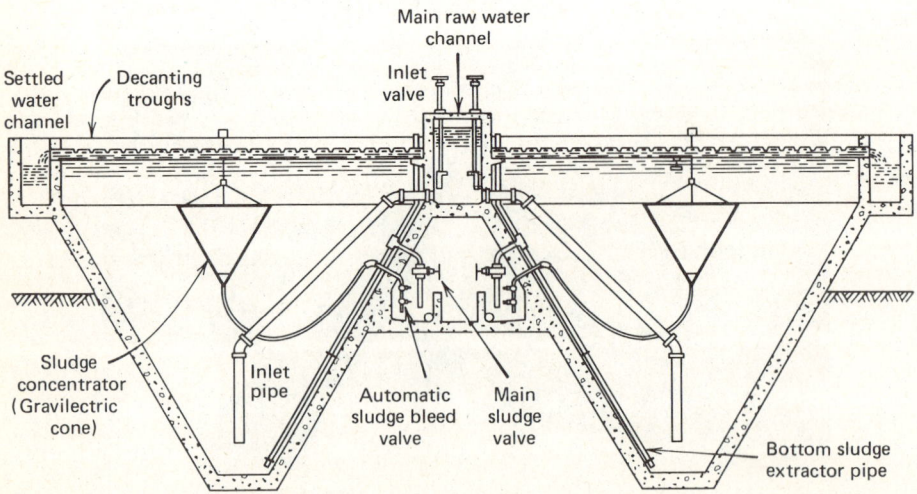

Figure 7.14. Upflow sludge-blanket clarifier with hopper bottoms. *Source:* Smethurst, 1979.

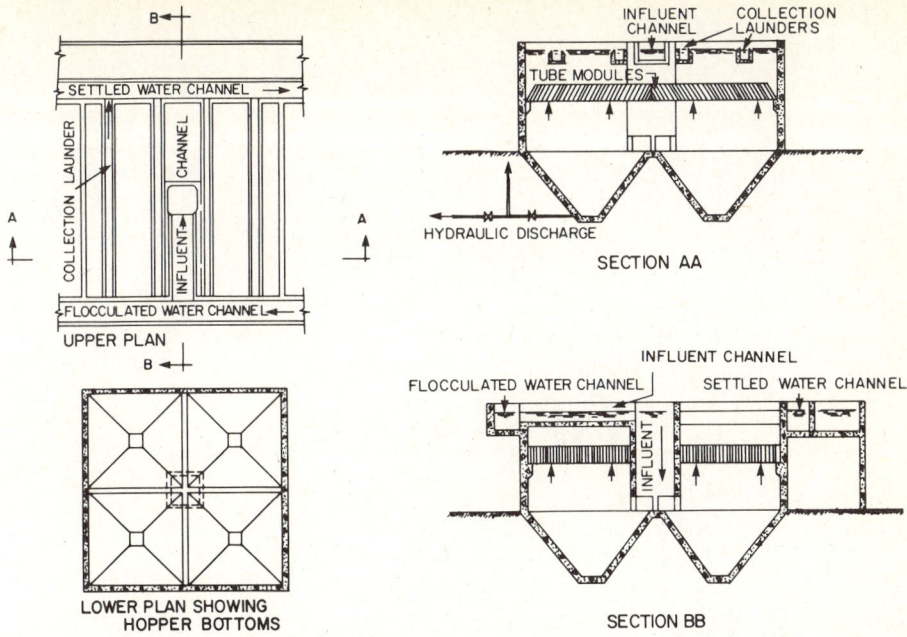

Figure 7.15. Concrete upflow clarifier with tube modules constructed in Brazil. *Source:* Azevedo-Netto, 1977.

bled off continuously or intermittently by taking the unit out of service to allow quiescent settling, in which case sludges of higher concentration are produced.

Such a plant was built in Chittagong, Bangladesh, of 76000 m³/day capacity, incorporating 24 units with a detention time of 1.2 hr and an overflow rate of 70 m/day, primarily because " . . . they are simple to operate, can be operated manually, and do not incorporate a lot of imported equipment requiring extensive maintenance" (Burlingame). Tube settlers can be used in upflow units as shown in Figure 7.15.

Despite the availability of many proprietary designs for upflow clarifiers, units designed and built locally are much to be preferred, because they will cost less and will incorporate local hardware, such as valves, thereby minimizing dependence on spare parts.

Upflow sedimentation is appropriate for modular treatment plant designs and package plants, mainly because of the relatively small area required, especially if inclined-plate or tube settlers are also used. Typical designs that employ upflow sedimentation are shown in Chapter Eight, under "Upflow-Downflow Filtration", and in Chapter Nine.

chapter eight

FILTRATION

Filtration is a physical, chemical, and (in some instances) biological process for separating suspended impurities from water by passage through porous media. Two general types of filters are commonly used in water treatment: the slow sand filter and the rapid filter. A slow sand filter consists of a layer of ungraded, fine sand through which water is filtered at a low rate; the filter is cleaned by periodically scraping a thin layer of dirty sand from the surface at intervals of several weeks to months. The sand is washed and after several scrapings, returned to the filter. The low rate of filtration allows the formation of an active layer of microorganisms called the *schmutzdecke* on top of the sand bed, which provides biological treatment. This layer is particularly effective in the removal of microorganisms, including pathogens, from water. A rapid filter, on the other hand, consists of a layer of graded sand, or in some instances, a layer of coarser filter media (e.g., anthracite) placed on top of a layer of sand, through which water is filtered at much higher rates; the filter is cleaned by backwashing with water.

Because of the higher filtration rates, the space requirement for a rapid filtration plant is about 20% of that required for slow sand filters; although the latter usually do not require pretreatment steps (i.e., chemical treatment, rapid mixing, flocculation, and sedimentation). Table 8.1 summarizes the design criteria, and Figure 8.1 shows simplified drawings for each type of filter. Another type of filtration scheme that has definite applications in developing countries, utilizes upflow filtration followed by a downflow bed, and can be an economical alternative to conventional flocculation-sedimentation-filtration schemes.

Rapid filtration plants are ubiquitous in the United States and most industrialized countries, although some countries (e.g. England and Switzerland) still use slow sand filters. Modern rapid filters are generally the most complex and costly structures

146

to construct, operate, and maintain in water treatment plants; they are often fully automated to reduce labor costs. Filter operation is generally controlled from an operating table located directly in front of the filter (as shown in Figure 8.2), which is often automatically operated, with pushbutton standby, and the option for operation from a central control room. Such equipment can shut off a filter at a predetermined head loss, backwash the filter, and put it back into service, all automatically, although provision is made for an operator to override the equipment.

In developing countries such labor-saving automation is neither necessary nor desirable. Simple rapid filter designs that employ manual controls, such as the hand-operated valve shown in Figure 8.3, and that eliminate, to the extent possible, unessential mechanical equipment and instrumentation are much to be preferred. Also, when conditions are suitable, slow sand filters or upflow-downflow type filters can provide simple solutions at low cost.

This chapter emphasizes the design and operation of simple types of rapid filters, including the upflow-downflow types. Direct filtration, which eliminates conventional coagulation and sedimentation, is also presented. Slow sand filtration, which may be the most practical technology for treating some surface water supplies in developing countries, is examined to a lesser extent, because other manuals and publications on this subject are widely available.

RAPID FILTRATION

Rapid filters can be classified in various ways. They may be classified according to (1) the type of filter media employed, (2) the type of filter rate-control system employed, (3) the direction of flow through the bed, or (4) whether they operate under gravity (free-surface) or pressure. In general, pressure filters are not well-suited for developing countries because they generally need to be imported. Furthermore, they require skilled operation and high-quality maintenance because the filter media cannot be monitored.

Constant-rate filtration and declining-rate filtration are two basic types of control systems. Constant-rate filters are equipped with a rate-control device in the influent or, preferably, the effluent line, the latter providing an adjustable resistance to the water flow, and compensating for the increasing head loss in the filter media. On the other hand, declining-rate filters do not use any rate controllers, allowing the rate to decline in each unit as the head loss increases. The design of both systems is discussed in this chapter.

The design variables for rapid filtration include (1) filter media, (2) filter bottoms and underdrains, (3) backwashing arrangements, (4) auxiliary scour wash systems, and (5) rate-control systems. These are described below. Interfilter-washing filtration units and direct filtration are also reviewed for their suitability in developing countries. Additional, more detailed information on the design of conventional rapid filters may be found in several standard references (American Water Works Association, 1971; Arboleda, 1973; Fair, Geyer, and Okun, 1968; Hudson, 1981; Sanks, 1979).

Table 8.1. General Features of Construction and Operation of Conventional Slow and Rapid Sand Filters

	Slow Sand Filters	Rapid Sand Filters
Rate of filtration	0.1 to 0.2 to 0.4 m/hr	4 to 5 to 21 m/hr
Size of bed	Large, 2000 m^2	Small, 40 to 400 m^2
Depth of bed	30 cm of gravel, 90 to 110 cm of sand, usually reduced to no less than 50–80 cm by scraping	30 to 45 cm of gravel, 60 to 70 cm of sand; not reduced by washing
Size of sand	Effective size 0.25 to 0.3 mm; uniformity coefficient 2 to 2.5 to 3	Effective sizes 0.55 mm and higher; uniformity coefficient 1.5 and lower, depending on underdrainage system.
Grain size distribution of sand in filter	Unstratified	Stratified with smallest or lightest grains at top and coarsest or heaviest at bottom.
Underdrainage system	1. Split tile laterals laid in coarse stone and discharging into tile or concrete main drains. 2. Perforated pipe laterals discharging into pipe mains.	1. Perforated pipe laterals discharging into pipe mains; 2. False floor type, with orifices 3. Many others, generally proprietary

Loss of head	6 cm initial to 120 cm final	30 cm initial to 240 or 275 cm final
Length of run between cleanings	20 to 30 to 60 days	12 to 24 to 72 hr
Penetration of suspended matter	Superficial	Deep, particularly with dual or mixed media
Method of cleaning	Scraping off surface layer of sand and washing and storing cleaned sand for periodic resanding of bed.	Dislodging and removing suspended matter by upward flow or backwashing which fluidizes the bed. Possible use of auxiliary scour systems.
Amount of water used in cleaning sand	0.2 to 0.6% of water filtered	1 to 4 to 6% of water filtered
Preparatory treatment of water	Generally none when raw water turbidity <50 NTU	Coagulation, flocculation, and sedimentation
Supplementary treatment of water	Chlorination	Chlorination
Cost of construction	Relatively low	Relatively high
Cost of operation	Relatively low where sand is cleaned in place	Relatively high
Depreciation cost	Relatively low	Relatively high

Source: Adapted from Fair, Geyer, and Okun, 1968.

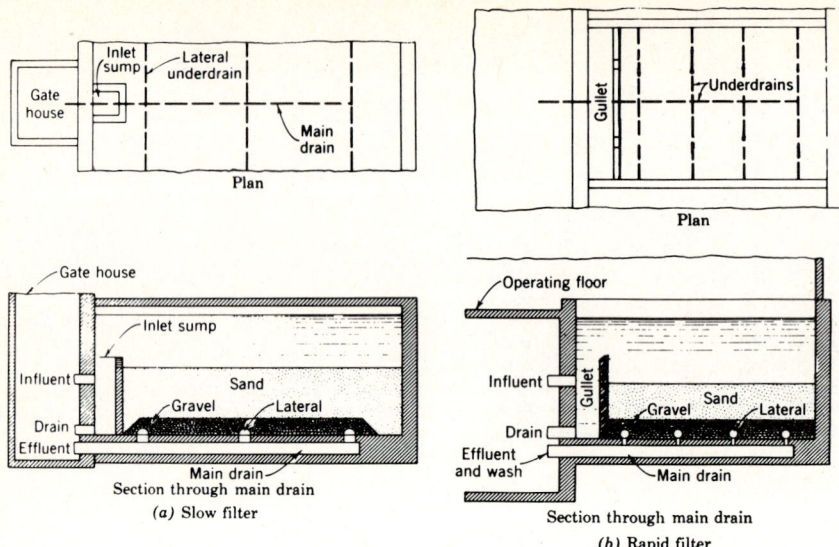

Figure 8.1. Simplified drawings of slow and rapid filters. *Source:* Adapted from Fair, Geyer, and Okun, 1968.

Figure 8.2. "Labor-saving" filter operating table at a large water treatment plant in Asia. *Source:* Okun.

Figure 8.3.   Hand-operated valve for washing a filter at a plant in India. *Source:* Okun.

Dual-media Filters

Sand has been used traditionally as the filter medium in water treatment plants because of its wide availability, low cost, and the satisfactory results that it has given. Sand filters remain the predominant method of filtration in developing countries. However, the grading of the sand that occurs in backwashing of rapid filters, leaving the finer sand on top, restricts the capacity of conventional rapid sand filters. The floc particles removed in filtration concentrate in the topmost layers of the filter and most of the depth of the filter is unused. This led to the introduction of dual- and multi-media filters, where lighter media of larger size occupy the upper layers of the filter, allowing greater penetration of the floc. Floc that escapes the upper layer is caught in the finer sand at the bottom of the filter.

Sand is still the medium of choice for filters in developing countries. Because they are so widely used throughout the world, and because information concerning conventional rapid sand filters is widely disseminated, primary attention in this volume is given to modifications in conventional practice that are appropriate to developing countries. Alternative indigenous media are discussed in their role as the coarse medium in dual-media filters and as the filter medium in single-media filters where suitable sand is not available.

Dual-media beds have gradually replaced sand alone in rapid filters in industrialized countries over the past 20 years (Culp and Culp, 1974). The dual-media filter is generally composed of a coarse coal upper layer (specific gravity of 1.45 to

1.55; effective size of 1.0 to 1.6 mm) on top of a lower sand layer (specific gravity of 2.65; effective size of 0.45 to 0.8 mm). Because of the different specific gravities of the two materials, the two layers retain their relative positions after backwashing, although the coarse material mixes with sand near the interface in "mixed media" filters.

Dual-media filters possess several distinct advantages over conventional sand filters: (1) higher filtration rates (10 to 15 m/hr) than for conventional filters, resulting in a reduction in the total filter area and cost for a given design capacity; (2) longer filter runs at any given loading; and (3) the capacity of existing sand filters can be easily increased at low cost, by their conversion to dual-media beds.

This latter advantage may be exceedingly beneficial to those communities in developing countries that are burdened with overloaded and inefficient treatment plants. Dual-media beds can be incorporated into an existing filtration system without changes in plant structure or method of operation if the hydraulic capacity of the influent and effluent piping is adequate. Also, a wide variety of unconventional filter media, such as crushed coconut shells or bituminous coal, which are indigenous and low in cost, are suitable as coarse material for the upper layer. The cost for converting plants from single-media sand filters to high-rate dual-media filters was estimated in India at only \$0.80 per m^3/day of plant capacity, or \$0.40 per m^3/day of increased capacity (1982 U.S.\$; Ranade and Gadgil, 1981). These estimates included the cost of modification in the influent and underdrain systems, and the cost of placing new media.

Experiments conducted in Wisconsin (Nelson, 1969) compared "coal-capped" sand filters (i.e., where 15 cm of sand is removed from a filter bed and replaced by 15 cm of anthracite coal) and sand alone under various raw water conditions. The capped filters were operated at 7.2 m/hr and the sand at 4.8 m/hr. The result showed that the coal-capped filters performed better and resulted in substantially longer filter runs for the worst raw water conditions.

The terms *effective size* and *uniformity coefficient* are used in defining filter media. The effective size (*ES*; P_{10}) is the particle size, in millimeters, such that 10% of the particles by weight are smaller and 90% are larger. The size distribution is characterized by the uniformity coefficient (*UC*), the ratio of the P_{60} to P_{10} sizes. These two parameters are determined for a particular filter material by standard sieve analyses (Cox, 1964; Fair, Geyer, and Okun, 1968). Most dual-media filters in use today in the United States contain sand with an effective size of 0.45 to 0.8 mm as the under layer, topped by a layer of coarse material with an effective size of 1 to 1.6 mm (IRC, 1981b). The uniformity coefficient for each layer is usually between 1.3 and 1.7. The depth of the filter bed should be large enough to prevent a significant amount of impurities from reaching the filtered water outlet, and is best determined by pilot-filter tests, especially if unconventional filter media with largely unknown physical properties are being considered.

Typically, dual-media filters consist of coarse media at a depth of 40 to 75 cm and a sand layer about 15 to 30 cm deep. Figure 8.4 shows a typical cross section through an anthracite-sand filter bed with graded layers for each filter medium and a reverse-gradation scheme for the supporting gravel.

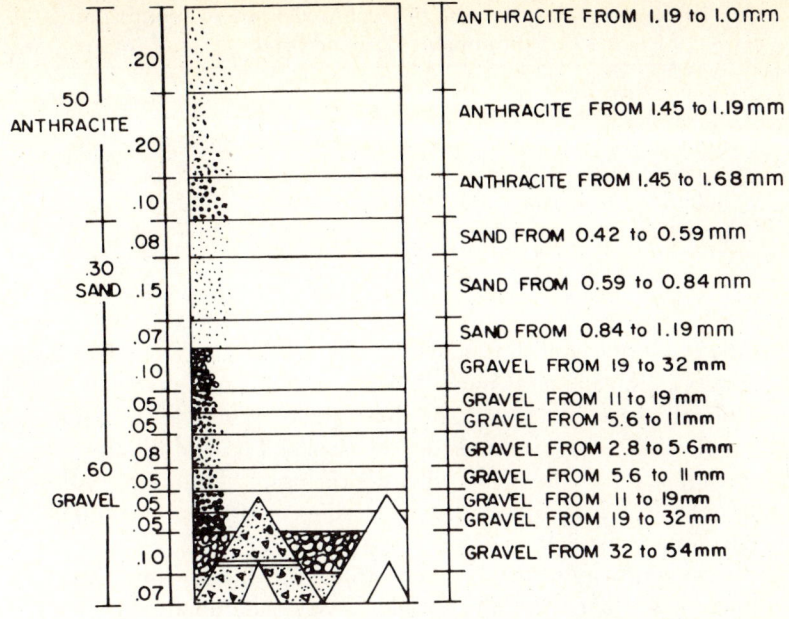

Figure 8.4. Typical dual-media filter bed. *Source:* Adapted from CEPIS, vol. 2, plan no. 37.

In places where sand cannot be purchased by specification and locally available sand is not properly sized, a simple procedure for grading filter sand may be used (Cox, 1964). The sand is screened through a coarse screen to remove foreign matter. It is then placed in a filter box and backwashed at higher than normal rates, allowing the undesirable fine sand to be wasted. The remaining fine sand that is undesirable is removed from the surface using hand tools.

Unconventional Coarse Filter Media   Indigenous coals and other unconventional coarse filter media (e.g., crushed coconut shells, pea gravel) are available and have been used in many countries, such as Brazil, Chile, Colombia, India, and the Philippines. They are now under investigation in Korea, and have been used in Japan. Bituminous coal with a specific gravity of 1.45 has performed well in Brazil. The continuing losses are greater than with anthracite, but its use has been quite satisfactory.

Pilot- and full-scale plant studies on unconventional coarse filter media have been conducted in India (Ranade and Gagdil, 1981). The lack of anthracite coal in India, which is commonly used in the United States as the coarse layer in dual-media filters, has prompted researchers to investigate various other materials such as high-grade bituminous coal (Paramasivam et al., 1973), crushed coconut shell

Table 8.2.   Characteristics of Dual-Media Filter Consisting of
Bituminous Coal and Sand

Item	Coal	Sand
Size range	0.85 to 1.6 mm	0.55 to 0.9 mm
Effective size	1.0 mm	0.6 mm
Uniformity coefficient	1.3 to 1.5	1.3 to 1.5
Specific gravity	1.4	2.65

Source: Adapted from Ranade and Gadgil, 1981.

(Kardile, 1972), berry seeds (Bhole and Nashikkar, 1974), and kernels of stone fruits such as apricots, (Ranade and Agrawal, 1974). All of these materials were found to be suitable as coarse filter media, but high-grade bituminous coal possessed the best overall characteristics in regards to cost, availability, and filtration properties. The specifications for a dual-media filter consisting of bituminous coal over sand are given in Table 8.2 and for a filter consisting of crushed coconut shells over sand are given in Table 8.3.

Several treatment plants have been constructed in India with crushed coconut shells as the coarse filter medium (Kardile, 1981). The dual-media filters in the Ramtek plant (see section on "Upflow-Downflow Filters") consist of a 30-cm layer of coconut shells (average size = 1 to 2 mm) over 50-cm layer of fine sand (E.S. = 0.45 to 0.55 mm), and have been operating for nearly a decade without any deterioration in filtrate quality.

A dual-cum-mixed media filter (composed of a coarse layer of crushed coconut shells and mixed media of 30% sand and 70% boiler clinker) was tested against a dual-media filter (composed of crushed coconut shells and sand) to compare filtration performance (Bhole and Rahate, 1977). After a 24-hr run, it was found that the dual-cum-mixed media filter reached only half the head loss of the dual-media filter, while still maintaining comparable turbidity removal. The characteristics of the three types of media (sand, coconut shells, and boiler clinker) are presented in Table 8.4.

Table 8.3.   Characteristics of Dual-media Filter Consisting of
Crushed Coconut Shells and Sand

Item	Crushed Coconut Shell	Sand
Size range	0.81 to 2.1 mm	0.5 to 0.81 mm
Effective size	0.80 mm	0.52 mm
Uniformity coefficient	1.2	1.3
Specific gravity	1.4	2.65
Depth	37.5 cm	37.5 cm

Source: Adapted from Nashikkar, Bhole, and Paramasivam, 1976.

Table 8.4. Characteristics of Mixed-media Filter Consisting of Crushed Coconut Shells, Boiler-Clinker and Sand

Item	Crushed Coconut Shell	Boiler-Clinker	Sand
Size range	1 to 2 mm	1 to 2 mm	0.5 to 0.85 mm
Effective size	1.1 mm	1.2 mm	0.54 mm
Uniformity coefficient	1.4	1.5	1.3
Specific gravity	1.4	1.9	2.65
Depth	20 cm	20 cm[a]	20 cm[a]

Source: Adapted from Bhole and Rahate, 1977.
[a]Sand 30%, boiler-clinker 70%.

Unconventional Fine Filter Media   A study conducted over a period of one year in India (Rao, 1981) has shown that selected crushed stone can be used as a filter medium instead of sand. Crushed stone is easily prepared from stone dust which is a waste product at quarries using stone crushers. Both fine grain (ES of 0.47 mm and a UC of 1.5) and coarse grain (ES of 0.7 mm and a UC of 1.3) crushed stone filter media were tested against a sand filter under filtration rates ranging from 4.7 m/hr to 9.8 m/hr. With both grain sizes, the performance of the crushed stone medium was better than that of the sand medium with respect to (1) turbidity removal, (2) bacterial removal, and (3) length of filter run. In places where good quality sand is not available, or must be transported from long distances at high cost, crushed stone may provide an economical alternative; but crushed lime-stone should not be used, as it may dissolve.

A laboratory study conducted in Australia (Barnes and Mampitiyarachichi, 1983) has shown rice hull ash to be an effective replacement for filter sand at low filtration rates. The rice hull, which is a waste product from rice production, constitutes about 20% of the harvested paddy mass. Combustion of rice hulls yields an ash which contains over 90% silica. A pilot filter containing rice hull ash (ES of 0.135mm and UC of 2.96) was tested against a sand filter. At filtration rates intermediate between slow and rapid filtration (0.25 to 2 m/hr) the performance of the rice hull ash medium was superior to that of the sand medium with respect to the following parameters when treating a water with an influent turbidity of 40 to 60 NTU:

Parameter	Rice Hull Ash Medium	Sand Medium
Effluent turbidity (NTU)	2.5–2.7	5–10
Length of filter run (hr) (time to reach head loss of 1.0 m)	460	186
Percentage removal of E. coli	90–99	60–96

At higher filtration rates (4 m/hr) the rice hull ash medium was unstable, because of its porosity.

Engineers should assess the potential of local materials that may serve as filter media. The design criteria for filter media are based on sizing with standard sieves, which have been calibrated by measuring the clear dimensions of a representative number of screen openings. In developing countries, special care should be taken in carrying out such tests, as uncalibrated sieves bought locally may result in incorrect media leading to inefficient hydraulic design and operation.

Filter gravel supports the filter media and aids in the distribution of backwash flow from the underdrain system. Gravel should consist of rounded silica stones with an average specific gravity of not less than 2.5. It should be free from clay, sand, and organic impurities. The depth and grading of gravel are related to the type of filter underdrain system used. A reverse gravel gradation, such as that used for the "teepee"-type filter bottom in Figure 8.4, has been found to be safe against movement of the top gravel layer (Hudson, 1981).

Filter Bottom and Underdrains

The two major requirements of the filter underdrainage system are the support of the filter bed without loss of media and the uniform distribution of the washwater across the entire filter bed. For the design of interfilter-washing units, however, the head loss is only 20 to 30 cm, possibly resulting in some sacrifice in uniformity of washwater distribution. However, the velocities are low enough in the plenum so that the slight pressure variations should not unduly affect backwash performance.

In many instances, bottoms can be either locally produced, reinforced concrete slabs with plastic or glass-tube orifices, or simple perforated-pipe lateral systems. A locally precast reinforced concrete filter bottom with a low orifice head loss system called the "teepee" has been adapted from California for use in interfilter-washing filtration systems in Latin America and in the Philippines (Arboleda, 1973). Because of its angular shape, this system was named after the American Indian teepee, which was a cone-shaped tent. The teepee-type filter bottom is illustrated in Figures 8.5 and 8.6. The angle-shaped beams that comprise the filter bottom are supported at each end by the side walls of the filter box. Plastic tubes of 6 to 20 mm diameter are inserted along the concrete beams at 10 to 25 cm on center to form the orifices. The beams are set with their bottom edges against one another, and then they are joined together with mortar to prevent loss of filter media and to waterproof the joints. Adequate space is provided between the beams so that three rows of 4 cm gravel below a graded layer of 2.5 to 1 cm gravel can be placed there. Porcelain balls or hollow plastic spheres filled with mortar may replace gravel in order to provide more uniform flow distribution (similar to the Wheeler underdrain system, which uses porcelain balls in recessed inverted pyramids underneath a smaller gravel layer).

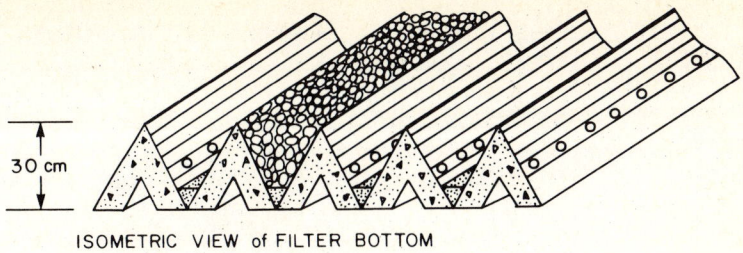

ISOMETRIC VIEW of FILTER BOTTOM

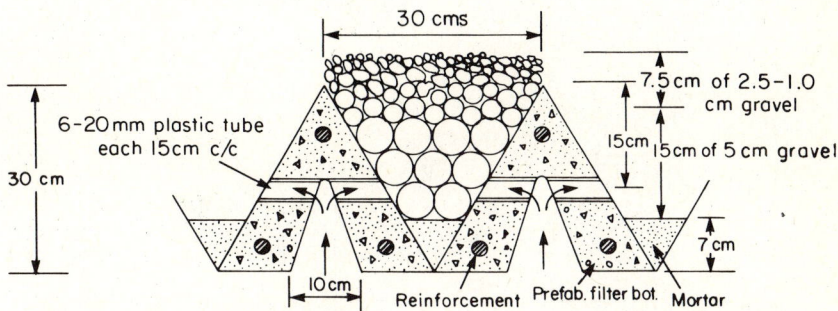

Figure 8.5. "Teepee-type" filter bottom. *Source:* Adapted from Arboleda, 1973.

The spacing of the orifices along the concrete beams dictates the head loss in the system, as indicated in the graph of Figure 8.7, which may be used for design purposes. For example, a washwater flow rate of 36 m/hr (or 600 l/m²/min, as shown on the graph) and an orifice spacing of 15 cm would produce a head loss of 25 cm with this type of underdrain system. The curves were obtained by experimentation using filter bottoms with the orifice dimensions listed within the figure; hence this graph would not apply to filter bottoms that are designed differently.

The perforated pipe lateral system consists of a central manifold pipe to which are attached a series of lateral pipes with orifices to distribute washwater or to collect the filtered water, as shown in Figure 8.8. The losses through the orifices are kept comparatively high (about 1 to 3 m) to maintain uniform distribution of backwash water. Fair, Geyer, and Okun (1968) suggest the following guidelines to pipe-lateral and underdrain design.

1. Ratio of area of orifice to area of bed served: .0015:1 to .005:1.
2. Ratio of area of lateral to area of orifices served: 2:1 to 4:1.
3. Ratio of area of manifold to area of laterals served: 1.5:1 to 3:1.
4. Diameter of orifices: 0.6 to 2 cm.
5. Spacing of orifices: 7.5 to 30 cm on centers.
6. Spacing of laterals: about the same as spacing of orifices.

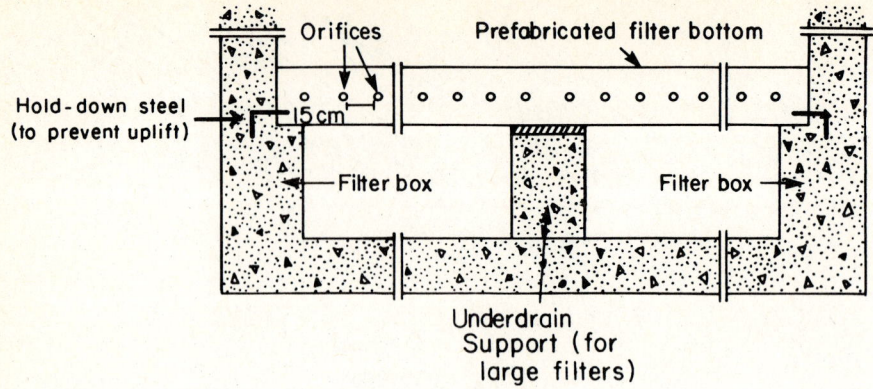

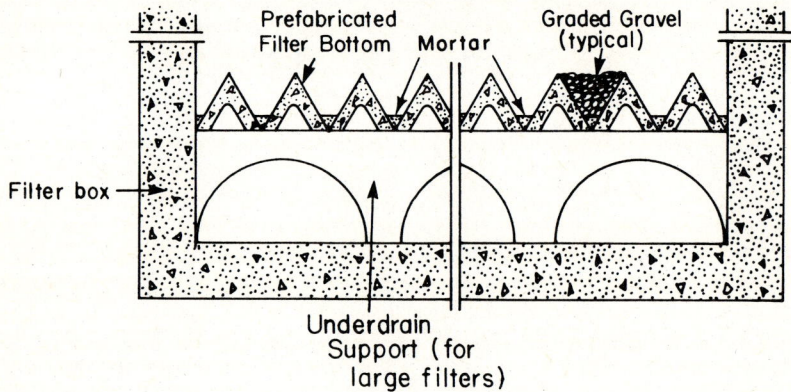

Figure 8.6. "Teepee-type" filter bottom placed in the filter cell. *Source:* Adapted from Arboleda, 1973.

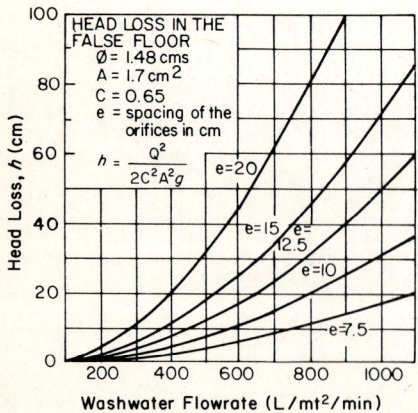

Figure 8.7. Head loss in the "Teepee-type" filter bottom for different flow rates. *Source:* Arboleda, 1973.

158

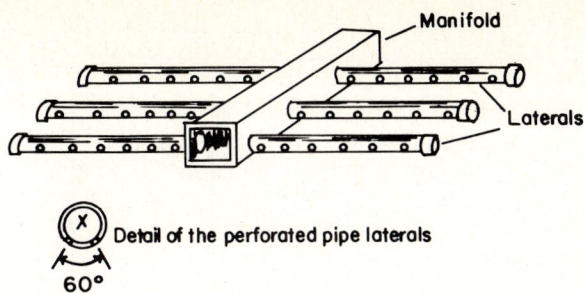

Figure 8.8. Main and lateral underdrain system. *Source:* Arboleda, 1973.

A number of different types of proprietary underdrain systems are available in the industrialized countries, which would be suitable for use in larger plants in developing countries if they can be manufactured within the country. The Leopold bottom consists of vitrified-clay or extruded plastic channels, with orifices that distribute water uniformly along the entire length of the channel. Types of proprietary filter bottoms that call for strainers or false bottoms or porous plates generally are not recommended because of clogging problems and breaking of the strainers, and ruptures of the falsefloor.

Backwashing Arrangements

The purpose of backwashing is to remove the suspended material that has been deposited in the filter bed during the filtration cycle. When a filter is backwashed, an upward flow is introduced at a rate sufficient to fluidize the filter media and to allow the accumulated contaminants to be carried away by the washwater to waste. Rates of backwash need to be high enough to fluidize all the filter media, but no higher. The percent expansion that accompanies any rate is a function of the size and specific gravity of the media, and the temperature of the water. Although optimum rates can be calculated, they are easily determined in the field through the use of a rod that can reach the top of the unexpanded sand or gravel. The backwash rate required to expand the entire bed can then be measured directly. Excessive rates should be avoided because they waste valuable water, may disturb the supporting gravel, are less effective in washing because the sand grains are separated further than necessary, and filter media may be washed overboard in the washwater gutters.

Figure 8.9 is useful for sizing the height of washwater gullets and indicates the percent of expansion that can be expected for different sizes of sand and anthracite when using a given flow rate (or velocity) of washwater at a temperature of 14°C. The minimum expansion that accompanies complete fluidization of the bed should be used. Table 8.5 can be used to adjust the values obtained from Figure 8.9 for any washwater temperature. The required rate of washwater flow is 35% greater at a temperature of 26°C than at 14°C. The time required for backwashing varies from 3 to 15 min, but may be substantially reduced by incorporating auxiliary surface

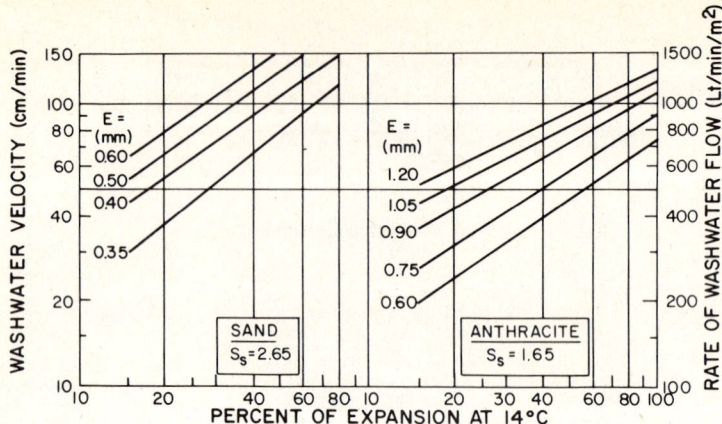

Figure 8.9. Backwash velocities and flow rates for sand and anthracite for different expansion rates at 14°C. *Source:* Arboleda, 1973.

wash systems in the filter units to provide faster and more thorough cleaning of the filter media.*

In order to determine the total head required to provide the design washwater rate, it is necessary to compute the head losses in the system during backwashing; including the losses attributed to the filter media, supporting gravel, underdrain system, and appurtenances, (e.g., washwater pipelines, fittings, and backwash-rate controllers). Most of the head loss occurs in the underdrain system (1 to 3 m), although interfilter-washing units are designed with relatively low head losses in the underdrains (20 to 30 cm). A graph for determining head losses in the "teepee"-type prefabricated concrete underdrain systems is given in Figure 8.7. Head loss

Table 8.5. Rates of Required Backwash for Equal Expansion at Various Temperatures Compared with rate at 14°C

Temperature °C	4	6	8	10	12	14	16
Backwash rate	0.75	0.80	0.85	0.90	0.95	1.00	1.05
Temperature °C	18	20	22	24	26		
Backwash rate	1.05	1.11	1.16	1.22	1.28	1.35	

Source: Arboleda, 1973.

*European practice calls for larger media (on the order of 1.0 mm), which would require much higher rates for effective backwashing. Accordingly, an air-water wash is used to reduce washwater requirements. Such systems require air compressors and specially designed underdrain systems, neither of which are appropriate for developing countries.

through the gravel is small, generally less than 8 cm. Head loss through the fluidized filter bed may be calculated from the following equation:

$$h = D (1 - f) (p - 1) \tag{8.1}$$

where

h = head loss across the fluidized bed (m)

D = unexpanded bed depth (m)

f = porosity of unexpanded bed (dimensionless)

p = specific gravity of the filter medium (dimensionless)

The head loss is the weight of the expanded media in the water, which is represented by equation (8.1). When the entire bed is expanded, the head loss becomes independent of the percent expansion, so that any washwater rate greater than that necessary to fluidize the bed separates the grains to no useful purpose. A useful guideline is to establish the rate that will assure complete bed expansion. This does not vary much where temperature variations are small.

Three types of backwash arrangements that are suitable for developing countries are (1) elevated washwater tanks; (2) taking washwater from the high-pressure distribution system tank, which is wasteful of energy but lower in installation costs; and (3) interfilter-washing units (i.e., an arrangement whereby one filter is backwashed with the effluent from other units).

Washwater pumps that take suction from the clear well must be sized to supply a backwash rate of at lease 36 m/hr; they are not generally recommended because they are more costly in small plants and must have motors with horsepower ratings substantially higher than any other motors in the plant.

An elevated washwater tank should have sufficient capacity to wash at least two filters for at least 10 min each at the maximum backwash flow rate without refilling. Small pumps are used to fill the washwater tank during intervals between successive backwashings. Two pumps are usually provided, with one serving as a reserve unit. The capacity of each pump should be based on expected frequency of washing, with 10% of the washwater rate generally being adequate (IRC, 1981b). The bottom of the tank should be high enough above the washwater gullet to provide the desired washwater flow rate, as determined from an analysis of the head losses in the system. This distance normally ranges from 4 to 6 m (IRC, 1981b). Washwater tanks should each be equipped with an overflow pipe, drain valve. air vent, vortex-breaking baffle, and manually operated washwater regulating valve, as shown in Figure 8.10. Tanks in developing countries are constructed of concrete or masonry, although steel prefabricated tanks are satisfactory where they are locally available.

When the high-lift pumps, transmission main, or distribution system (with high pressure) are near the treatment plant, washwater may be taken from them. Such

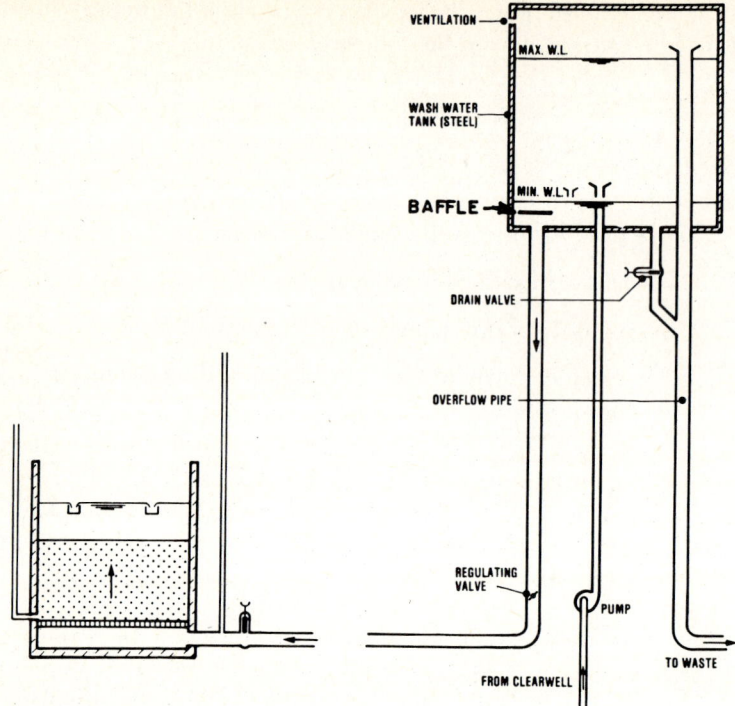

Figure 8.10.   Washwater tank arrangement. *Source:* Adapted from IRC, 1981b.

an arrangement eliminates the need for a separate washwater tank and pumps to fill the tank. However, it may be necessary to reduce the pressure, because distribution system pressures are generally higher than those needed for effective backwashing and, if not controlled, may blow out the filter. A pressure-reducing valve in the washwater line will serve—although it would have to be imported. A preferable option is to draw water from the distribution system into an elevated washwater storage tank, whence water can flow by gravity to the filters when backwashing is needed.

Interfilter-washing systems are virtually free from ancillary backwash equipment such as washwater tanks, pumps, pipe galleries, and washwater rate controllers. The washwater and pressure head for backwashing a cell are obtained from companion cells that are connected in parallel through a common underdrain system, as shown in Figure 8.11. A cell is backwashed by closing the inlet and opening the drainage outlet of the cell. The water level in the cell is thus lowered, creating a positive head (h_b), which reverses the direction of flow through the filter bed and initiates the backwash cycle. After washing, the drain is closed and the inlet opened. The cell then resumes its filtration cycle.

The available head for backwashing, h_b, is the difference in elevation between the effluent weir and the gullet lip in the filter cell. The required value of h_b to expand the filter bed is the sum of the head loss in the underdrain and pipe system

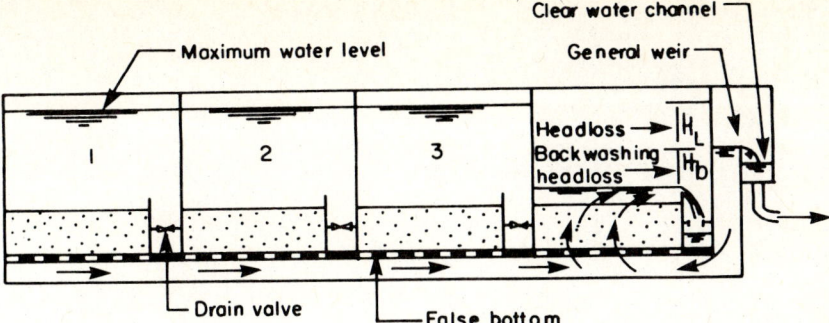

Figure 8.11.   Backwashing of one filter with the flow of the others. *Source:* Arboleda, 1973.

and the head required to keep the filter media in suspension. By increasing the depth of the water over the filter beds (about 1.5 to 2.5 m), limiting the head loss in the underdrain system (about 20 to 30 cm), interconnecting the underdrain systems, and using dual-media filter beds, the backwashing head (h_b) will be sufficient to produce the desired expansion rates.

A 60 m³/day pilot unit was tested at the National Hydraulic Laboratory in Lima, Peru (Arboleda, 1974). It consisted of four 20 × 30-cm Plexiglas units, each unit containing 15 cm of sand (effective size 0.55 mm) below 45 cm of anthracite (effective size 1.0 mm) and a "teepee"-type underdrain system. The average filtration rate was 10 m/hr, and the maximum water depth above the filter bed was 90 cm. When filter 1 was washed, the flow in filter 2, 3, and 4 increased slowly during the 20-min backwash period to a peak about 20% greater than the original flow. After the wash, when all the filters returned to normal operation, filter 2, 3, and 4 produced about 10% less than they produced before the backwashing, because part of their flow was taken by newly cleaned filter 1. In actual practice, backwashing periods are normally about 10 min, and the peak flows for the operating filters during this

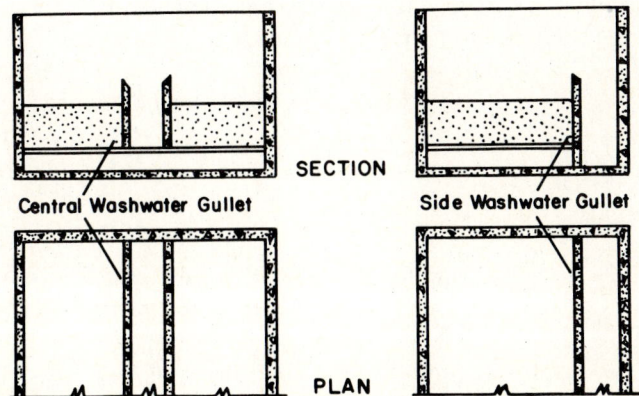

Figure 8.12.   Arrangements for washwater gullets. *Source:* Arboleda, 1973.

A. Submerged Discharge

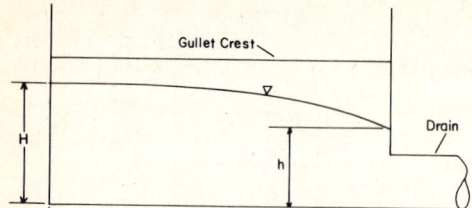

B. Free Discharge

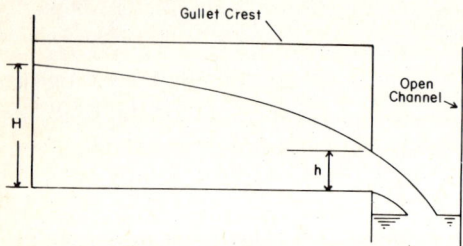

Figure 8.13. Nomenclature diagrams for side weir or gullet design. *Source:* Adapted from Hudson, 1981.

period would be somewhat less. The design and operation of interfilter washing units are discussed below.

Washwater may be collected and removed from the filter cell by either (1) a system of troughs and gullets, or (2) only gullets. Although washwater troughs are used extensively in the United States (mainly because of tradition), there has been little evidence to indicate that they measurably improve the backwashing process. In fact, gullets have performed admirably as the sole washwater collection system in a number of plants in the United States (Hudson, 1981). Center and side gullet designs are shown in Figure 8.12. A photograph of a filter box with center gullet is shown in Figure 8.15. In constructing a gullet wall it is convenient to bevel the edge so that the flat top surface is narrow and can be leveled accurately by grinding. Flat-bottomed gullets or troughs may be designed with the help of the following formulas and the nomenclature diagrams presented in Figure 8.13:

$$H = [h^2 + (2Q^2/gb^2h)]^{1/2} \quad \text{for submerged discharge} \tag{8.2}$$

$$H = 1.73h_c \qquad\qquad\qquad \text{for free discharge} \tag{8.3}$$

where

H = depth at upstream end (m)

h = depth at downstream end (m)

h_c = critical depth for free discharge − $(Q^2/gb^2)^{1/3}$

Q = rate of discharge (m³/s)

g = gravity constant (9.81 m/s^2)

b = width of channel (m)

Free discharge is the preferable design, with an allowance for some freeboard in the channel above H.

Auxiliary-scour Wash Systems

Auxiliary scour is used to assist in cleaning the filter media and to prevent mud ball formation and filter cracking. Air scouring and surface wash are two basic types of auxiliary scour systems. The fixed-grid types of surface-wash systems are best suited for developing countries because of their simplicity in design and lack of moving parts. It involves the installation of a system of pipes located over the filter media in a manner to project jets of water under considerable pressure into the upper portion of the filter media while it is being washed. Baylis (1935) designed such a system, which consisted of a manifold passing along the center of the filter with laterals taken off approximately 60 cm apart. The pipes may be supported by supports at the top of the filter box. At about 60-cm intervals, 2.5-cm pipes project downwards to within 10 cm of the surface of the filter media. Plastic caps with five 6-mm holes are screwed over the ends of the pipes. Water pressures of 345 to 500 KPa are used. Details are shown in Figure 8.14.

A second type of surface-wash system consists of horizontal pipes located about 5 cm above the top of the filter media with sand-proof orifice assemblies pointing downward at an angle of 30° below the horizontal (Hudson, 1981). These units are designed for pressures of 500 KPa and apply a flow rate of 2 to 5 m/hr to the bed surface. The orifices are spaced so that they provide complete coverage of the filter bed, as shown in Figure 8.15. The influence of the water jets emitted from the orifices has been shown in practice to carry up to 45 cm laterally.

In surface-wash systems, the piping provides a direct cross-connection between the washwater and unfiltered water; if the surface washwater is drawn from the clear well or the distribution system, the surface-wash header should be above the filter box, and fitted with a vacuum-beaker to prevent back siphonage.

Filter Control Systems

For any given rate of filtration, the head loss through the filter bed increases during the filter run, because impurities from the applied water gradually clog the bed. Where the head loss through the filter is fixed, clogging gradually reduces the rate of filtration. Conventional filter designs compensate for increasing head loss in the filter, and thereby maintain uniform flow by incorporating a mechanical rate controller in the filter outlet pipe to provide a decreasing resistance to the flow, so that the total head loss across the filter unit (and hence the filtration rate) remains constant. Another type of design obtains a constant output by allowing the water level above the filter to rise as the run proceeds. These types of control systems are classified as *constant-rate* filtration systems. On the other hand, *declining-rate* filtration systems

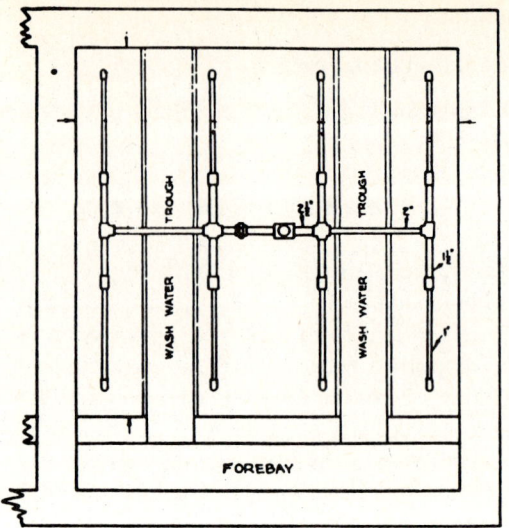

PLAN OF SURFACE-WASH PIPING IN PLANT FILTER.

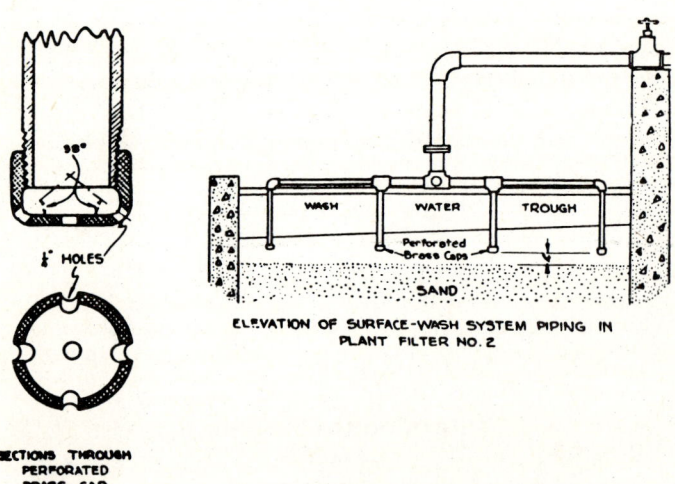

Figure 8.14.   Details of Baylis surface-wash piping. *Source: Baylis, 1935.*

do not use rate controllers, and the filtration rate declines in each unit as the head loss in the filter medium increases.

A comparison of head losses, flow rates, and water levels as a function of filtration time for constant-rate and declining-rate filters is presented in Figure 8.16. The total head loss (H_T) for each filter control system is comprised of several components: (1) the head loss in the clean bed and in the underdrains, valves, pipes, and fittings

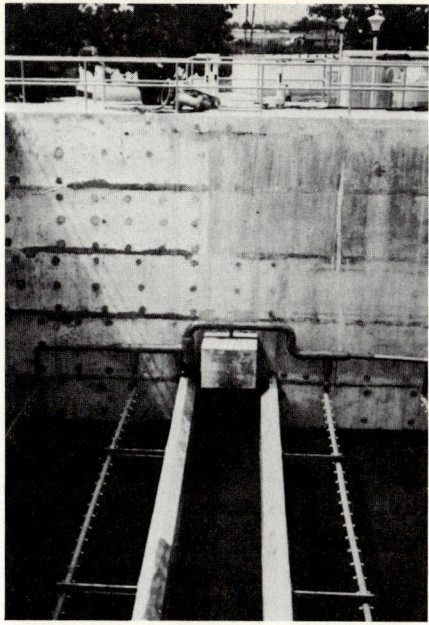

Figure 8.15. Fixed-grid surface-wash system at a plant in Cali, Colombia. *Source:* Hudson.

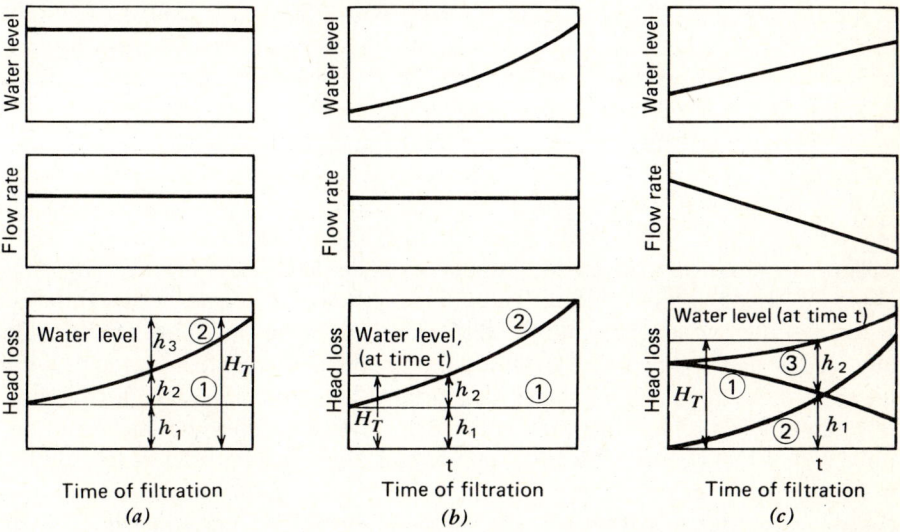

Figure 8.16. Head losses, flow rates, and water levels in filter control systems: (a) constant-rate filtration with rate controllers; (b) constant-rate filtration with increasing water level; (c) declining-rate filtration. One, h_1 = head loss due to clean bed, underdrains, valves, pipes and fittings; two, h_2 = head loss due to clogging of the filter bed; three, H_T = total head loss; h_3 = excess head (expended in rate controller or valve).

(h_1); (2) head loss due to clogging of the filter bed (h_2); and (3) the excess head that has to be expended by a rate controller or valve (h_3).

Constant-rate filters have been traditionally employed in the United States, and consequently the procedures for their design are well-documented (Fair, Geyer, and Okun, 1968; Sanks, 1978). The intent of the following section is to describe several simple methods for achieving constant-rate filtration without using sophisticated designs and/or imported equipment.

Declining-rate filters, on the other hand, although perhaps less popular in the United States, are particularly suitable for developing countries, because they may lead to simpler and less-expensive designs. The design of declining-rate filters, including detailed examples and hydraulic calculations, is provided by Hudson (1981). The design of hydraulic control systems for declining-rate filters is covered by Arboleda (1974), who includes procedures and typical layouts for minimizing the use of equipment and simplifying filter operations. The discussion presented in the section on declining-rate filtration summarizes design procedures set out in these two publications.

Constant-Rate Filtration

A constant-rate filter requires some type of filtration-rate control system to compensate for the increased head loss in the filter bed as it becomes clogged with impurities. There are several types of rate controllers: inlet rate control devices (equal distribution of the constant plant flow to the filters, often by overflow weirs, called "flow splitting"), and outlet rate control devices (throttling valves tied to flow-measuring or differential pressure or water level devices). The following types are relatively simple to build, operate, and maintain in developing countries.

Filter control systems using an even distribution of the raw water ("flow splitting") over the filter units (see Figure 8.17) are perhaps the simplest, because they do not require any mechanical equipment. The incoming raw water is divided equally among the filters by means of a weir at each filter inlet. The influent conduit feeding the filters is generously sized so that the water level does not vary significantly along the length of the conduit. Thus, the overflow rate at each weir will be the same, and the raw water feed to the filter units will be equally split. The filtration rate is controlled jointly for all the filter units by the raw water feeding rate. During filtration the water surface in each filter unit will rise to compensate for the head loss buildup in the filter bed. When the water surface reaches the maximum permissible level above the filter bed, such as, for example, the level of the influent weir, the filter is taken out of service for backwashing. With this type of inlet rate control system there will be considerable variations of the raw water level in the filters (known as "variable-head" filtration); hence the filter box must be deeper than in filters with rate controllers.

The new 65,000 m³/day direct filtration plant at Oceanside, California (see Chapter Three, under "Baffled Channel Flocculators") has eight influent flow-splitting filters with variable head and interfilter-backwashing capabilities (Ainsworth,

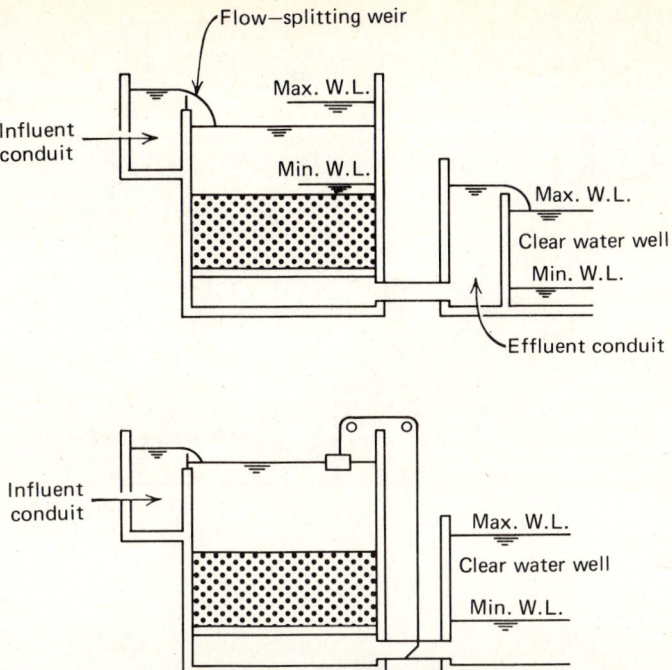

Figure 8.17. Constant-rate filter control arrangements: (a) influent flow splitting with variable head; (b) float-operated outlet control system with constant head. *Source:* IRC, 1981b.

1984). Each filter contains conventional dual media and operates at a constant design filtration rate of 30 m/hr. A schematic of a typical filter cell showing flows during filtration is presented in Figure 8.18. Water levels in the filter are typically 1.2 m above the media at the beginning of a filter run and rise to the full available depth of 3.2 m (to flood the influent control weir) before backwashing becomes necessary.

When it is desirable to maintain a constant water surface in the filters, a rate controller must be provided in the filter outlet pipe. A very simple control consists of a float connected by a small cable running over sheaves to a butterfly valve on the filter-outlet pipe (see Figure 8.17b). The weight of the float must be adequate to insure operation of the valve. When the filter is first placed in operation after backwashing, the water surface will tend to drop, causing the butterfly valve to close. As the filter head loss builds up, the filter surface will tend to rise, causing the valve to open. When the valve becomes wide open, it can no longer control the water surface in the filter, and the operator, by observing the position of the float, knows that it is time to backwash the filter. This is a very simple means of achieving constant-rate filtration and is best adapted to relatively small filters where the outlet pipe is 300 mm in diameter or less.

An 86,000 m³/day plant in Surabaya, Indonesia, with 96 small float-operated filters, has been in successful operation for about 50 years. The outlet pipe in each

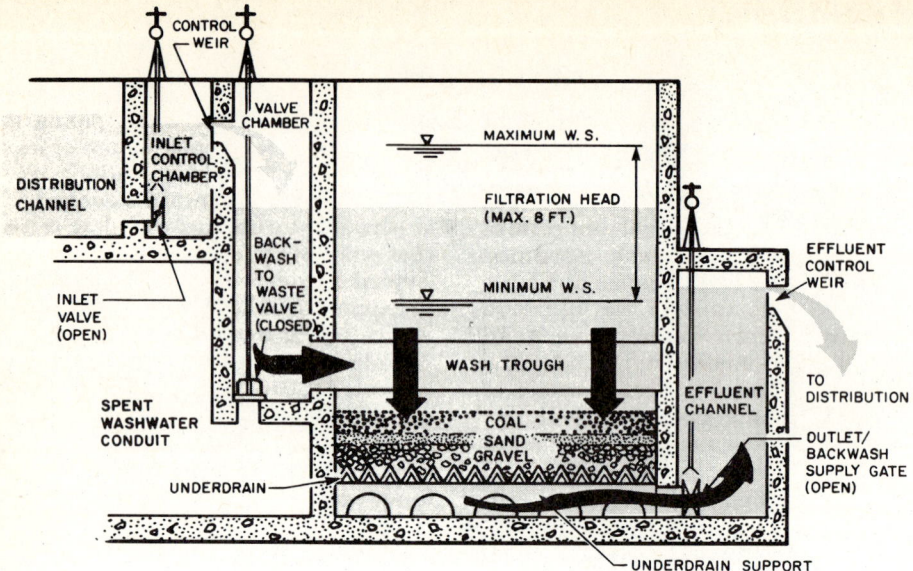

Figure 8.18. Constant-rate filter with influent flow-splitting at direct filtration plant in Oceanside, California. *Source:* Ainsworth, 1984.

Figure 8.19. Filter outlet piping at large treatment plant in Surabaya, Indonesia. Each outlet pipe is equipped with a four inch butterfly valve and a counter weighted lever (painted white), which maintain the flow at a constant rate. A thin cable connects the lever to a float inside the filter box. *Source:* Okun.

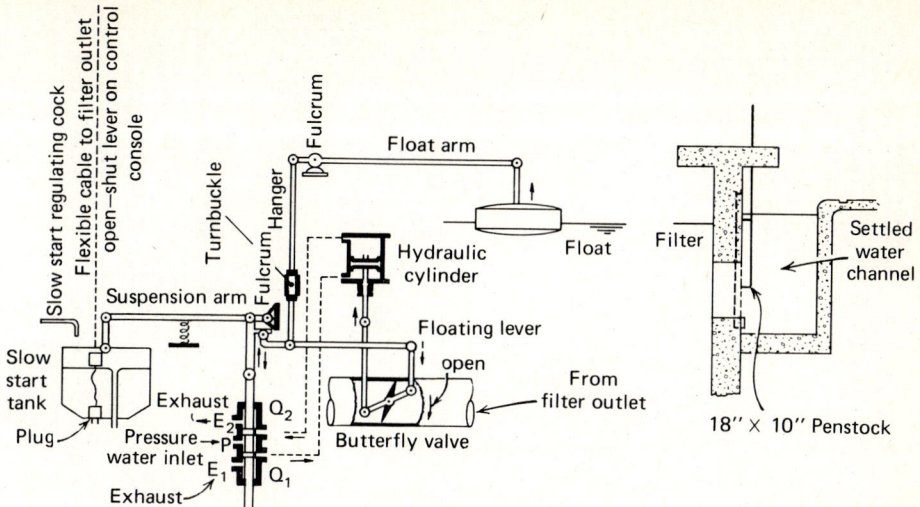

Figure 8.20. Rate-controller at large plant in Penang, Malaysia. *Source:* Ching, 1979.

filter is equipped with a 4-in. butterfly valve, as shown in Figure 8.19. A similar float-operated control system was installed in a 140,000 m³/day plant in Penang, Malaysia (Ching, 1979). The system consists mainly of a hydraulic cylinder, a pilot valve, and a small water tank connected to a filter outlet butterfly valve, and a float valve that is inside the filter cell. A schematic of the system is shown in Figure 8.20. The float inside the filter cell responds to changes in the water level, which causes the pilot valve to operate the hydraulic cylinder to open the outlet butterfly valve. The slow-start water tank allows the filter outlet to be closed or opened slowly by draining or filling the tank with a steady flow of water. This feature eliminates the frequent adjustment of outlet controllers practiced in other types of constant-rate filter plants.

Declining-Rate Filtration

When no rate controllers are used and the filter inlet is placed below the minimum water level, filtration will take place at a declining rate. One of the advantages of declining rate filtration is that breakthrough of particles or turbidity is less likely at the end of a run than with constant-rate filters.

The graph of head loss versus filtration velocity and the accompanying diagram presented in Figure 8.21 show the basic hydraulic elements in the design of declining-rate filters. If the water level at the beginning of the filter run (line 1) is known, a vertical line can be drawn passing through the maximum safe flow rate. The initial head loss at that point (h_1) is comprised of the friction head loss in both the clean filter bed and the filter appurtenances. The excess head (h_3) is the difference between the initial head loss at that point h_1 and the initial water level in the filter. When

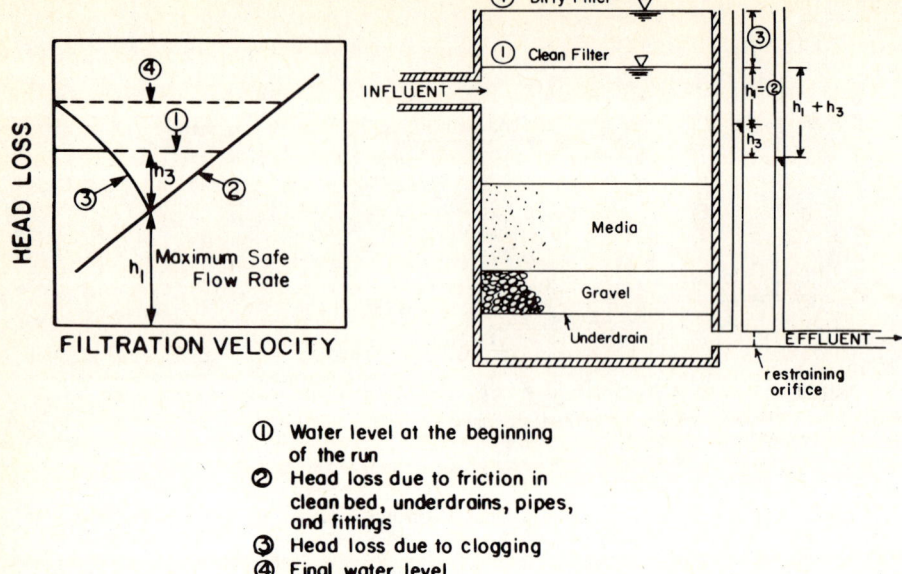

- ① Water level at the beginning of the run
- ② Head loss due to friction in clean bed, underdrains, pipes, and fittings
- ③ Head loss due to clogging
- ④ Final water level
- h_1 Total initial head loss
- h_3 Excess head

Figure 8.21. Heads and water levels in declining-rate filtration systems. *Source:* Adapted from Arboleda, 1974.

a filter has just been cleaned, it may be necessary to dissipate this excess head through some type of restraining orifice or adjustable gate, in order to limit filtration rates to the maximum rate of flow permitted. For example, if a rapid filter is designed for a total available head ($h_1 + h_3$) of 3 m, and the frictional head loss in the clean sand bed and appurtenances are calculated to be 2.50 m, then the restraining orifice should be sized to produce a head loss of 0.5 m. The equation relating flow and head loss through an orifice is

$$Q = cA \, (2gh)^{1/2} \tag{8.4}$$

where

$$Q = \text{flow (m}^3\text{/s)}$$

$$c = \text{orifice coefficient, generally } 0.6$$

$$A = \text{area of orifice (m}^2\text{)}$$

$$g = \text{gravity constant (9.81 m/s}^2\text{)}$$

$$h = \text{head loss or differential (m)}$$

For a filter bed surface area of 25 m² and an initial filtration rate of 15 m/hr, the flow rate onto the filter bed would be 9000 m³/day or 0.10 m³/s; hence, from equation 8.4, the area of the orifice opening to produce a head loss of 0.5 m would be

$$A = 0.10/0.6 \ (2 \times 9.81 \times 0.5)^{1/2} = 0.053 \ m^3$$

Friction head losses in the filter appurtenances have been approximated by Hudson (1981) for a flow rate of 36 m/hr:

Filter inlet and piping	0.60 m
Filter gravel	0.05 m
Filter underdrains	1.0 m
Outlet piping, filter to equalizing chamber	0.60 m
Total	2.25 m

The Carmen-Kozeny equation (see Chapter Six, eq. 6.8) is used to calculate head losses in the filter media, which take place under laminar flow conditions.

The total available head, represented by $h_1 + h_3$ in Figure 8.21, is usually equal to the difference between the water level in the common influent header preceding the filters and the minimum water level in the filtered water outlet chamber. Hudson recommends designing for a total available head of 3 m, unless there is assurance that filter runs will be of reasonable length under lower heads. Declining-rate filters are generally designed to operate from 150 to 50% of the average filtration rate, which ranges from 5 to 7 m/hr. However, filtration rates two to three times greater than this rate may be possible when using dual-media filters.

Design and Operation of Interfilter-washing Units

Interfilter-washing filtration units are an appropriate filtration system for developing countries. An early design for such a system was proposed by Greenleaf (1964). Studies conducted in Latin America (Sperandio and Perez, 1976) have demonstrated the high efficiency of this type of filter when operated at a declining rate. Moreover, such filters are easier to build, operate, and maintain than conventional filters. For example, only two butterfly valves or sluice gates are needed for filter control (three valves, if using a surface-wash system), and the entire system may be designed with concrete channels or box conduits. It is also possible to eliminate completely pipe galleries containing elaborate piping, valves, and controlling systems that are common to conventional filtration schemes and which represent a major portion of the filtration complexity and cost. Other advantages of interfilter-washing filtration systems are enumerated below (Arboleda, 1974).

1. Backwashing is automatically controlled by the level of the effluent weir.
2. The backwashing operation starts slowly with the descending level in the filter.

3. If the filters are not backwashed at the proper time, the plant flow decreases and there is a "backwater" effect, which forces the operator to act immediately.

4. There is less possibility of producing a negative head within the filter.

The major disadvantage is that the filter boxes need to be deeper than in conventional filters—not an important constraint in developing countries.

Interfilter-washing units have been planned or are operating successfully in large plants in Latin America, including those for the cities of Monterey, Mexico (1.04 million m³/day); Mexico City (1.08 million m³/day); Rio Grande, Brazil (518,000 m³/day); Santo Domingo, Dominican Republic (691,000 m³/day); Cali, Colombia (259,000 m³/day); and Guatemala City, Guatemala (112,300 m³/day), as well as in at least 100 smaller plants. The hydraulic characteristics of these filters have been described by Arboleda (1973, 1974), who has also developed several innovative and simple filter layouts. The material presented is based largely on his published works.

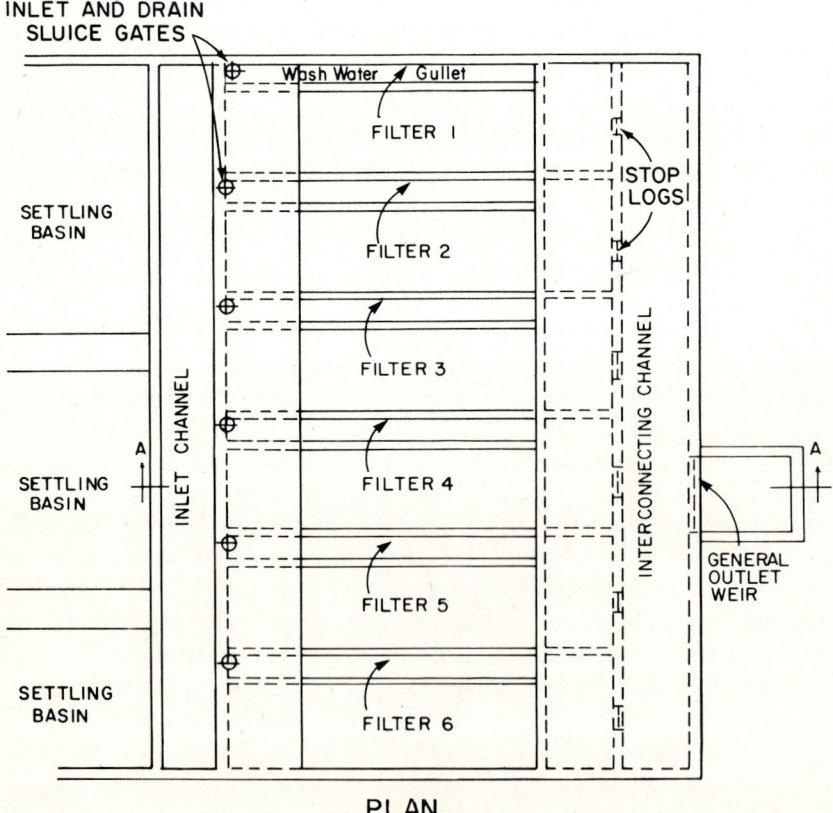

Figure 8.22.   Battery of interfilter-washing cells at the plant in Cochabamba, Bolivia. *Source:* Arboleda, 1974.

Interfilter-washing units can be designed with either unrestricted or restricted declining flow rate, or at constant rate. The filtration system for the plant in Cochabamba, Bolivia, a plan of which is shown in Figure 8.22, employs unrestricted declining flow rate. A section of a typical filter cell in the plant, which gives the water levels during filtration, is shown in Figure 8.23. At the start of the filter cycle, the minimum water level is established slightly above the surface of the water flowing over the general outlet weir, in accordance with the initial head loss. The water rises slowly in the filter box as head loss builds up during filtration, until the maximum allowable head loss is reached (normally 1.5 to 2.5 m), at which point the filter must be backwashed. To initiate the backwash cycle only one sluice gate need be manipulated. In Figure 8.23, gate A slides upward, opening the drain C and closing the inlet D. The rate of backwashing is controlled by adjusting the level of the general outlet weir. In order that newly cleaned filters do not take an excessive load when put back into operation, all filter units are kept reasonably clean, so that the filtration rate in any filter does not exceed 30 m/hr at the beginning of the run. This is best achieved by washing the filters in succession on a time schedule.

The second type of design introduces a constriction at the outlet side of the filter by means of a sluice gate, which can be adjusted to control the filter velocity after washing. A plan view of a battery of filters that employ restricted declining flow rate is shown in Figure 8.24; section views that show the water levels during filtration and backwashing are shown in Figures 8.25a and b. During the filtration cycle, the head losses due to the filter bed and underdrain system, h_f, dictate the water level in the clear-water channel which under normal operation is the same for all

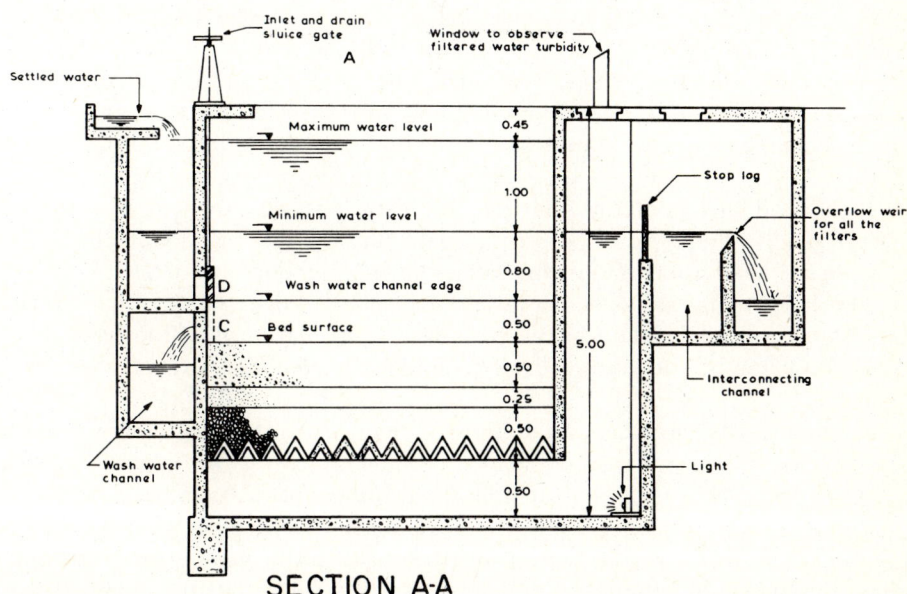

SECTION A-A

Figure 8.23. Typical filter cell at the Cochabamba plant, showing water levels during filtration. *Source:* Arboleda, 1974. (Dimensions in meters.)

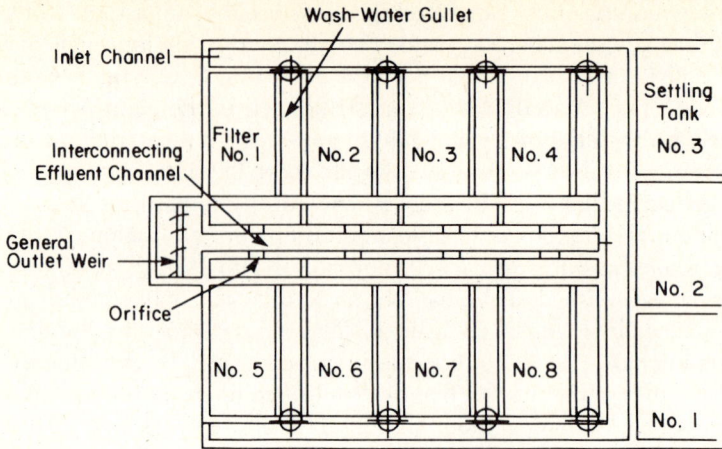

Figure 8.24. Battery of interfilter washing cells with outlet-gate control (plan). *Source:* Arboleda, 1973.

filter units. The difference in water level, Δ_h, between the clear-water channel and the interconnecting flume is controlled by the outlet sluice gate, and dictates the rate of filtration.

During the backwash cycle (Figure 8.23), the same gate is used to control the head available for backwashing, h_b, and thus the rate of rise of washwater. To change from a filtration to backwash mode, and vice versa, only two gates are used: (1) the gate located at the inlet side of the filter which controls the opening and closing of the inlet channel and drain; and (2) the gate located at the outlet side of the filter, which regulates the filtration and washwater rates. The water level difference, Δh, controlled by the outlet sluice gate, can be recorded to measure flow rates through the constriction. From the position of the gate, the opening between the two chambers would be known, and the flow can be calculated from equation 8.4.

In both of the preceding filter designs, care must be exercised in selecting the size and type of gates to be used; in particular the gate that admits the washwater must be capable of opening fast enough to produce the required rate of backwash before the available washwater is depleted. For instance, a hand-operated sluice gate of standard design may take up to 300 turns of the handwheel and several minutes to open fully. Accordingly, a butterfly valve may be more desirable.

Hudson (1981) has designed a declining-rate interfilter-washing filtration system whereby the flow is restricted on the inlet side and butterfly valves are used in place of sluice gates. This design was used for the treatment plant in Cali, Colombia. A schematic of a typical filter cell in the plant is shown in Figure 8.26. The filters receive water from a relatively deep inlet channel. From this channel the water flows through influent pipes, each with a butterfly valve and restraining orifice.

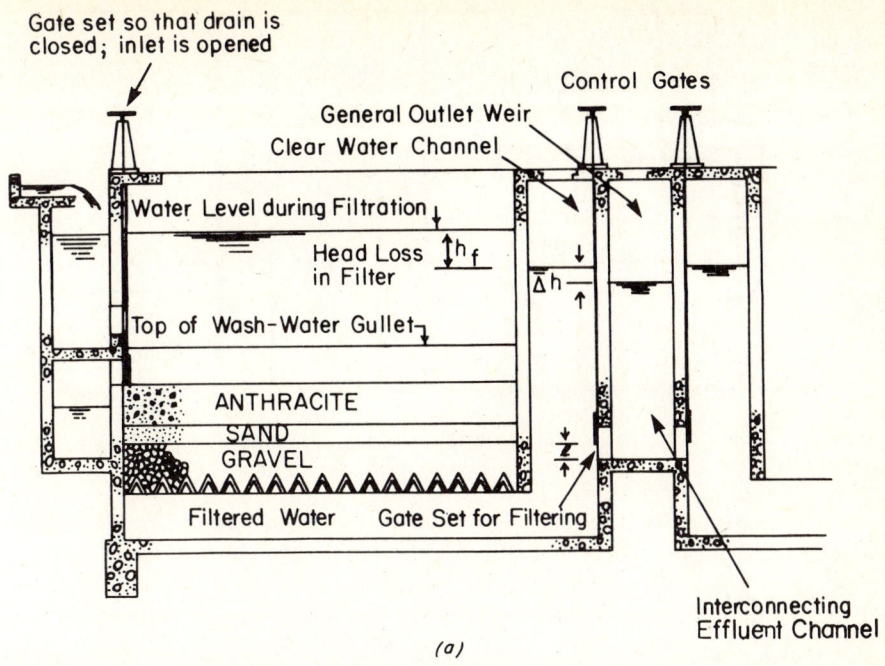

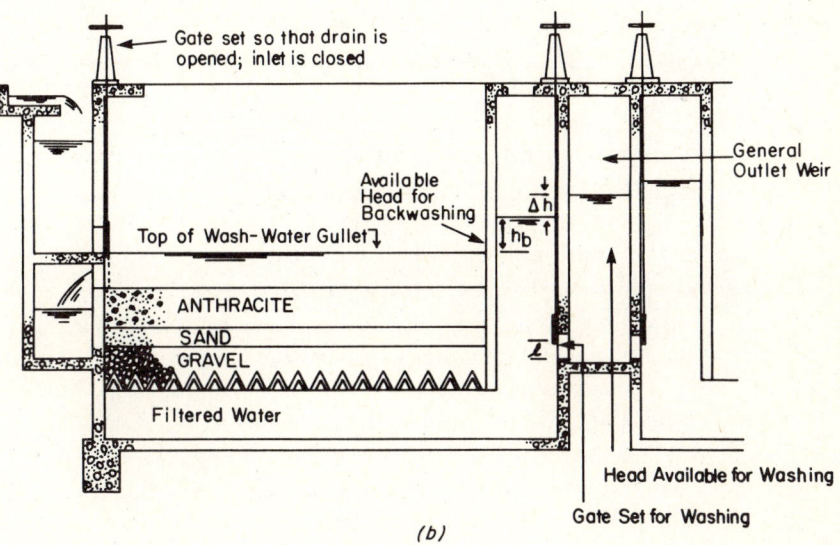

Figure 8.25. Typical filter cell with outlet-gate control showing water levels during (a) filtration and (b) backwashing. *Source:* Arboleda, 1973.

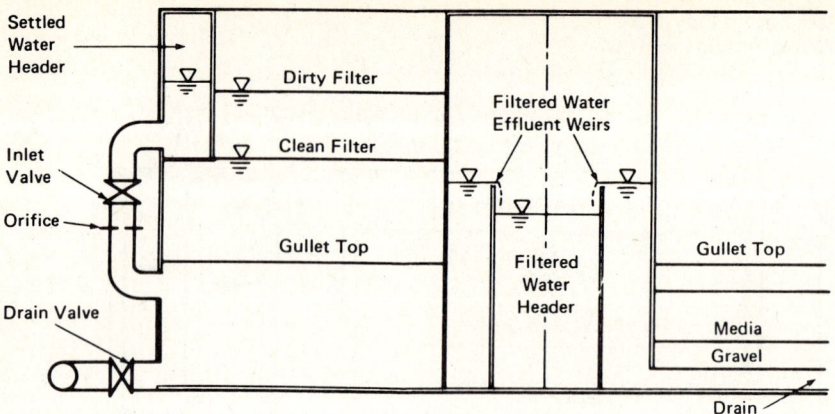

Figure 8.26. Influent-controlled, declining-rate filter system for the plant in Cali, Colombia. *Source:* Hudson 1981.

Each cell also has a drain valve. The filtered water discharges over a common effluent weir. Hudson claims the following advantages for this type of system.

1. The inlet constriction allows a changing flow rate to be applied to the system, and also provides a declining flow rate through each filter as the filtration progresses.
2. Valve structure is required only at the inlet end to regulate the flow supply or the backwashing operation.
3. Butterfly valves are simpler and faster to operate than sluice gates and are easily maintained. Butterfly valves are also generally less expensive.

Several design guidelines that should be taken into account when designing interfilter-washing filtration systems are summarized below.

1. The capacity of and the flow through the plant must be at least equal to the washwater flow needed to clean one filter.
2. A minimum of four filters, each capable of operating at one-third higher rate, is necessary to operate at design capacity when one unit is out of service for washing. The larger the number of filters, the smaller the total filter area and the greater the flexibility in operation.
3. The filters must be so designed that any one may be taken out of service for repairs without interruption of the normal operation of the others. This may be done by the insertion of slide gates.
4. The underdrain system must be especially designed to produce low head loss.
5. The influent channel should be able to carry the flow to any filter unit with a low loss of head.

Direct Filtration

Direct filtration of raw waters low in turbidity and color is a comparatively low-cost option that has distinct advantages for developing countries. The direct filtration process subjects raw water to rapid mixing of coagulants, and often flocculation, followed directly by filtration. Figure 8.27 shows separate flow sheets for a conventional filtration plant and a direct filtration plant, the latter consisting of the addition of alum and a polymer to the rapid mix influent, followed by dual-media filters. The chief advantage of direct filtration for developing countries lies in the potential for reduced chemical consumption. Plants designed with direct filtration also have lower capital construction costs than conventional plants; up to 30% under certain conditions, which results from the elimination of settling basin structures, sludge removal equipment and, sometimes, flocculation structures and equipment. On the other hand, washwater usage in direct filtration is greater than for conventional treatment, ranging as high as 6% as compared with 3 to 4% for conventional plants, because of shorter filter runs.

Conventional filtration plants with horizontal-flow settling basins may be operated in the direct-filtration mode by reducing the coagulant dosage in the rapid mix, thereby reducing sedimentation in the settling basin. Of course, the plant always has the option of reverting back to the conventional mode if, for example, rains increase the raw water turbidity above acceptable levels for direct filtration.

The application of direct filtration is generally limited to raw waters where turbidity and color are each less than 25 units. Waters higher in color or turbidity may be treatable by direct filtration, but laboratory and/or pilot-plant studies would need to be performed in each case. Experience has demonstrated that full-scale plants perform better in direct filtration than most pilot plants, and substantially better than bench-scale tests.

Wagner and Hudson (1982) have developed a simple bench-scale test that evaluates the possibility of using direct filtration. Basically, the method utilizes jar testing to sort out the variables of best coagulant, the most effective polymer, optimum

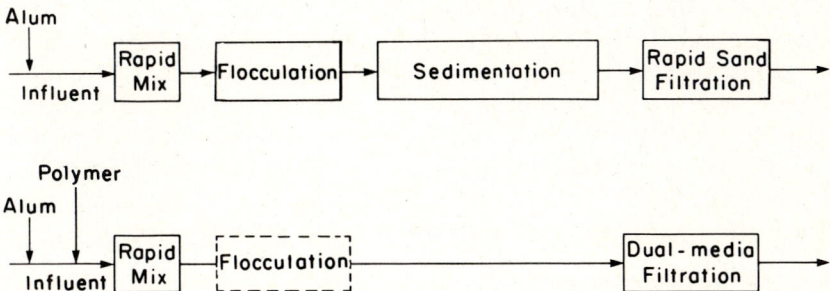

Figure 8.27. Flow sheets comparing conventional filtration (using alum) with direct filtration (using alum and a polymer). *Source:* Adapted from Culp, 1977.

dosages, stirring intensities and time. Samples are taken from each of the jars and then filtered through standard laboratory filter paper. The filter test run takes no more than 2 or 3 min. When the coagulant dose required to produce a low filtered water turbidity is less than 6 to 7 mg/l with the addition of a small dose of polymer, the raw water has the potential to be treated by direct filtration. When higher doses are required, say 15 mg/l, then treatment by direct filtration may not be appropriate.

Filter clogging is related directly to the floc volume that is loaded onto the filter, and the floc volume is a function of coagulant dose. Hence, where low coagulant doses are found to be effective, the potential for direct filtration is increased. Positive results obtained from bench-scale testing justify undertaking more conclusive pilot-plant investigations to determine the plant-scale design parameters.

Because there is no sedimentation step in the direct filtration process, all impurities in the raw water are removed entirely by the filters. Hence, filter units should generally be designed with large storage capacities in the beds. This is most readily accomplished by deepening the filter and by using dual media with larger sizes to permit greater penetration of the floc, so that the full depth of the bed can be used for storage of the floc. The following bench-scale and pilot-plant investigations conducted in Brazil and Jordan by Wagner and Hudson (1982) provide evidence that there may be good potential for direct filtration in developing countries.

In Brasilia, bench-scale testing was carried out during the dry season (raw water turbidity of 12 NTU), which confirmed that direct filtration was feasible and that pilot-plant investigations were justified. Pilot-plant investigations were carried out during the rainy season (raw water turbidity averaged 25 NTU, with levels as high as 45 NTU) with several pilot filters. The results of the bench-scale and pilot-plant studies are presented in Table 8.6. The most efficient filter design was a dual-media filter comprised of 25 cm of sand (0.5 mm effective size and 1.35 UC) and 45 cm of indigenous bituminous coal (1.1 mm effective size and 1.45 UC). The average filter run was 15 hr at a rate of 18 m/hr. Based on these results, a 520,000 m³/day direct filtration plant has been built in Brasilia.

In Amman, Jordan, pilot filter tests were run on settled (but uncoagulated) raw water with turbidities that averaged 62 NTU. Data from these filter tests are shown in Table 8.7. The most effective pilot filter used a filter medium of 85 cm of anthracite (1.4 mm effective size and 1.57 UC) over 26 cm of sand (0.67 mm effective size and 1.05 UC). In spite of the high raw water turbidity, the filtration rate could be kept high (11.5 m/hr) without causing unduly short filter runs. Interestingly, only a small amount of cationic polymer (0.5 mg/l) was needed for coagulation. In Jordan, where the price of alum is $300 to $350 per metric ton, direct filtration would cut chemical consumption costs by 70 to 80%.

A comparative study of the efficiencies of direct and conventional filtration was conducted in a plant serving the city of Linhares, Brazil (Sperandio and Perez, 1976). The plant contained a rapid mixer followed by two independent treatment processes operated in parallel: conventional rapid filtration working with declining rate and direct filtration using upflow filters. The raw water turbidity entering the plant averaged 11 NTU and the color averaged 23 color units. The upflow filter contained a sand bed 2 m deep having an effective size of 0.6 mm and uniformity coefficient of 2.5. In general, the direct-filtration upflow unit (or contact unit) was

Table 8.6. Summary of Bench and Pilot Filter Tests at Brasilia, Brazil

Source	Turbidity (NTU)			Dosages (mg/l)			Filter Rate (m/hr)	Filter Run (hr)	Terminal Head Loss (m)	Number of Runs
	Raw	Pilot Filter	Paper Filter	Alum	Lime	Polymer				
Rio Descoberto Reservoir	12		0.95	3						
			0.40	3	1.0					
			0.61	2.5	0.2					
			0.82	2.5	0.8					
			1.0	2.0						
			20.0							
	25									
	12	0.32		4.0	1.0		18	15	1.2	13
	25	0.55		3.0	1.0		18	15	1.2	7
	45	0.65		4.5		0.5	18	15	1.2	11

Source: Wagner and Hudson, 1982.

Table 8.7. Summary of Bench and Pilot Filter Testing at Amman, Jordan

Turbidity (NTU)				Dosages (mg/l)		Filter Run (hr)	Average Filtration Rate (m/hr)
Raw	Settled	Paper Filter	Pilot Filter	Alum	Polymer		
Bench-Scale Testing							
114		0.6		1.0	1.0		
114		1.6		1.0	0.6		
114		1.8		1.0	0.4		
	12.5	1.4			0.5		
	12.5	1.1			0.5		
Pilot Filter Testing							
125	64		0.5		0.5	22	6.8
120	64		0.63		0.5	27	7.6
120	61		0.5		0.5	12 to 13	9.4
120	64		0.5		0.5	10	11.5
125	64		0.39		0.5	17	11.5

Source: Wagner and Hudson, 1982.

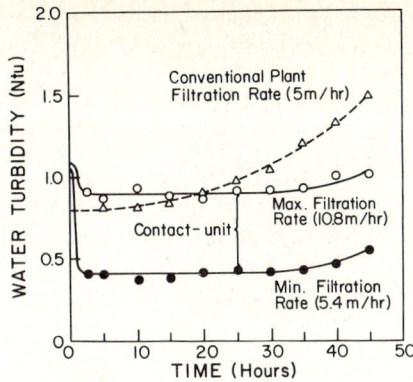

Figure 8.28. Comparative efficiencies of a conventional plant and contact unit at a plant in Linhares, Brazil. *Source:* Sperandio and Perez, 1976.

more efficient, as shown in the comparative graph in Figure 8.28. Two filtration rates were used in the upflow unit: a minimum rate of 5.4 m/hr and a maximum rate of 10.8 m/hr. The filter runs averaged about 30 hr. The filters were backwashed at an upflow velocity of 36 m/hr for a period of 20 min, owing to the great depth of the sand bed (twice the normal depth). The coagulant doses were reported to be much lower than those for the conventional filters.

In conclusion, because direct filtration omits sedimentation, it is not generally appropriate where raw water quality is highly variable. In any event, before any major commitment is made to direct filtration, pilot studies are warranted.

UPFLOW-DOWNFLOW FILTRATION*

Upflow-downflow filtration is a simple and economic treatment method for smaller water treatment plants in developing countries. In this type of system a battery of upflow roughing filters (called contact clarifiers) replaces the conventional arrangements for mixing, flocculation, and sedimentation used in rapid filtration plants. This can result in reduced construction and operation costs; the latter because the coagulant dosage used with this type of design is said to be smaller than that used for conventional treatment (Azevedo-Netto, 1977). Upflow-downflow type filters may be designed to treat turbid waters as well as relatively clean waters with the typical designs described in this section. Where sudden variations in the raw water quality are expected, however, the time available for adjusting the coagulant dosage is less than in conventional treatment. An important application of upflow-downflow filtration is in the design of simple modular and package water treatment plants. All of the designs described herein can be so adapted for fabrication of modular plants, which will further reduce their cost and enable prefabricated units to be transported to remote areas where on-site construction is impracticable (see Chapter 9).

---

*Upflow filtration alone is not recommended for potable water treatment as it incorporates a cross connection, in that contaminated washwater occupies the same space above the filter as the filter effluent; there is always the danger that the contaminated wastewater will flow into the filtered water clearwell.

The filter medium of the upflow unit may range from coarse sand having an effective size of 0.7 to 2.0 mm, up to graded gravel ranging in size from about 10 to 60 mm. The depth of the unit should be between 1.5 and 3 m. For coarse sand beds, filtration rates as high as 12 to 16 m/hr may be used, whereas those for gravel beds are limited to 4 to 8 m/hr. The choice of media for the contact clarifier should be based on pilot-filter studies of the water to be treated. The design parameters for the downflow (or polishing) filter are analogous to those used for rapid filters. Dual-media filter beds may be used to allow for higher filtration rates and longer filter runs. Backwash arrangements are necessary for both the contact clarifier and polishing filter. Water pressure for backwashing may be provided by an elevated washwater tank.

A simple upflow-downflow filtration scheme involving two units, generally concrete pipes, called "superfiltration," is shown in Figure 8.29 (Azevedo-Netto, 1977). Superfilters, based on a Soviet Union design, have been tested and installed in a number of small communities in Brazil. Table 8.8 indicates the required structural dimensions and washwater flow rates for superfilters having capacities ranging from 100 to 450 m^3/day. For larger installations, the capacity may be increased by placing a battery of superfilters in parallel. For efficient treatment with superfiltration, raw water turbidity should not exceed 160 NTU and normally be less than 50 NTU, whereas the color should not exceed 80 color units.

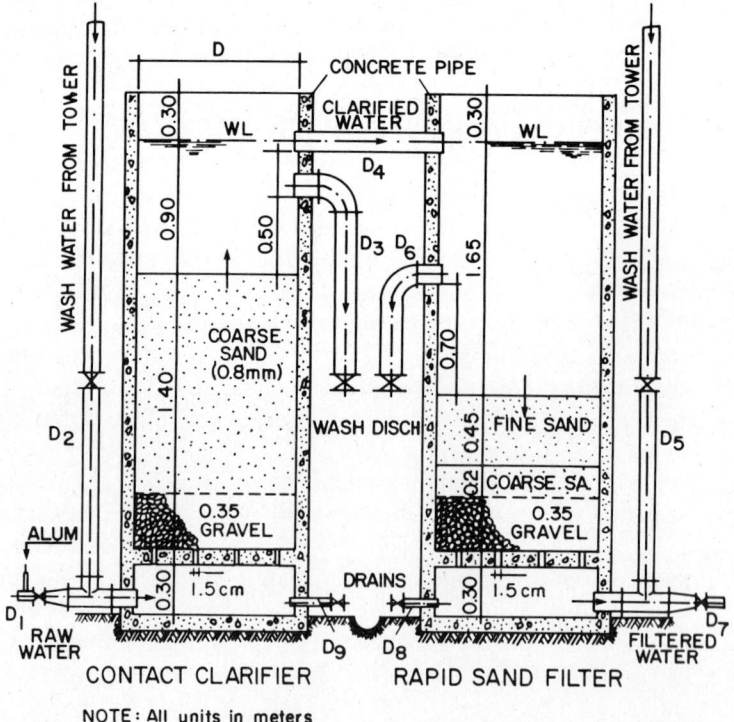

Figure 8.29. Gravity "superfilter"—Brazil. *Source:* Azevedo-Netto.

Table 8.8. Design Guidelines for Upflow-Downflow Filtration Units in Brazil[a]

Flow (m³/day)	$D_1 = D_7$ (cm)	$D_2 = D_5$ (cm)	$D_3 = D_6$ (cm)	D_4 (cm)	$D_8 = D_9$ (cm)	Diameter D (cm)	Washing (l/sec)
100	5	10	15	15	4	100	12
150	5	10	15	20	4	120	17
250	7.5	10	15	25	4	150	30
450	7.5	15	20	30	5	200	50

Source: Adapted from Azevedo-Netto, 1981, personal communication.
[a]For the significance of the various symbols, see Figure 8-29.

Field data on the performance of superfilters are limited; available data have shown mixed results. Two horizontal superfilters, having capacities of 1800 and 2400 m³/day, respectively, were installed at the Naval Base in Bahia, Brazil (Azevedo-Netto). The upflow contact clarifiers work with a filtration rate of about 5 m/hr, whereas the downflow filters work with a rate of about 7.5 m/hr. The turbidity and color of the raw water influent to the superfilters range from 20 to 60 NTU and 100 to 180 mg/l Pt, respectively, whereas the turbidity and color of the treated effluent are reduced to less than one NTU and 5 mg/l Pt, respectively. Filter runs averaged less than 24 hr, with effluent turbidities of 0.4 NTU for 17 hr, increasing to about 0.7 NTU at 23 hr. In Brazil, construction costs of superfiltration plants are about 50 to 60% of the cost of conventional installations.

Superfilters were also installed recently in the city of Colon, Costa Rica, and two smaller communities in that country (Institute of Water Supply and Sewerage, Costa Rica). Several faults in the design and construction of these units have led to numerous operational problems, the most important being inefficient distribution of washwater and excessive backwash rates that have resulted in loss of anthracite filter media and clogging of the inlet pipes to both filters. Furthermore, the overall performance of the superfilters was generally poor, with about 50 to 90% turbidity removal when treating raw water with a turbidity less than 50 NTU. Filter runs averaged typically about 10.5 to 12 hr. The difficulties were attributed primarily to improper hydraulic design of the filters and ancillary piping.

Plans and specifications for two types of upflow-downflow treatment plants have been published by CEPIS (1982), for capacities ranging from 800 to 1600 m³/day. In both designs, raw and backwash waters are delivered to a battery of three upflow filters from an elevated storage tank via either a single or dual-pipe system. In the single-pipe system, when an upflow filter has to be washed, operation of the remaining units must be interrupted by closing appropriate valves in order to provide sufficient flow of washwater to the dirty filter. In the dual-pipe system, the storage tank has two compartments; one for water to be filtered and one for backwash water. Each tank is provided with a pipe system connected to the filters; hence the flow of raw water to the upflow filters need not be interrupted while one of the filters is being backwashed. The battery of four downflow filters that follow the upflow filters are characterized by declining-rate filtration, dual-media beds, and interfilter-washing capabilities. Wherever possible, concrete distribution channels have been used in place of pipes. Raw water quality restrictions are the same as those for superfiltration. Several plants similar to the designs developed by CEPIS have been installed in Brazil, but data on their performance or cost were not available.

Kardile (1981) has developed three upflow-downflow filtration designs for rural communities in India. A 2400 m³/day plant for the community of Ramtek (population of 20,000) was designed for the treatment of low turbidity waters, whereas a 1000 m³/day plant for Chandori (population of 15,000) and a 4200 m³/day plant for Varangaon (population of 35,000) were designed for the treatment of turbid waters. The Ramtek plant is comprised of a gravel-bed pretreatment unit followed by dual-media filters; the Chandori plant is comprised of a gravel bed-cum-tube settler pretreatment unit followed by dual-media filters; and the Varangaon plant

PREFILTER   DUAL MEDIA FILTER

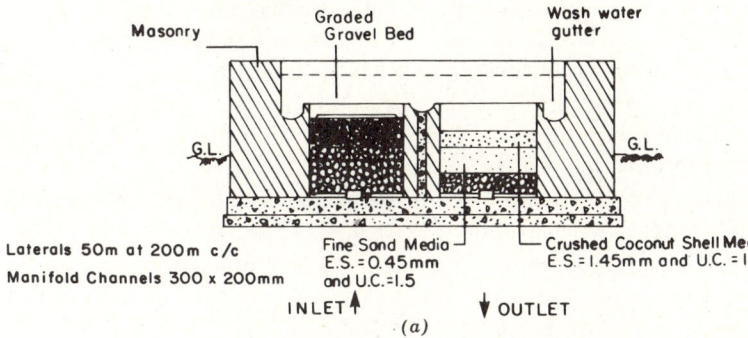

Masonry
Graded Gravel Bed
Wash water gutter

G.L.
G.L.

Laterals 50m at 200m c/c

Manifold Channels 300 x 200mm

Fine Sand Media
E.S.=0.45mm
and U.C.=1.5

Crushed Coconut Shell Media
E.S.=1.45mm and U.C.=1.47

INLET ↑   ↓ OUTLET

(a)

DUAL MEDIA FILTER   PRETREATOR

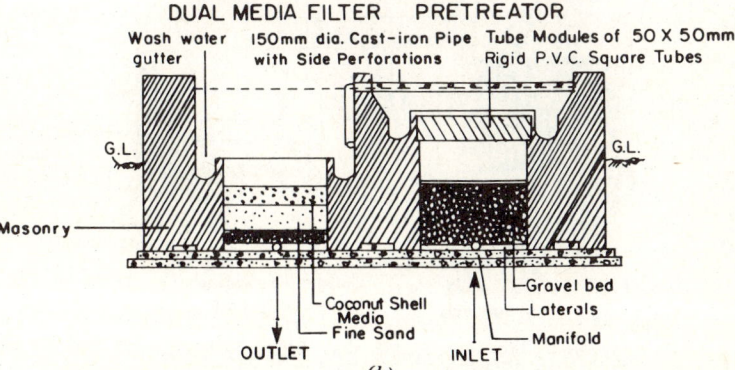

Wash water gutter
150mm dia. Cast-iron Pipe with Side Perforations
Tube Modules of 50 X 50mm Rigid P.V.C. Square Tubes

G.L.
G.L.

Masonry

Coconut Shell Media
Fine Sand

Gravel bed
Laterals
Manifold

OUTLET   INLET

(b)

Figure 8.30.   Upflow-downflow filtration plants in India (a) flow diagrams of Ramtek filter and aerial view of plant, (b) flow diagram and aerial view of Chandori treatment plant. *Source:* Kardile, 1981.

187

has two treatment units in parallel consisting of one gravel bed flocculator and tube-settling tank, which are followed by three dual-media filters. Flow diagrams and aerial views for the Ramtek and Chandori plants are shown in Figure 8.30; the Varangaon plant is shown in Figure 8.31. Recommended design criteria for each of these plants are listed in Table 8.9. The design of gravel-bed flocculators, which are common to all three plants, is discussed in Chapter 6 (under "Gravel-Bed Flocculators"). The dual-media filters for each of the plants consist of a layer of crushed coconut shells (specific gravity 1.4) placed on top of layers of sand and gravel. The underdrain system is a simple perforated pipe and manifold design that is manufactured locally. The thick masonry sidewalls (clearly outlined in Figure 8.30) for the Ramtek and Chandori plants, were used in place of reinforced concrete to take advantage of local materials and unskilled labor found in the villages. They also served to support the rapid-mixing channels and walkways. The construction costs for the three plants were between 30 to 50% of the construction costs for the same capacity conventional rapid filtration plants (see Chapter 10).

An evaluation of the performance of the Varangaon treatment plant was conducted over a three-year period, from 1978 to 1981 (Tasgaonkar, 1981). Performance criteria included (1) reductions in turbidity, (2) washwater consumption, and (3)

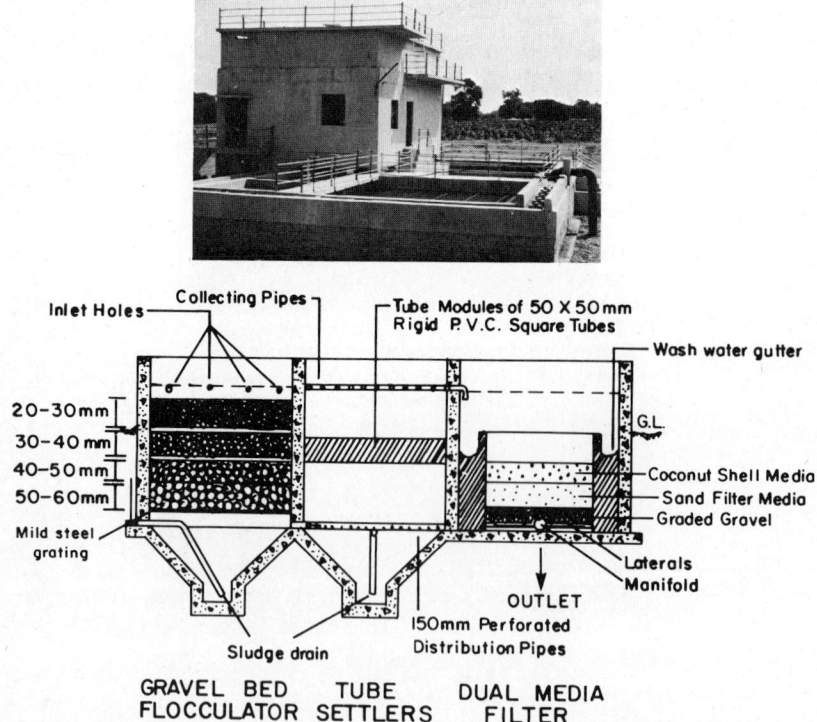

Figure 8.31. Flow diagram and ground-level view of upflow-downflow plant in Varangaon, India. *Source:* Kardile, 1981.

Table 8.9. Recommended Design Criteria for the Indian Upflow-Downflow Treatment Plants

	Design Criteria	Ramtek Plant (2400 m³/day)	Varangaon Plant (4200 m³/day)	Chandori Plant (1000 m³/day)
		I. Raw Water Turbidity		
i.	General Recommendations	For low turbidity sources	For high turbidity sources	For moderate turbidity sources
ii.	Average range in NTU	10 to 30	30 to 100	30 to 100
iii.	Maximum range in NTU	300 to 500	1000 to 5000	1000 to 2000
		II. Pretreatment		
A.	Mixing unit	Mixing channel	Mixing channel	Mixing channel
	i. Type of gravel bed units	Prefilter	Flocculator	Pretreator
	ii. Direction of flow	Upward	Downward	Upward
	iii. Surface loading in m/hr	4 to 7	4 to 7	4 to 8
	iv. Depth of the gravel bed in m	1.5 to 2.0	2.5 to 3.0	1.5 to 2.0
B.	Tube settling tank	Not adopted	Tube settler	Gravel bed-cum-tube
	i. Surface loading in m/hr	—	5 to 10	4 to 8
	ii. Detention period in minutes	—	30 to 50	30 to 50
	iii. Depth of the tank in m	—	3 m above hopper	3.5 to 4.0
	iv. Direction of Flow	—	Upward	Upward
	v. Size of PVC square tubes	—	50 mm x 50 mm	50 mm x 50 mm
	vi. Depth of tube settler	—	0.5 to 0.6 m	0.5 to 0.6 m
		III. Dual Media Filter Bed		
i.	Surface loading in m/hr	4 to 7	5 to 10	4 to 8
ii.	Dual media details			
	a. Coconut shell media depth	30 to 40 cm	30 to 50 cm	30 to 50 cm
	Average size in mm	1.0 to 2.0	1.0 to 2.0	1.0 to 2.0
	b. Fine sand media depth	50 to 60 cm	50 to 60 cm	50 to 60 cm
	Effective size in mm	0.45 to 0.55	0.45 to 0.55	0.45 to 0.55
	Uniformity coefficient	Below 1.5	Below 1.5	Below 1.5
iii.	Back wash method	Hard wash	Hard wash	Hard wash

Source: Adapted from Kardile, 1981.

Table 8.10.   Turbidity of Raw, Settled, and Filtered Water
for the Varangaon Plant—India

Date	Turbidity of Water (NTU)			Alum Dose (mg/l)
	Raw	Settled	Filtered	
August 1978	5100	22	1.8	144
August 1978	9610	75	2.1	196
September 1978	2810	20	1.8	112
October 1978	28	13	1.5	10.6
March 1979	30	17	1.6	8.0
August 1979	10,200	29	2.2	208
August 1979	4700	24	2.0	120

Source: Tasgaon-kar, 1982.

length of filter run. Turbidity readings for raw, settled, and filtered waters, together with corresponding alum doses, are presented in Table 8.10 for several randomly selected dates. Despite some periods of extremely high raw water turbidities (e.g., 10,200 NTU was recorded during a monsoon on March 29, 1979), the plant was still able to produce filtered water with turbidities below the current Indian standard of 2.5 NTU. Alum consumption varied directly with raw water turbidity, and in some instances was quite high (208 mg/l on March 29, 1979). The washwater consumption as a percentage of total filtered water varied from 1.4 to 2.9%. The average length of filter runs was 45 hr.

SLOW SAND FILTRATION

The flow rates for slow sand filters are about 20 to 50 times slower than for rapid filters. Because the filter is cleaned by manually removing the dirty top sand rather than backwashing, the sand is not stratified and its hydraulic characteristics are governed by the finer portions of the sand. Another distinguishing feature of slow sand filters is the presence of a thin layer, called the *schmutzdecke,* which forms on the surface of the sand bed and includes a large variety of biologically-active microorganisms. These break down organic matter and also fill the interstices of the sand so that solid matter is retained quite effectively. The impurities present in the raw water are removed almost entirely in the upper 0.5 to 2 cm of the filter bed.

Cleaning of the filter bed is carried out by scraping off this top layer when it becomes too clogged with impurities. Unless the water being treated is excessively turbid or has high algal concentrations, slow sand filters may run continuously for a period of several months before cleaning is necessary. The filter-cleaning operation may be carried out by unskilled laborers using hand tools, and completed in one or two days. Figure 8.32 shows the manual cleaning of a slow sand filter in India.

Figure 8.32. Manually cleaned slow sand filter in India. *Source:* IRC, 1978.

After cleaning, some time (up to one week) is required to ripen the *schmutzdecke* and return the filter effluent quality to its former level.

The principal use of slow sand filtration is in the removal of organic matter and pathogenic organisms from raw waters of relatively low turbidity. The biological treatment that takes place in the *schmutzdecke* of the filter is capable of reducing the total bacteria count by a factor of 10^3 to 10^4 and E. coli count by a factor of 10^2 to 10^3 (IRC, 1981b). Accordingly, considerable savings can be realized in the quantities of chlorine required for disinfection. Such an advantage is particularly important in rural areas of developing countries where chlorination practices have proven to be very unreliable, and where slow sand filtration can provide a more reliable safety barrier than, for example, rapid filters that require uninterrupted chlorination to assure safety.

Slow sand filters are most practical in the treatment of water with turbidity below 50 NTU, although higher turbidities can be tolerated for a few days. The best purification occurs when the turbidity is below 10 NTU (Huisman and Wood, 1974). When higher turbidities are expected, slow sand filters should be preceded by some type of pretreatment (see Chapter Three). Lest these units be thought to be outmoded, London continues to build such plants, with roughing filters for pretreatment.

Slow sand filters provide a number of distinct advantages for developing countries which are summarized here (Feachem, McGarry, and Mara, 1977).

1. The cost of construction is low.
2. Simplicity of design and operation means that filters can be built and used with limited technical supervision. Little special pipework, equipment, or instrumentation is needed.

3. The labor required for maintenance can be unskilled because the major job is cleaning the beds, which can be done by hand.

4. Imports of material and equipment can be negligible and no chemicals are required.

5. Power is not required if gravity head is available, and there are no moving parts or requirements for compressed air or high-pressure water.

6. Variations in raw water quality and temperature can be accommodated provided turbidity does not become excessive; overloading for short periods does no harm.

7. Water is saved—an important matter in many areas—because large quantities of washwater are not required.

The factors that weigh against the use of slow sand filtration and lead to the choice of rapid filters as the more appropriate treatment method apply mainly in the industrialized countries: (1) large land requirements (about five times that required for rapid filtration plants); (2) higher construction costs in countries where construction methods are largely mechanized and labor is expensive; (3) higher costs for cleaning the filters in countries where manual labor is expensive; (4) need to cover the filters in freezing climates (it also is difficult to find people who will work at cleaning in cold weather); (5) working of the biological layer (i.e., *schmutzdecke*) may be upset by certain types of toxic industrial wastes or heavy concentrations of colloids; and (6) certain types of algae may interfere with the working of the filters, usually choking the filter bed, which calls for frequent cleaning.

Interestingly, these limitations (with the exception of the last) generally do not apply in developing countries, and this problem may be ameliorated by covering the filter or using an algicide to inhibit algal growths (see Chapter Three; under "Control of Microorganisms"). Under suitable circumstances, then, slow sand filtration is the cheapest, simplest, and most efficient method of water treatment for low turbidity surface waters in developing countries. However, it should not be considered as a panacea for the treatment of all surface waters in developing countries, even when preceded by pretreatment units.

In the state of Maharashtra, India, slow sand filter plants have been installed with coagulation and settling tanks ahead of the filters in order to treat high turbidity waters (50 to 500 NTU) during the rainy season. Observations on these plants by Kardile have shown that pretreatment has not generally been effective in reducing floc carryover to the filters sufficiently to prevent rapid clogging. In some cases the entire sand bed had to be removed and cleaned because of clogging. One filter was anaerobic, which resulted in taste and odor problems in the effluent. This was attributed to intermittent operation of the filters, which allowed stagnant, turbid water to remain on top of the filters for extended periods of time, depriving microorganisms in the sand bed of oxygen.

Design of Slow Sand Filters

The essential parts of a slow sand filter are shown in Figure 8.33. Slow sand filters, because they are not backwashed, are much simpler in design than rapid filters. Pertinent design criteria for the design of slow sand filters are summarized below (adapted from Huisman and Wood, 1974).

1.   Rate of Filtration.   The traditional rate of filtration used for normal operation is 0.1 m/hr, although it is possible to produce safe water at rates as high as 0.4 m/hr. At higher filtration rates, the intervals between filter cleanings are shortened, but the quality of the treated water does not deteriorate. Higher rates of filtration can be used during periods when a filter is out of service for cleaning.

2.   Supernatant Water Layer.   The depth of water should provide a head sufficient to overcome the resistance of the filter bed and prevent air binding. In practice a head of 1.0 to 1.5 m is usually selected.

3.   Filter Bed.   The sand bed thickness varies between 1.0 and 1.4 m. This thickness should be reduced to not less than 0.5 to 0.8 m after removing the upper sand layers during filter cleaning. Filter sand should have an effective size between 0.15 and 0.35 mm and a uniformity coefficient between 1.5 and 3, although a coefficient of less than 2 is desirable. The careful selection and grading of sand are not as critical as in rapid filters. Use of builder grade or locally available sand can reduce costs.

4.   Filter Gravel.   The filter gravel should be so graded that the sand does not penetrate the underdrain system, yet provides free flow of water when a limited number of underdrains are provided. For example, when using a filter bottom composed of stacked bricks with open joints (10 mm wide) four layers of gravel are normally used with the following size ranges: 0.4 to 0.6 mm; 1.5 to 2 mm; 5 to 8 mm; and 15 to 25 mm; each layer about 10 cm thick (IRC, 1981b). A photograph depicting this graded-gravel scheme is shown in Figure 8.34.

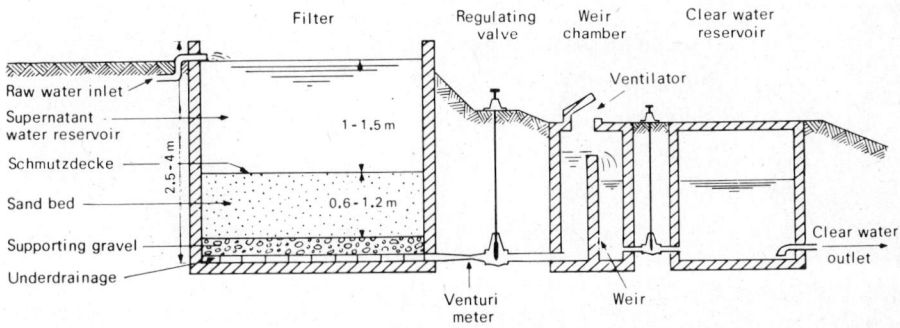

Figure 8.33.   Diagram of a slow sand filter. *Source:* Huisman and Wood, 1974.

Figure 8.34.   Graded-gravel scheme for slow sand filter. *Source*: Hisman and Wood, 1974.

   5.   *Underdrain System.*   A simple method of underdrainage consists of a system of main and lateral drains made from perforated pipes of asbestos cement or plastic (see Figure 8.8). A filter bottom of stacked bricks, concrete slabs, or porous concrete may also be used, as shown in Figure 8.35. Bricks should be placed so that (1) the clear spacing between adjacent bricks is smaller than the size of the supporting media, and (2) rows about 10 cm apart are formed to obtain lateral drainage conduits that drain into a large collector. The collector is usually connected to the filter control chamber through an orifice or small pipe.
   6.   *Depth of Filter Box.*   The minimum depth of the filter box is determined from the following elements (Paramasivam and Mhaisalkar, 1981).

Freeboard above supernatant level	0.20 m
Supernatant water	1.00 m
Filter medium (initially)	1.00 m
Four-layer gravel support	0.30 m
Brick filter bottom	0.20 m
Total	2.70 m

It is general practice to use a filter box 3 to 4 m deep, but a depth of 2.70 m will reduce construction cost without sacrificing filter efficiency.
   7.   *Number of Filter Beds.*   At least two filter units should always be built, and reserve units should be provided for large treatment plants. Table 8.11 gives

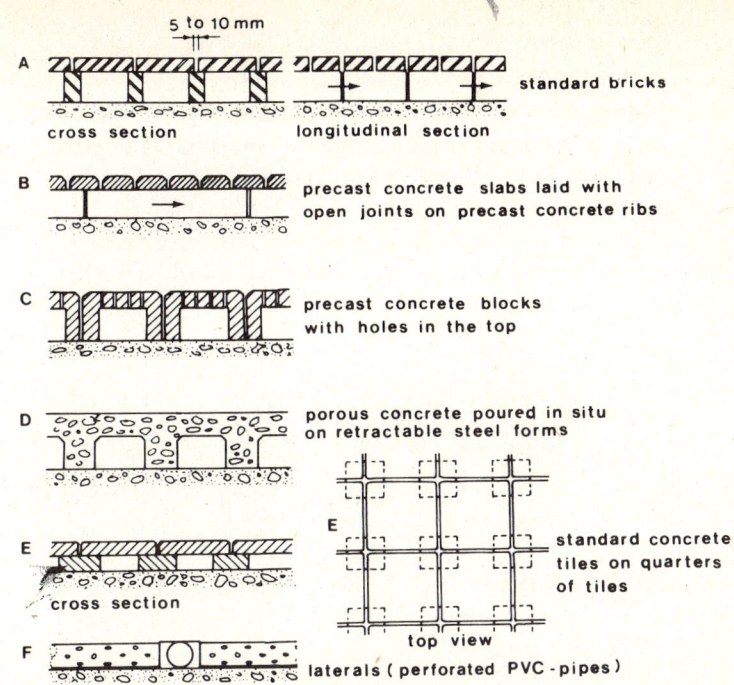

Figure 8.35. Different types of underdrain systems for slow sand filters. *Source:* Adapted from Thanh and Hettiaratchi, 1982.

some rough guidelines for determining the number of filter units for a given design population (Arboleda, 1973).

    8. Filter Control. Slow sand filters are operated conventionally at a constant rate. The rate is controlled by maintaining a constant head loss across the filter. A hand-operated valve preceded by a venturi meter can be used to regulate the filtration rate and depth of water over the filter (Figure 8.33). The normal range of head loss from clean to clogged conditions in the sand bed and filter appurtenances is 0.6 to 1.2 m. An effluent weir, such as that shown in Figure 8.33, is a valuable device to prevent air binding. Another practical method for controlling the rate of filtration

Table 8.11. General Guidelines for Determining the Number of Slow Sand Filters Required for Different-sized Communities

Population	Total Number of Units	Reserve Capacity
>2000	2	100%
2,000 to 10,000	3	50%
10,000 to 60,000	4	33%
60,000 to 100,000	5	25%

Source: Arboleda, 1973.

is by installing a float valve on the filter effluent line, as shown in Figure 8.36*b*. The valve is activated by the water level in a float chamber, which is constructed to maintain a reasonably constant head over an orifice in the float chamber. The level of the orifice or filter outlet must be above the top of the sand bed.

Telescopic piping outlet structures (Figure 8.36*c*) are commonly used in dynamic filters. The lower pipe is fixed vertically and connected to the filtered water effluent line that leads to the clear well; the higher pipe is adjustable and moves vertically inside the fixed pipe, responding to fluctuations in the water level by means of a float attached to the end of the pipe. A circular weir, supported by the float, maintains

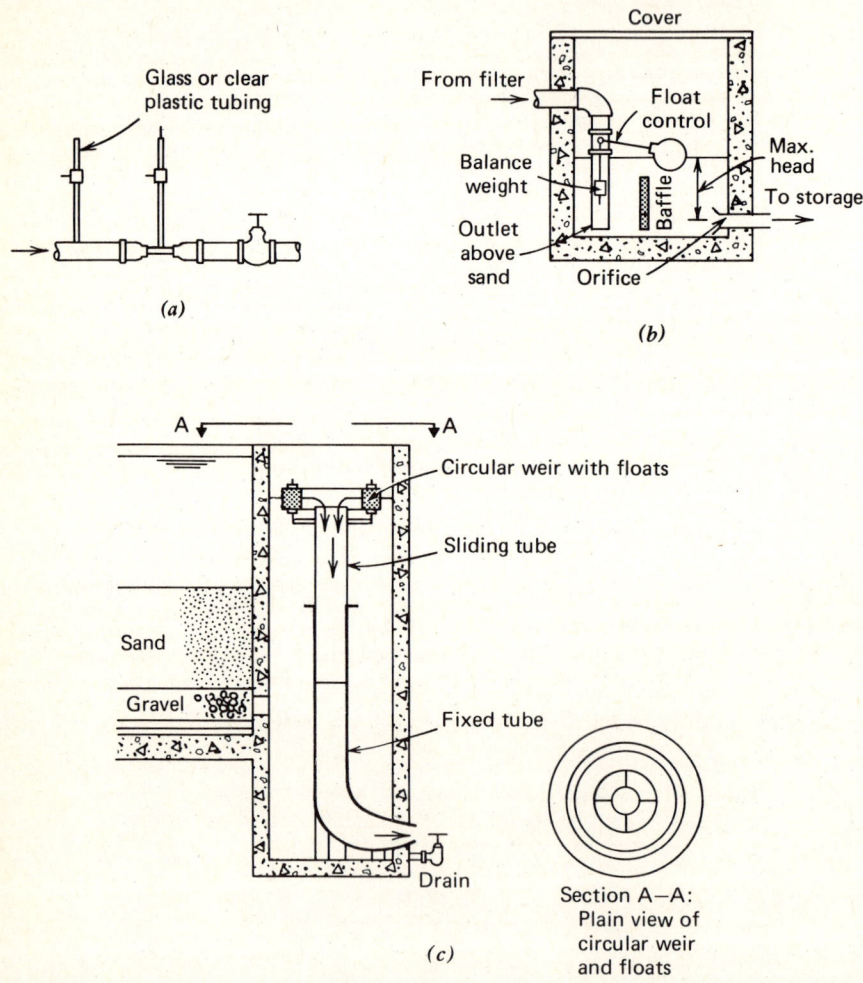

Figure 8.36. Typical devices for the control of the rate of flow in slow sand filters: (a) simple operation and control using common plumbing fittings calibrated before installation. Outlet elevation is above sand surface; (b) float control and orifice outlet structure; (c) telescopic piping outlet structure. *Source:* (a, b) Salvato, 1982; (c) Arboleda, 1973.

a constant flow of water to the clear well. The rate of flow can be changed by adjusting the level of the weir on the sliding pipe. The height of the float is limited by chains, so that filtration will stop after reaching the maximum allowable head loss.

Consideration should be given to the possibility of operating the filter at a declining rate, where the raw water inlet is closed with the filter outlet valve open. The supernatant drains through the filter at a declining rate. The effluent weir should be set at least 0.2 m above the top of the filter bed, to prevent drying of the *schmutz-decke* at the end of the filtration period. A sufficient quantity of water is required above the filter bed for storage.

Criteria for the design of slow sand filters are summarized in Table 8.12.

In areas where sand is expensive or difficult to obtain, the surface scrapings from a slow sand filter may be washed, stored, and reused at a later date. The scrapings must be washed immediately however, otherwise the material may go anaerobic, yielding taste- and odor-producing substances that are nearly impossible to remove by washing.

The hydraulically operated device shown in Figure 8.37a can be used to wash sand removed from a filter. In this unit sand and dirt are separated by water jetting through a pipe into a chamber where the sand settles but the dirt is carried away by the water. The rate of overflow of the washer should not exceed the settling velocity of the smallest sand particle to be retained. However, in practice, turbulence and sand concentration reduce the desired rate of overflow appreciably, so that the rate of flow of the incoming sand-water solution is generally sufficient to effect separation without supplementary water. The sand or grit that settles to the bottom is ejected hydraulically or can be removed by means of a shear gate. Several such sand washers are usually connected in series. The clean sand is finally jetted into an open sand storage bin. Pipes carrying sand-water solutions should be sized for velocities of 1.5 m/s or higher. About 8 m^3 of sand per hour can be washed per square meter of washer surface area (Fair, Geyer, and Okun, 1968).

For small installations, the sand washer (equipped with a shear gate) can serve as a storage area for cleaned sand; for larger installations a sand separator (Figure

Table 8.12.   General Design Criteria for Slow Sand Filters

Parameter	Range	Preferred Value
Filtration velocity (m/h)	0.1 to 0.2	0.1
Depth of filter bed (m)	1. to 1.4	1.0
Area per filter bed (m^2)	10 to 100	—
Height of supernatant water (m)	1 to 1.5	1.0
Depth of system of underdrains (m)	0.3 to 0.5	0.4
Specifications of filter bed	UC = 1.5 to 3.0	
	E.S. = 0.15 to 0.35 mm	
Number of filters	minimum of 2	

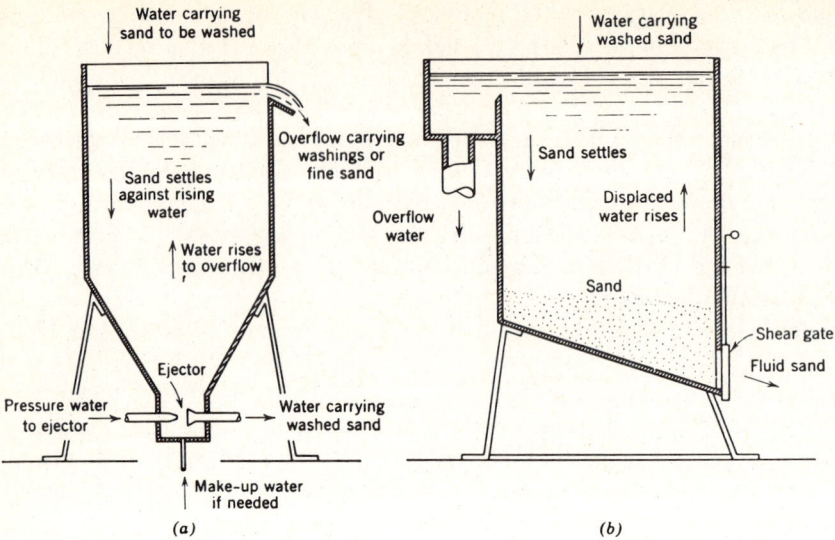

Figure 8.37. (a) Hydraulically operated sand washer (b) Gravity-operated sand separator. *Source:* Fair, Geyer, and Okun, 1968, vol. 2.

8.37b) can be used to effect separation of the sand from the washing water and for storage. An adaptation of the foregoing sand washers has been used successfully for years at the Madras waterworks in India.

In smaller plants, where hydraulically operated sand washers are not practicable, sand may be washed entirely by hand as illustrated in Figure 8.38. The sand is agitated in a box with water running through it at a low velocity so as not to wash out the fine particles. This process continues until the washing water clears, indicating that the sand is clean. The sand can then be stored and is ready for replacement on the filter.

When the bed reaches a minimum thickness of 0.5 to 0.8 m, the bed should be resanded. The clean sand should be placed on top of the gravel, and the older sand should be placed on top to provide seeding with microorganisms to form the *schmutzdecke* more rapidly and to assure that all the sand will be cleaned from time to time.

A two-year study by Imperial College, London, for the Oxfam relief agency is looking at the feasibility of using mats to cover the sand beds to permit easy cleaning (Nigel Graham). The mat is made of Bondina, a synthetic material normally used in air conditioning, which is characterized by small interstices that provide a suitable environment for the growth of microorganisms to form the *schmutzdecke*. When the mat becomes clogged, it can be removed manually from the sand bed, cleaned by pressurized water, and replaced on the bed. This procedure replaces scraping of the top 1 to 3 cm of sand from the bed. Also, the ripening period that is required for the formation of the *schmutzdecke* is thought to be shorter with synthetic mats

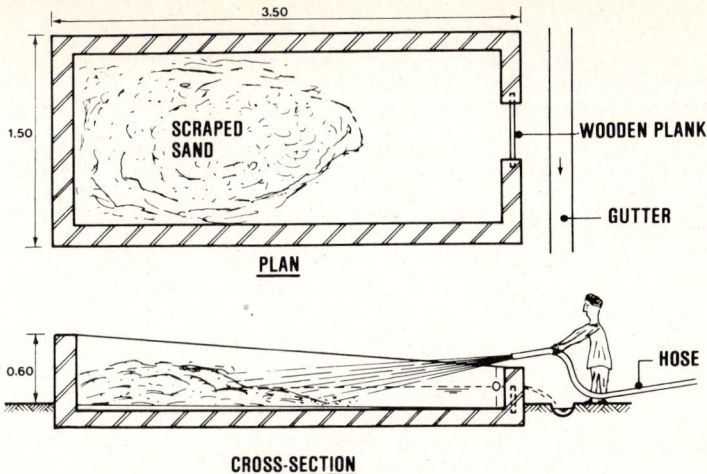

Figure 8.38.   Washing platform for manual cleaning of sand. *Source:* IRC, 1981b.

as compared with that for the sand bed alone. In large filters several rectangular mats can be placed on top of the sand beds, so that their edges overlap with one another, covering the entire surface area of the filter. This facilitates removal and cleaning operations. Field testing of this new technology is underway in Peru.

Dynamic Filtration

A unique type of slow sand filtration system, called dynamic filtration, originally developed in the Soviet Union, has been used in rural areas of Argentina and Ecuador. The basis and design of dynamic filters have been studied in Latin America (Perez, 1980; Rodriguez, 1977). The filter consists of a shallow channel, about one meter in depth, with a sand bed and underdrain system similar to those used in conventional slow sand filters (see Figure 8.39). The raw water flows over the bed surface in a thin fluid layer, about 1 to 3 cm deep, and then over a weir into an overflow channel to waste. Part of the flow (about 10% of the total) percolates through the sand bed into the underdrain system, and is conveyed to a clear well at about the same rate as in a conventional slow sand filter.

The main advantages of dynamic filters in relation to conventional slow sand filters are: (1) the low construction costs, because the filter walls can be very low, requiring less excavation, and can be built of unreinforced concrete or brick masonry; (2) raw water turbidities as high as 50 NTU can be applied regularly, and even higher turbidities accepted for short periods, such as during storms; and (3) cleaning of the sand bed is simpler, because the surface is raked daily to prevent excessive clogging, rather than physically removing the top few centimeters of dirty sand for cleaning.

The main disadvantages are (1) a much greater volume of water is needed to

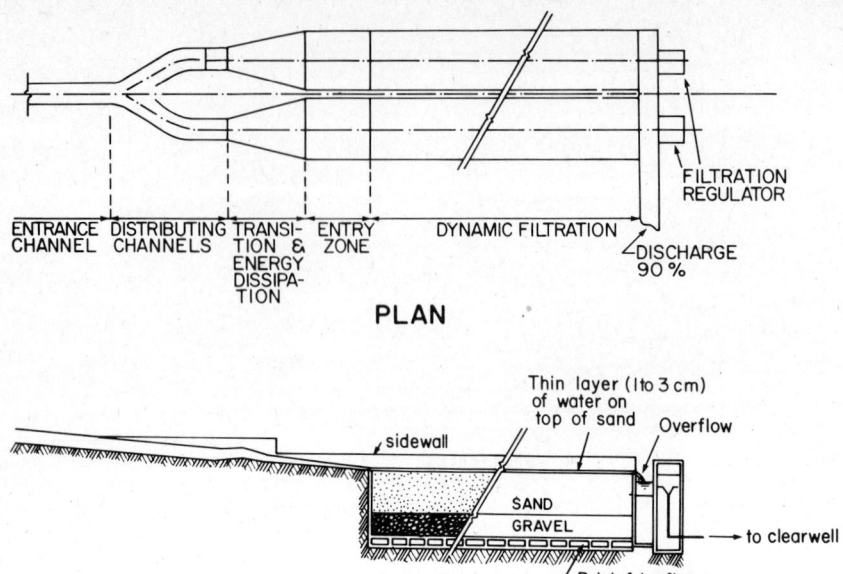

ENTRANCE DISTRIBUTING TRANSI- ENTRY   DYNAMIC FILTRATION
CHANNEL   CHANNELS   TION &   ZONE
                      ENERGY
                      DISSIPA-
                      TION

FILTRATION
REGULATOR

DISCHARGE
90 %

PLAN

Thin layer (1 to 3 cm)
of water on
top of sand   Overflow

sidewall

SAND
GRAVEL

Brick false floor

to clearwell

LONGITUDINAL SECTION

Figure 8.39.   Diagram of a dynamic filter—Argentina. *Source:* Arboleda, 1973.

serve an equivalent population (the source of supply must be at least 10 times the capacity needed), and (2) the *schmutzdecke* formed in these filters is likely to be less effective than in conventional slow sand filters, although there have been no conclusive studies conducted to verify this assumption. Furthermore, because of the large areas required, these filters are suitable only for smaller facilities. Design criteria and descriptions of system components for dynamic filters are given below. Filtration velocities, underdrain systems, and filter control are the same as those for conventional slow sand filters.

Shape of Filter   The filter is rectangular, with a length to width ratio of approximately 5:1. A minimum of two units should be built, each designed for no less than 65% of the required capacity, so that one unit can be taken out of service without overloading the operating unit(s) excessively.

The raw water flows over the bed surface with a velocity of 0.2 to 0.35 m/s, forming a thin layer about 1 to 3 cm deep, which assures that there is no erosion of filter media or sedimentation of larger organic particles transported by the water. Hence, a presettling basin is not normally required, because much of the material that would precipitate on the filter surface is swept away by the water flowing over the surface of the sand.

The following filter media sizes are recommended:

Media	Effective Size (mm)	Coefficient of Uniformity	Thickness (cm)
Filter sand	0.25 to 1.0	2 to 3	60 to 80
Upper supporting gravel layer	1.0 to 2.0	—	5
Lower supporting gravel layer	2.0 to 4.0	—	5

The unique design aspects of dynamic filtration are contained within the inlet and outlet structures. The inlet structure is comprised of a transition and energy dissipation channel, which is designed to reduce turbulence and scouring velocities as the raw water flows over the filter surface. A shallow water depth over the media combined with turbulent conditions can also encourage the formation of air bubbles that would be trapped in the filter media, thus reducing its efficiency. The transition and energy dissipation channel connects the raw water conduit to the entry zone, and consists of an expanding cross-sectional area and an inverted slope bottom which serve to dissipate the energy of the incoming raw water. The entry zone, located ahead of the filter box, should be constructed of concrete, because experience has shown that vortices are likely to form in this region, which can cause erosion problems. The lateral dimensions of the entry zone are the same as those of the filter unit.

The outlet structure consists of a rectangular overflow weir placed along the entire filter width, and an overflow channel that carries the nonfiltered flow back to the raw water source. The water elevation over the weir crest should be similar to that of the filter to allow full drainage of the excess flow without eroding the filter media. The overflow channel should be designed to carry about 90% of total raw water flow.

The most commonly used rate control system for dynamic filters are telescopic pipes. Other simple filter control systems, such as float-controlled butterfly valves, may also be used (see section on "Design of Slow Sand Filters" and Figure 8.36).

Operation and Maintenance   The filter is cleaned by scraping the top portion of the filter media (about 5 cm), so the retained particles are washed out by the surface current. The cleaning operation should be made when the head loss reaches a predetermined value. Filter runs of 24 to 48 hr are used in practice, depending on the quality of the raw water and rate of filtration. Rates less than 0.30 m/hr and turbidities higher than 30 NTU call for filter runs of 24 hr. When the lower layers of the filter media show signs of clogging, the filter media should be removed and replaced. This operation normally takes place every 6 to 12 months.

The performance of dynamic filters was studied in the laboratory at the University of Chile (Rodriguez, 1977). Rates of filtration at the beginning of runs were well over 0.4 m/hr, dropping to 0.25 and 0.1 m/hr after 60 and 80 hr of operation.

Effluent turbidities increased only slightly over the period, beginning about the twentieth hour of operation.

As a relatively new approach to filtration, field or pilot-plant demonstrations of its performance are warranted before dynamic filtration is widely adopted.

Information Sources on Slow Sand Filtration

Information on the design and operation of slow sand filters is contained in two comprehensive design manuals by Huisman and Wood (1974), and the IRC (1978). The former describes the design, construction, and operation of modern slow sand filters, the theory of biological filtration, and the various methods of cleaning filters, which range from simple manual techniques to advanced mechanical or hydraulic systems. The latter focuses on simple designs for developing countries, and includes guidelines for the design and construction of small slow sand filters, suitable treatment methods for the removal of turbidity, as well as four typical designs with capacities between 25 to 1000 m^3/day. Complete plans and specifications for these designs are also included.

Arboleda (1973) discusses pertinent design criteria and simple flow rate controllers for three types of slow sand filters, namely conventional, upflow, and dynamic filters. The IRC design manual, *Small Community Water Supplies* (IRC, 1981*b*), devotes an entire chapter to the design and construction of slow sand filters for small communities in developing countries, and includes illustrations of several simple types of designs.

Economic considerations in the design of slow sand filters, together with basic design criteria, are given by Paramasivan and Mhaisalkar (1981). Mathematical models are developed which optimize the number and dimensions of filters in order to minimize filter costs (see Chapter Ten, under "Construction Costs of Water Treatment Plants").

A comprehensive annotated bibliography on the subject of slow sand filtration has been published by the IRC (1977c). The selected references deal mainly with the technical aspects of the process. An author and key word index is provided as well as a list of institutions and organizations that can provide further information on the subject.

chapter nine

MODULAR AND PACKAGE DESIGNS FOR STANDARDIZED WATER TREATMENT PLANTS

The conventional engineering approach for water treatment plant design involves planning on an individual community basis, or regional planning when several communities are to be served by a single project, followed by preliminary engineering and detailed engineering design. This approach, however, does not lend itself to the swift construction of a large number of small projects, which is the situation commonly encountered in the developing countries, particularly those committed to the International Water Supply and Sanitation Decade. One alternative is to adopt standardized procedures for the planning, design, and construction of water supply elements in order to decrease the time needed for the design of projects, reduce the number of experienced designers needed, and lower the overall cost. Other advantages that accrue from the reasonable use of standard designs are listed below (Brown and Okun, 1968; IRC, 1981a).

1. Expansion of the productivity of the skilled engineering designer.
2. Simplification of the design problem for the experienced engineer or technician.
3. Reduction in the cost of detailed design, thereby allowing more to be spent on preliminary studies which are often slighted.
4. Reduction in construction costs if standards are based on the use of local materials.

5. Improvement in construction quality.
6. Simpler operation training.
7. Lower maintenance costs as spare parts could be more easily stocked.
8. Promotion of local industry and expertise in the manufacture of equipment at low cost.

The main disadvantage in adopting standard designs is that they may become rigid and inflexible, thereby tending to inhibit the imaginative engineer and hence stifle improvement. This could be a serious problem if standards do not permit or encourage innovation.

Standard design manuals that are written for a particular country should reflect that country's unique conditions and needs, although the experiences of other countries or the work of international agencies may be helpful. For example, the design manual on slow sand filtration published by the IRC (1978), the manual on modular water treatment plants produced by CEPIS (1982), or the standard designs presented in this manual (e.g., upflow-downflow filters described in Chapter Eight), could readily be incorporated into a country-specific design manual. In general, standard designs and specifications should be kept simple, keeping in mind the need for quick installations at low cost, and ease of construction, operation and maintenance, while still providing the minimum acceptable level of service.

In Latin America, a "systems" approach has been developed for the promotion, design, construction, and operation of water and sanitation projects for small communities (Donaldson, 1976). Under this approach, projects are broken down into their component parts, and each is studied for its effect upon the others. These elements are then coordinated to yield the lowest cost solution that meets the desired goals, that is, the implementation of the greatest number of systems in the shortest time. For example, the technical aspects of a project are designed using existing maps or aerial photographs, standardized design criteria (e.g., 200 l per capita per day, etc.), predesigned elements (modular treatment units, pump houses, etc.), and standardized equipment lists. The materials are gathered in a central place and sent to the community as a package, together with any necessary tools or equipment not available locally. Also, professionals are available to involve the local community in the project, including the training and supervision of workers. Using this approach it is possible to delegate a considerable amount of work to intermediate level technicians and local workers.

Two types of standardized designs that have been used for treating water in small communities are package plant and modular plant designs. For the purposes of this text, package plant designs refer to compact treatment units, generally made of steel, manufactured entirely in the factory, and transportable to the site. Modular plant designs refer to compact treatment units, generally made of concrete or masonry, and assembled either partly or entirely on-site without large or complicated equipment. The compact nature of both types of plants may be attributed in part to technological advances, such as gravel-bed and heliocoidal-type flocculators, inclined-plate settling, and dual-media filters which, when properly designed, can greatly reduce the size of the treatment units. However, using such technologically advanced units at higher loadings requires skilled operation. Often, simple standardized units should be used at low loadings to assure proper operation and reliability. The merits of package and modular types of designs for developing countries, as well as some typical plant layouts and design criteria, are presented below.

PACKAGE WATER TREATMENT PLANTS

The popularity of package plants in the industrialized countries has grown in recent years, stimulated by rising construction and labor costs of custom-designed treatment facilities, particularly for smaller installations. Such plants are often automatic and designed for virtually unattended operation, offering savings over conventional plants. Package plants are preassembled in factories, where costs can be more carefully controlled than in the field. In most designs, on-site assembly and installation requirements are kept to a minimum. Accordingly, package plants have become a practical and economical solution for water treatment in small communities in North America and Europe.

On the other hand, these package water treatment plants are not well-suited for small communities in developing countries as contrasted with facilities constructed wholly or partly on site, for the following reasons.

1. Low-cost local labor is available in most communities and hence on-site construction costs are lower than those encountered in the industrialized countries.

2. On-site construction can provide additional jobs for the local community, and concomitantly instill a sense of ownership to those that contribute their time and effort toward the project. This encourages better operation and maintenance than is provided when package units are installed by outside contractors.

3. The use of steel package plants in humid tropical countries in conjunction with corrosive chemicals (e.g., alum, hypochlorites) requires special attention to preventive maintenance.

4. Some developing countries do not have the technical capability or supporting infrastructure to manufacture and maintain package plants. The economies of scale resulting from manufacturing package plants demand a large-scale operation in order to procure the necessary materials, manufacture the several types of plants, and transport and erect the plants. The importation of package plants from foreign proprietors is not likely to be economically feasible, because such units are expensive and overly mechanized, and when repairs are necessary, the user is completely dependent upon spare parts from abroad.

Simple package water treatment plants manufactured inside the country may be practical in places where a large number of small treatment facilities are needed, or where local conditions are unfavorable for on-site construction (e.g., lack of construction materials or low-cost labor, poor terrain or soil conditions). Bhole (1981) suggests that package plants for rural areas in developing countries fulfill the following requirements. The plants should be (1) sturdy, (2) simple to operate and provide easy access to any of their parts, (3) reliable, (4) furnished with minimal mechanical equipment and running costs, (5) able to operate without electrical energy, (6) low cost, (7) easy to transport and install with minimum construction work at site, and (8) able to treat surface water.

Bhole has designed a simple package plant, taking into account the above mentioned criteria. A detailed plan and elevation of the plant is shown in Figure 9.1. The plant is comprised of the following: (1) an alum dosing unit, consisting of a large-size plastic bucket for storing alum solution; (2) a gravel-bed flocculator, consisting of sections of increasing cross-sectional areas to produce tapered velocity gradients (described in Chapter Six, under "Gravel-Bed Flocculators"); (3) an inclined-plate settling tank consisting of 26 plates located below a V-notched weir that conveys the settled water to the filter unit; (4) a filter unit, consisting of sand and supporting gravel media, and a perforated pipe underdrain system; and (5) a chlorination unit, similar to the alum unit but containing a solution of bleaching powder.

Raw water is pumped by a diesel-driven pump (P_1) to the elevated tank (ET), which provides the necessary head for gravity flow through the treatment unit. The

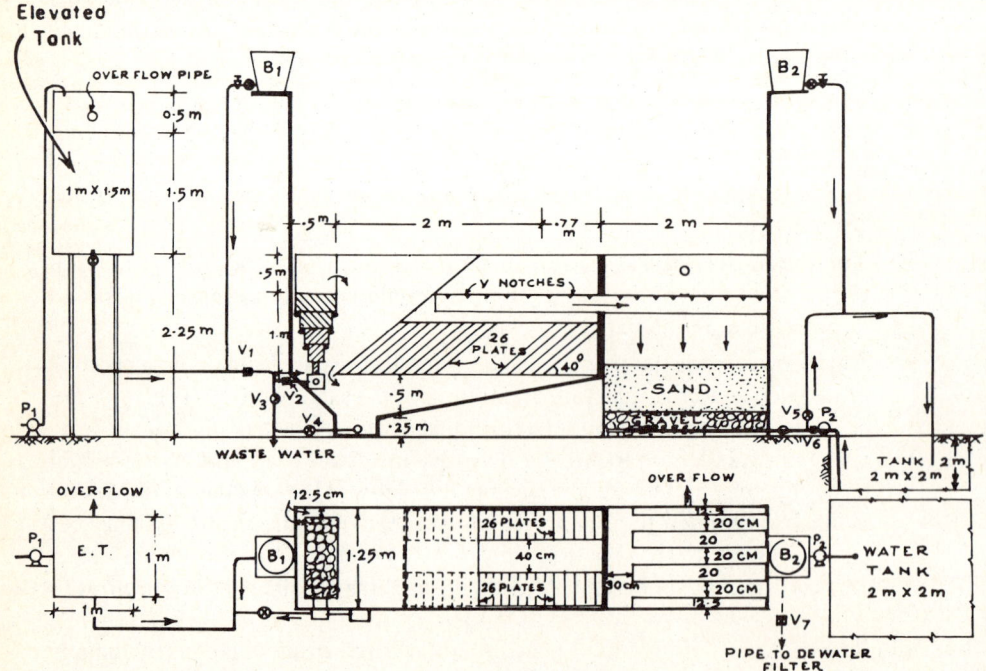

Figure 9.1.   Package water treatment plant—India. *Source:* Bhole, 1980.

second pump (P_2) is used for cleaning the filter and settling tank. During filtration, valves V_1, V_2, and V_5 are opened and valves V_3, V_4, and V_7 are closed. During backwashing, valves V_4 and V_6 are opened and valves V_1, V_2, V_3, V_5, and V_7 are closed, and pump P_2 is started. The wash water flows upward through the filter, collects in the troughs, and flows into the settling tanks where it is drained via a floor drain. To drain the flocculator unit, valves V_1 and V_4 are closed and valves V_2 and V_3 are opened.

This package plant is $5.3 \times 1.25 \times 1.25$ m, weighs 1.3 tons, and has a capacity of 270 m^3/day, which is sufficient for a population of 2000 people, assuming a per capita consumption of 45 l per day. Several package plants can be operated in parallel to meet larger water demands. The cost of the plant is about 20,000 Indian rupees ($2500; 1982 US$), but could be reduced if it is manufactured in large numbers. Figure 9.2 shows the "packaged" appearance of this plant, and pertinent installation criteria and process capabilities.

A steel package plant designed specifically by APS Technical Services Ltd. (1982) in England for developing country applications consists of a single module containing hydraulic flocculation, inclined-plate settling, and rapid sand filtration. The module itself is a standard 6-m shipping container which can be handled by any port or railroad that takes conventional containers, and it will fit any container ship, train, or truck. By making the container part of the plant, the problems and costs of packing for shipping are overcome, while permitting a low and readily determined shipping cost to almost any destination in the world. A flow diagram of the plant is shown in Figure 9.3.

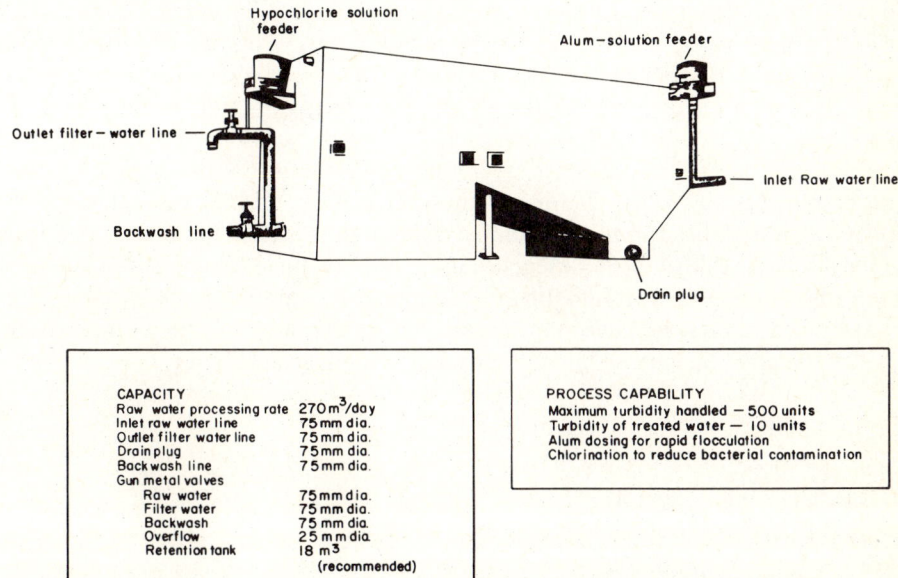

Figure 9.2. Isometric view of Indian package plant, together with installation requirements and process capabilities. *Source:* Bhole, 1980.

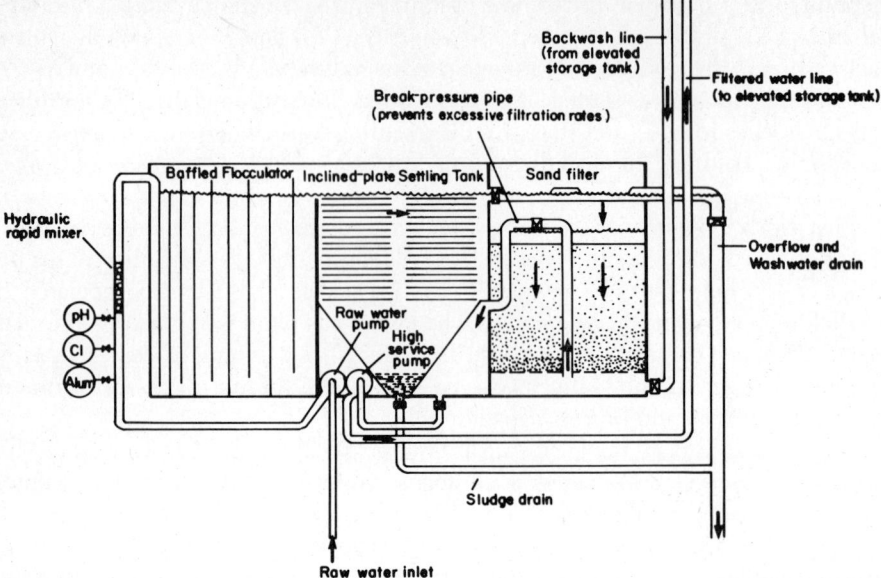

Figure 9.3. Flow diagram of steel package plant manufactured in England. *Source:* APS Technical Services, Ltd. 1982.

The rapid mixer consists of a short length of pipe enclosed by wire screening at both ends, and is filled with short pieces of small-diameter plastic pipes that serve to agitate the water passing around and through them; the flocculator is comprised of lightweight plastic baffles that can be removed easily for cleaning; the settling tank contains parallel plastic plates inclined at 60° from the horizontal; and the filter box contains a 60-cm layer of sand supported by gravel, and a main and lateral underdrain system. The underdrain system of the filter is connected to a backwash line and a break-pressure pipe, the latter used to avoid excessive filtration rates when the sand bed is clean. All chemicals (alum, chlorine, and lime) are added to the incoming water ahead of the rapid mixer, although it is possible to add chemicals at the discharge side of the high service pump. The chemical solutions are contained in flexible bags and drawn into the influent pipeline, in proportion to the flow rate, by the negative pressure created by an orifice placed upstream of the dosing points.

The technical specifications of the package plant are as follows:

1. *dimensions*: 6 m × 2.4 m × 2.4 m
2. *dry weight*: 13,000 kg
3. *operating weight*: 45,000 kg
4. *flow rate*: 360 to 530 m^3/day
5. *ground loading*: 75 kg/m^2
6. *tank construction*: lined carbon steel
7. *pumps*: electric motor or diesel

An elevated storage tank, which can be erected by four or five laborers without special equipment, is provided by the manufacturers for backwashing the filter as well as providing pressure for distribution.

MODULAR WATER TREATMENT PLANTS

Water treatment plants based on modular designs may be built more quickly than conventional plants, while still allowing contributions from the local community by way of materials, and involvement in construction. Standardized modular designs can reduce the type and number of plant devices that need to be ordered and stored, thereby facilitating a more efficient system of procurement of spare parts, training of operators, and ease of repairs. To further shorten the time span for project implementation, plants may be comprised of modular elements that are prefabricated and transportable to construction sites for final assembly. Although modular designs are amenable to either concrete or steel construction, concrete is generally preferred because of its wide availability in developing countries, comparatively low cost, and resistance to corrosion. Moreover, most of the skilled and unskilled workers employed in developing countries are more familiar and proficient with concrete construction than with steel. Two types of modular water treatment plants developed in Latin America and Indonesia are described below. In addition, the upflow-downflow filtration plants described in Chapter Eight and the various plants described in the CEPIS manual (1982) are suitable modular units for developing countries.

The water treatment plant serving the city of Prudentopolis, Brazil (population 7500), consists of a modular unit 4 m square in plan, having a capacity of 1000 m^3/day (Arboleda, 1976; Sperandio and Perez, 1976). A plan and section of the plant are shown in Figures 9.4 and 9.5, respectively. The plant consists of a hopper bottom square tank with four 1×1 m dual-media filters located at the corners, four 1×2 m inclined-plate settling tanks placed near the outside walls, and, at the center, a flocculation chamber with four compartments. The raw water enters a rapid mix chamber at one corner of the tank, where alum and lime are added. Agitation is caused by the discharge of the raw water over a circular weir. The water then enters a distribution channel, flows over one of three triangular weirs, and is conveyed via a cast-iron pipe to the flocculation chamber, which is comprised of four vertical compartments. Tapered mixing is provided by wooden paddles of different cross-sectional area that are driven by a small 370-W (½ hp) electric motor.

The flocculated water leaves the bottom chamber via six cast-iron pipes that discharge into four upflow settling tanks equipped with a series of 1×1-m asbestos-cement parallel plates placed 5 cm apart. The settled water enters the filter via cast-iron pipes that are attached to a metal box located on the outside faces of the filter walls. The drainpipe is also attached to this metal box. Two butterfly valves, controlled by a single handle, can simultaneously close the filter influent pipe and open the drain pipe, or vice versa, to initiate either filtration or backwashing operations. The four filters are designed for interfilter backwashing, because they are all interconnected by a 300-mm cast-iron pipe (see Chapter Eight, under "Design

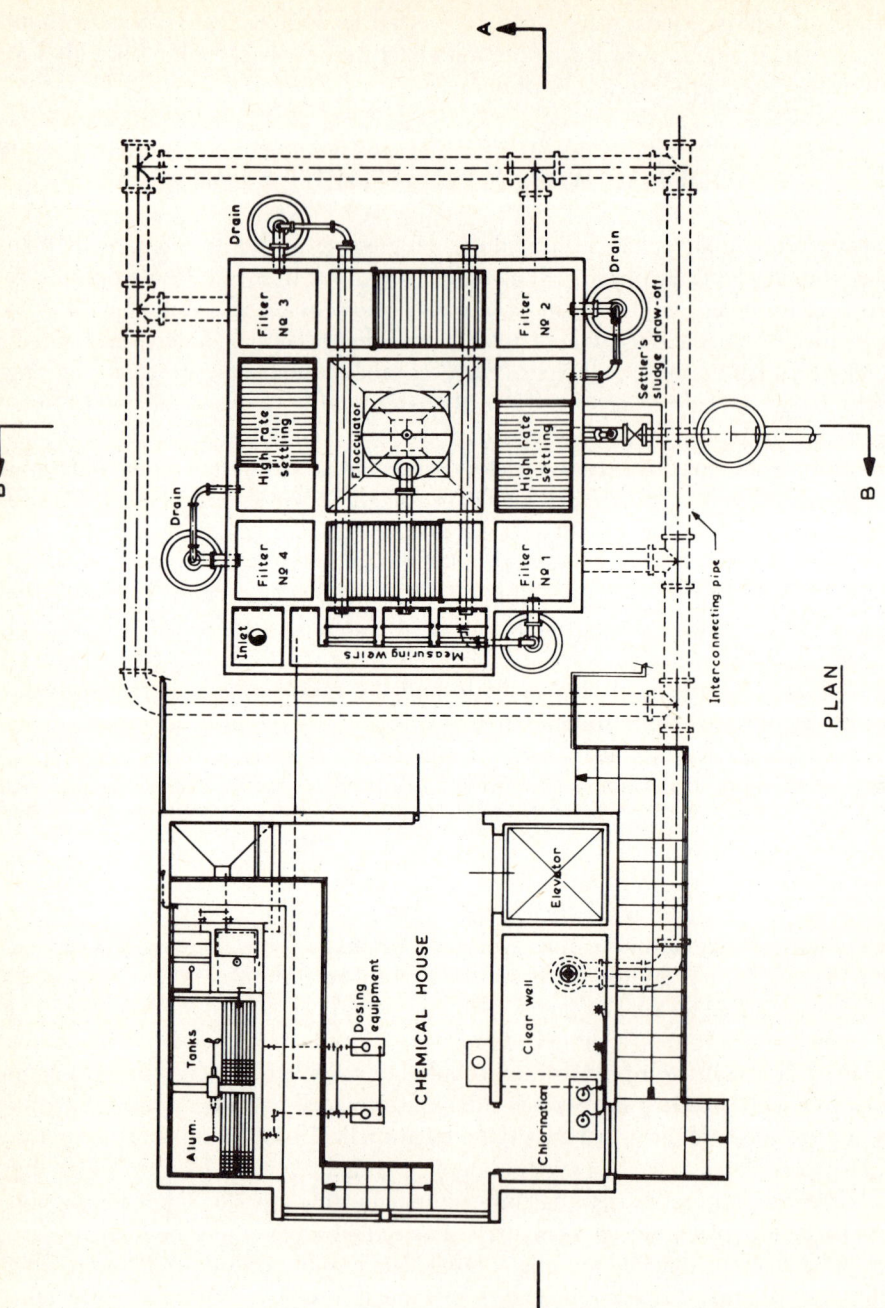

PLAN

Figure 9.4. Modular treatment plant in Prudentopolis, Brazil (plan). *Source:* Arboleda, 1976.

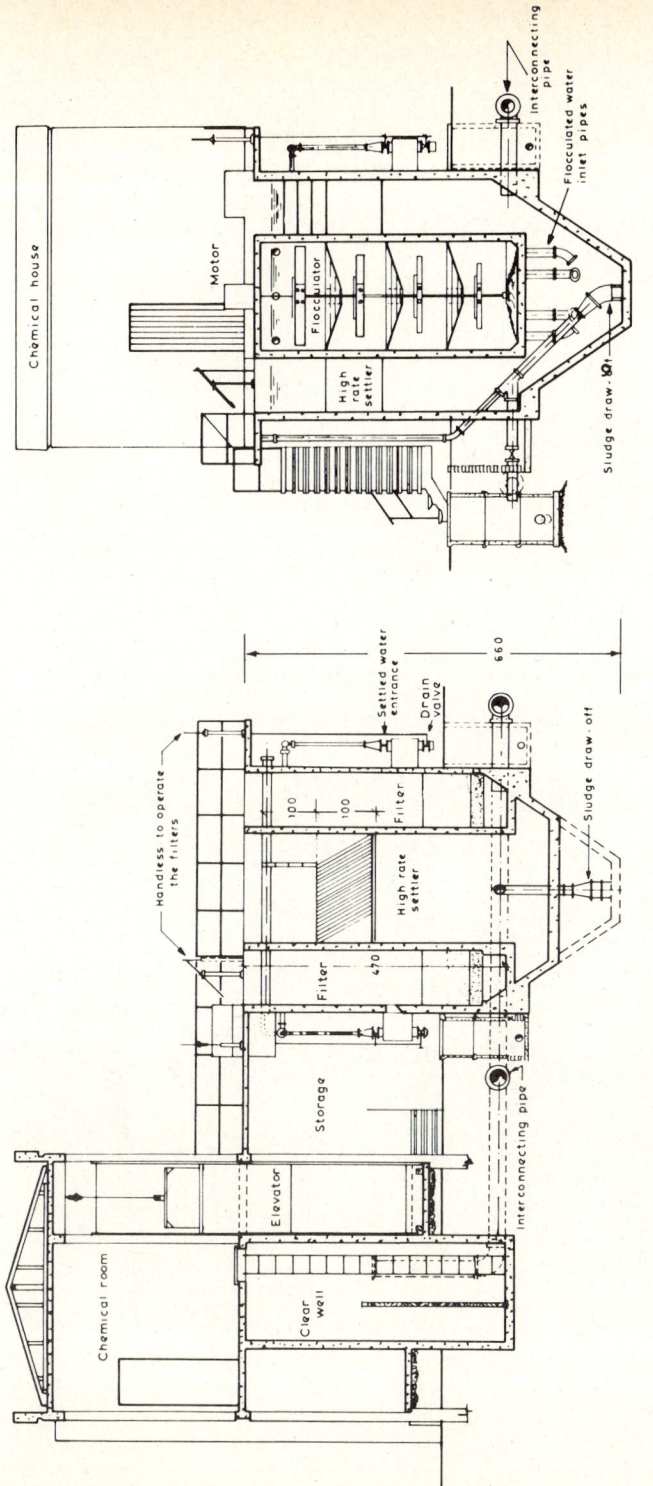

SECTION B-B

SECTION A-A

Figure 9.5. Modular treatment plant in Prudentopolis, Brazil (sections A-A; B-B). *Source:* Arboleda, 1976.

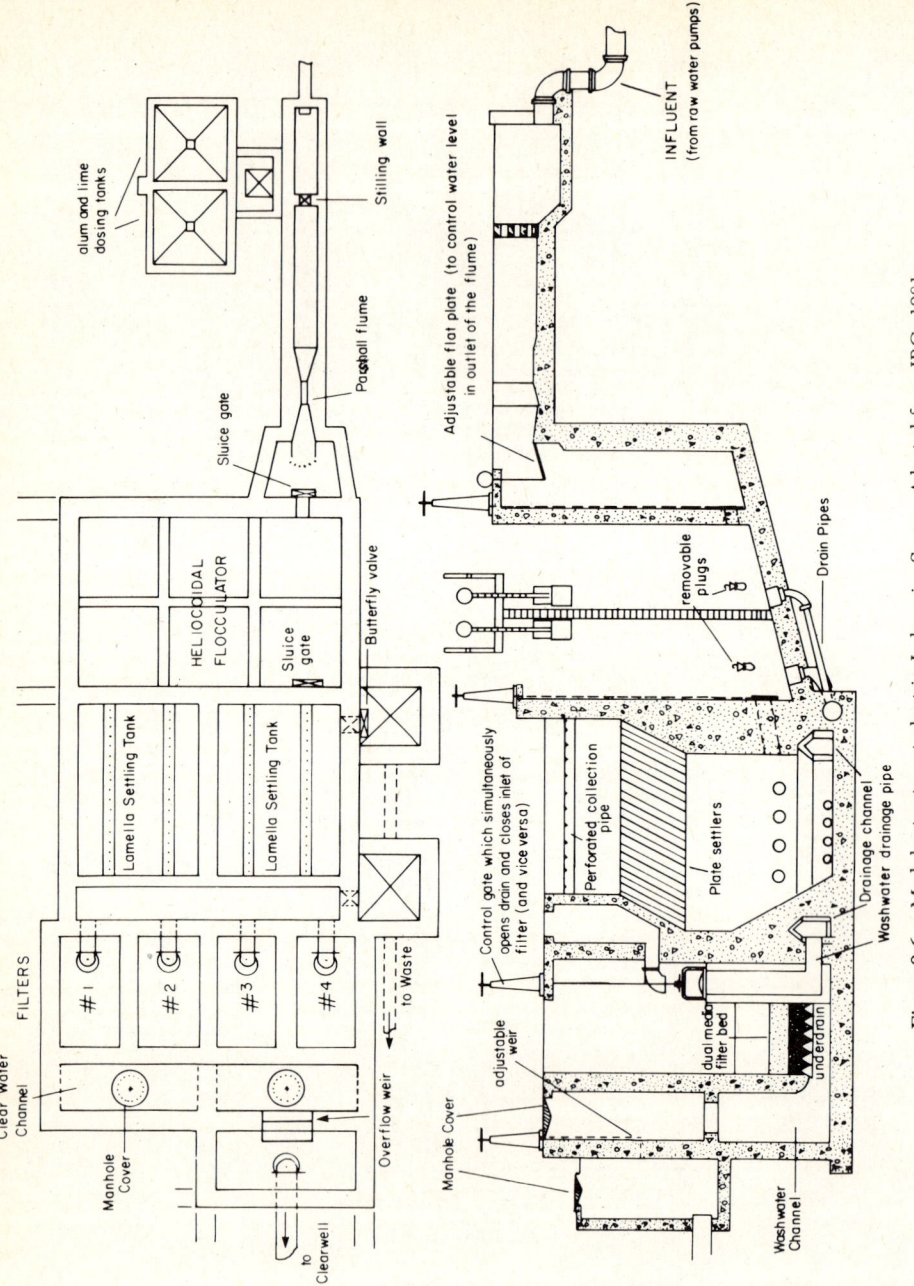

Figure 9.6. Modular treatment plant in Indonesia. *Source:* Adapted from IRC, 1981a.

alum and lime dosing tanks

Stilling wall

Sluice gate

Parshall flume

HELIOCOIDAL FLOCCULATOR

Sluice gate

Butterfly valve

Lamella Settling Tank

Lamella Settling Tank

Clear Water Channel

FILTERS

#1

#2

#3

#4

Manhole Cover

Overflow weir

to Waste

to Clearwell

Adjustable flat plate (to control water level in outlet of the flume)

INFLUENT (from raw water pumps)

removable plugs

Drain Pipes

Control gate which simultaneously opens drain and closes inlet of filter (and vice versa)

Perforated collection pipe

Plate settlers

Drainage channel

Washwater drainage pipe

dual media filter bed

underdrain

adjustable weir

Manhole Cover

Washwater Channel

and Operation of Interfilter Washing Units"). To regulate the backwash flow, a sliding pipe placed in the clear well can be raised or lowered to decrease or increase the backwash rate.

An important feature of the plant, which should be included in any type of design, is the potential for expanding plant capacity. In this case, when two or three modules are to be used, the raw water influent flow is split by means of three triangular weirs installed on the side of the distribution channel. The two outside weirs discharge directly into cast-iron pipes, which are used to convey the influent to two additional treatment modules. The total construction cost of the single-module Prudentopolis plant, including the chemical building, is $54,000 (1982 US$) as compared with a conventional plant of the same capacity (between $80,000 and $110,000). A complete description of this plant, including a comprehensive technical and economic evaluation, is given in a report issued by the U.S. Agency for International Development (Sperandio and Perez, 1976).

An extensive modular water supply program for rural communities in Indonesia was initiated in 1979 under the joint direction of the Indonesia Directorate of Sanitary Engineering and the IRC (IRC, 1981a). The purposes of the program were threefold: (1) to study a modular approach (i.e., using standard components for small water treatment plants in Indonesia); (2) to prepare criteria, specifications, and working plans for the planning and design of these modules for domestic manufacture; and (3) to study and to comment on existing designs in order to evaluate the economic aspects of the use of local building materials (concrete and steel) and to make recommendations accordingly.

Standard designs were developed for both concrete and steel plants having capacities of 1730, 3460, 5190, and 6920 m^3/day. Figure 9.6 shows a 1730 m^3/day concrete plant comprised of: (1) rapid mixing by means of a hydraulic jump formed immediately downstream of a Parshall flume, which is also used as a flow measuring device; (2) flocculation in six square chambers where flow alternates between upflow and downflow (heliocoidal-flow type); (3) inclined-plate sedimentation using asbestos-cement plates inclined at 60°; (4) filtration with dual-media filters having interfilter-washing capabilities; and (5) chemical feeding using simple constant-rate feeders for alum and hypochlorite dosing and lime saturation towers. The above-mentioned unit processes were employed in all of the concrete and steel plant designs, but the unit processes for the steel plants were designed particularly for prefabrication and transportability. Accordingly, the steel components were built no larger than 2 × 5 × 3 m, weighed no more than 4.5 tons, and could be readily transported and quickly assembled at the construction site.

An international seminar held in Indonesia on modular approaches for water supply programs (IRC, 1981a) recommended that the Indonesian designs be used as models in other developing countries with (if possible) intercountry field testing of the modular treatment plants. An unpublished report on the standard water treatment plants in Indonesia, which includes general and detailed design criteria, descriptions of the concrete and steel plant designs, and the application of the standard designs for various types of surface waters, is available from the IRC (1980).

chapter ten

COSTS OF WATER TREATMENT PLANTS IN DEVELOPING COUNTRIES

Preliminary planning of water supply projects, including the final selection of treatment components and arrangements for financing, must be based on reliable cost data. Such data are difficult to obtain in developing countries, particularly in areas where water treatment facilities are being built for the first time. In such instances, reasonable cost estimates for construction, operation, and maintenance may be obtained indirectly by using (1) cost data for similar plants that have been built in other areas or in another developing country with similar characteristics; (2) general cost curves that are based on the costs of a variety of plants constructed within the country; or (3) general predictive costs equations developed for similar situations. Although costs from one country usually are not directly applicable to other countries, the relationships among the experienced costs of the various types of treatment (e.g., conventional, direct filtration, package plants), and particularly the unit costs as a function of the size of the plant, are useful.

The purposes of the cost data and the predictive-cost equations presented in this chapter are: (1) to assist administrators, engineers, and public officials in developing countries, who are planning water treatment schemes, to assess the general level of capital and recurrent costs as a tool for planning; (2) to allow officials to check whether costs estimates are reasonable; and (3) to provide financial guidelines for making preliminary decisions on water supply schemes.

This chapter begins by discussing the general cost functions used widely by officials in the water supply fields; followed by sections on: (1) construction-cost curves, predictive equations, and tables specific to countries in Asia and Latin America, and comparisons among these data; and (2) operation and maintenance (O&M) costs and the inherent difficulty in estimating such costs.

Unless otherwise stated, the cost data in this text have been adjusted to the March 1982 Engineering News Record (ENR) Construction Cost Index of 3729 based on an ENR Index of 100 in 1907, so that all costs are on a common basis (*Engineering News Record*, March 1982). The ENR index is based upon the average cost, at a particular time, of constant quantities of structural steel, portland cement, lumber, and common labor in 20 cities in the United States.

Foreign currencies have been converted to U.S. dollars using the July 1982 exchange rates listed below:

1. Brazil (Cruzeiro)   170 Cr = U.S. $1.00
2. England (Pound)   0.58 £ = U.S. $1.00
3. India (Rupee)   8.8 Rs. = U.S. $1.00

THE GENERAL COST EQUATION

The relationship between plant capacity and construction costs is often represented by the following power function, which reflects the economies of scale present in large water projects:

$$C = aQ^b \tag{10.1}$$

where C = construction costs; Q = plant capacity; and a and b = constants. This relationship is the basis for most of the cost curves presented in this chapter.

The constant b determines the manner in which cost changes with plant capacity. Large economies of scale are associated with small values of b. For example, the EPA cost curves for conventional and direct filtration plants constructed in the United States have b-values of 0.70 and 0.48, respectively; therefore, the construction costs in both cases will increase with capacity, but the cost/unit capacity will decline. On the other hand, the construction cost equation for slow sand filters in India (eq. 10.2) has a comparatively high b-value of 0.86; hence, the economies of scale are not as great as those for U.S. filtration plants. Although optimum design values for new facilities depend upon many factors including expected rates of growth, discount rates, useful life of the facilities, and the ease of expansion, unless b-values are below 0.7, there is little economic incentive to overdesign plants (Paramasivam et al., 1981).

The constant a is equivalent to the construction cost of a plant with unit capacity, and is also a function of the ENR construction cost index. For example, the EPA costs in Figure 10.1 for U.S. plants, originally based on 1975 prices, have been adjusted to 1982 prices by multiplying each of the a-values of the respective cost equations by 3729 (March 1982 ENR cost index), and dividing by 2212 (1975 ENR cost index). The component b remains unchanged in this updating procedure.

Numerous factors affect the costs of water treatment, apart from plant capacity (Q) or basic construction costs (as reflected by the ENR construction cost index). Some of these factors include (1) the type of plant; (2) the local costs of materials and labor; (3) design criteria (conservative designs lead to larger components and higher costs); (4) geographical location; (5) transportation; (6) climatic conditions; (7) level of competition among building contractors; and (8) delivery time for critical items. A factor of great importance in developing countries is the cost of the equipment and materials that have to be imported. Obviously, it would not be feasible to incorporate all of these factors into a preliminary cost estimate for a particular project; nevertheless, any known conditions that would substantially affect the cost of a project should be considered, and appropriate adjustments made to the cost data.

CONSTRUCTION COSTS OF WATER TREATMENT PLANTS

This section contains cost curves, tables, and cost equations for the construction of rapid and slow sand filtration plants in developing countries. Brief descriptions and construction costs for plants described in this text, and for other plants designed simply and economically, are summarized in Tables 10.1 and 10.2, respectively. Plants are characterized and grouped generally as conventional, direct filtration, and package or modular. Table 10.2 does not show separate costs for the major treatment plant components, but the following guidelines may be used as a rough estimate (Sanks, 1979).

Component	% of Total Construction Cost
Earthwork, general site work, and yard piping	15 to 20
Sedimentation and flocculation basins	20 to 30
Filters and appurtenant systems	20 to 35
Operations and administration buildings	10 to 20
Miscellaneous chemical tanks, small structures	10 to 15

The costs of the clear wells are not included because of their highly variable capacities, which depend on local circumstances.

Cost data from Table 10.2 are plotted in Figure 10.1 for (1) conventional rapid filtration plants; (2) direct filtration plants; and (3) modular and package plants; all of which have been designed with practical, low-cost technologies in developing countries. Also, cost curves developed by the EPA for U.S. plants are shown in the figure for comparative purposes (EPA, 1978). These curves, originally based on 1975 cost data, have been adjusted to 1982 dollars. The costs for the simplified

Table 10.1. Description of Simplified Water Treatment Plants

Plant Location (reference)	Plant Type[a]	Year of Construction	Capacity (m³/day)	MGD	Plant Unit Processes[b]
Barranquilla, Colombia (Arboleda)	C	1982	86,000	23	WRM, MF, HFSIP, IFW/DMF
Becerril, Colombia (Arboleda)	C	1982	3,500	0.92	PFRM, HFF, HFSIP, IFW/DMF
Cali, Colombia* (Wagner, 1982)	C	1979	260,000	68.0	WRM, MF, HFSIP, RSF
Cali, Colombia* (Arboleda)	C	1982	86,000	23	WRM, MF, HFSIP, IFW/DMF
Cochabamba, Bolivia* (Arobleda, 1976)	C	1975	20,000	5.0	PFRM, BCF, HFSIP, IFW/DMF
La Paz, Colombia (Arboleda)	C	1982	12,000	3.2	PFM, PWF, HFSIP, IFW/DMF
Manaure, Colombia (Arboleda)	C	1982	2,200	0.57	PFRM, HFF, HFSIP, IFW/DMF
Manizales, Colombia (Arboleda)	C	1982	69,000	18	WRM, MF, HFSIP, IFW/DMF
Oceanside, California* (MacDonald & Streicher, 1977)	C	1977	65,000	17	MJRM, BCF, HFS, IFW/DMF
Pereira, Colombia (Arboleda)	C	1982	52,000	14	WRM, MF, HFSIP, IFW/DMF
Parana, Brazil (Wagner, 1982)	C	1978	87,000	23	PFRM, MF, UFS, DMF

Table 10.1. (Continued)

Plant Location (reference)	Plant Type[a]	Year of Construction	Capacity (m³/day)	Capacity MGD	Plant Unit Processes[b]
Piracicaba, Brazil (Azevedo-Netto)	C	1981	65,000	17	PFRM, MF, HFS, DMF
Sao Paulo, Brazil (Azevedo-Netto)	C	1981	1,000	0.27	OPRM, VFBCF, UFSIP, IFW/DMF
Sao Paulo, Brazil (Azevedo-Netto)	C	1981	2,200	0.57	OPRM, VFBCF, UFSIP, IFW/DMF
Murtizapur, India (Kardile)	C	1973	1,900	0.50	AF, HRM, BCF, UFS, RSF
Shegaon, India (Kardile)	C	1978	4,800	1.3	AF, HRM, BCF, HFS, RSF
Pusad Bloch, India (Kardile)	C	1975	4,300	1.1	HRM, BCF, HFS, RSF, WT
Linhares, Brazil* (Sperandio & Perez, 1976)	D	1974	5,200	1.4	PFRM, CUSF
Brasilia, Brazil* (Wagner, 1982)	D	1981	520,000	1.4	HRM, DF
Parana, Brazil (Azevedo-Netto)	SF	1975	1,300	0.34	UCC, DPF
Colon, Costa Rica* (Institute of Water Supply & Sewerage)	SF	1979	560	0.15	UCC, DPF
Ramtek, India* (Kardile, 1981)	UD	1973	2,400	0.63	GBF, DMF

218

Chandori, India* (Kardile, 1981)	UD	1977	1,000	0.26	GBF/TS, DMF
Varangaon, India* (Kardile, 1981)	UD	1977	4,200	1.1	GBF, UFSIP, DMF
Bhagpur, India (Khardile)	UD	1978	4,200	1.1	HRM, GBF/TS, DMF
India* (Bhole, 1981)	P	1981	270	0.07	GBRM, GBF, UFIP, DMF
Parana, Brazil (Wagner, 1982)	P	1980	830	0.23	GBRM, GBF, UFSIP, DMPF
Parana, Brazil (Wagner, 1982)	M	1979	43,000	11	OPRM, MF, UFSIP, IFW/DMF
Parana, Brazil (Wagner, 1982)	M	1979	17,000	4.5	OPRM, MF, UFSIP, IFW/DMF
Parana, Brazil (Wagner, 1982)	M	1979	4,400	1.2	OPRM, MF, UFSIP, IFW/DMF
Parana, Brazil (Wagner, 1982)	M	1980	830	0.23	GBRM, GBF, UFSIP, DMPF
Prudentopolis, Brazil* (Arboleda, 1976)	M	1975	1,000	0.26	OPRM, MF, UFSIP, DMF

*Plants described in this text.

[a]C = conventional rapid filtration; D = direct filtration; M = modular rapid filtration; P = package plant; SF = superfiltration; UD = upflow-downflow filtration.

[b]AF = aeration fountain; BCF = baffled channel flocculators; CUSF = contact upflow sand filter; DF = direct filtration; DMF = dual-media filters; DMPF = dual-media pressure filters; DPF = downflow polishing filter; GBF = gravel bed flocculator; GBRM = gravel bed rapid mixer; HFF = helicoidal flow flocculator; HFS = horizontal-flow settling tank; HRM = hydraulic rapid mixer; IFW = interfilter washing; IP = inclined plates; MF = mechanical flocculators; MJRM = multijet rapid mixer; OPRM = orifice plate rapid mixer; PFRM = Parshall flume rapid mixer; PWF = Pelton wheel flocculator; RSF = rapid sand filter; TS = tube settler; UCC = upflow contact clarifier; UFS = upflow settling tank; VF = vertical flow; WRM = weir rapid mixer; WT = washwater tank.

Table 10.2. Construction Costs of Simplified Water Treatment Plants

Location/Source of Plant (reference)	Year of Construction	Total Construction Costs[b]	1982 Construction Costs Per Unit Capacity[a]	
			(U.S. $/m³/day)	(U.S. $/MGD)
Barranquilla, Colombia (Arboleda)	1982	U.S. $2,985,000	34	130,000
Becerril, Colombia (Arboleda)	1982	U.S. $120,000	34	130,000
Cali, Colombia* (Wagner, 1982)	1979	U.S. $3,800,000	18	70,000
Cali, Colombia (Arboleda)	1982	U.S. $2,354,000	27	100,000
Cochabamba, Bolivia* (Arboleda, 1976)	1975	U.S. $260,000	22	83,000
La Paz, Colombia (Arboleda)	1982	U.S. $265,000	22	84,000
Manaure, Colombia (Arboleda)	1982	U.S. $109,000	50	190,000
Manizales, Colombia (Arboleda)	1982	U.S. $1,446,000	21	79,000
Oceanside, California* (MacDonald & Streicher, 1977)	1977	U.S. $3,700,000	82	310,000
Pereira, Colombia (Arboleda)	1982	U.S. $881,700	17	64,000
Parana, Brazil (Wagner, 1982)	1978	U.S. $1,000,000	16	59,000

Location (Source)	Year	Cost		
Piracicaba, Brazil (Azevedo-Netto)	1982	U.S.$5,000,000	77	290,000
São Paulo, Brazil (Azevedo-Netto)	1981	U.S.$79,000	80	300,000
São Paulo, Brazil (Azevedo-Netto)	1981	U.S.$97,000	47	180,000
Murtizapur, India (Kardile)	1973	Rs. 491,000	29	110,000
Shegaon, India (Kardile)	1978	Rs. 595,000	19	72,000
Pusad Bloch, India (Kardile)	1975	Rs. 498,000	22	83,000
Linhares, Brazil* (Sperandio & Perez, 1976)	1974	U.S.$69,000	24	91,000
Brasilia, Brazil* (Wagner, 1982)	1981	U.S.$5,300,000 (design estimate)	11	41,000
Parana, Brazil (Azevedo-Netto)	1975	U.S.$36,000	47	180,000
Colon, Costa Rica* (Institute of Water Supply and Sewerage)	1979	U.S.$10,000	22	83,000
Ramtek, India* (Kardile, 1981)	1973	U.S.$16,000	13	50,000
Chandori, India* (Kardile, 1982)	1980	U.S.$19,000	20	77,000
Varangaon, India* (Kardile, 1981)	1977	U.S.$50,000	17	65,000

Table 10.2. *(Continued)*

Location/Source of Plant (reference)	Year of Construction	Total Construction Costs[b]	1982 Construction Costs Per Unit Capacity[a]	
			(U.S. \$/m³/day)	(U.S. \$/MGD)
Bhagpur, India (Kardile)	1978	Rs. 190,000	16	61,000
India* (Bhole, 1981)	1982	Rs. 20,000	8.4	32,000
Parana, Brazil (Wagner, 1982)	1980	U.S.\$27,000	38	140,000
Parana, Brazil (Wagner, 1982)	1979	U.S.\$650,000	19	71,000
Parana, Brazil (Wagner, 1982)	1979	U.S.\$320,000	23	88,000
Parana, Brazil (Wagner, 1982)	1979	U.S.\$140,000	31	140,000
Parana, Brazil (Wagner, 1982)	1979	U.S.\$76,000	55	210,000
Prudentopolis, Brazil* (Arboleda, 1976)	1975	U.S.\$35,000	59	23,000

*Plants described in this text.
[a]ENR Cost Index for: 1973 = 1895; 1974 = 2020; 1975 = 2212; 1977 = 2577; 1978 = 2776; 1979 = 3003; 1980 = 3237, 1981 = 3467; 1982 = 3729.
[b]8.8 Rs. = U.S.\$1.00.

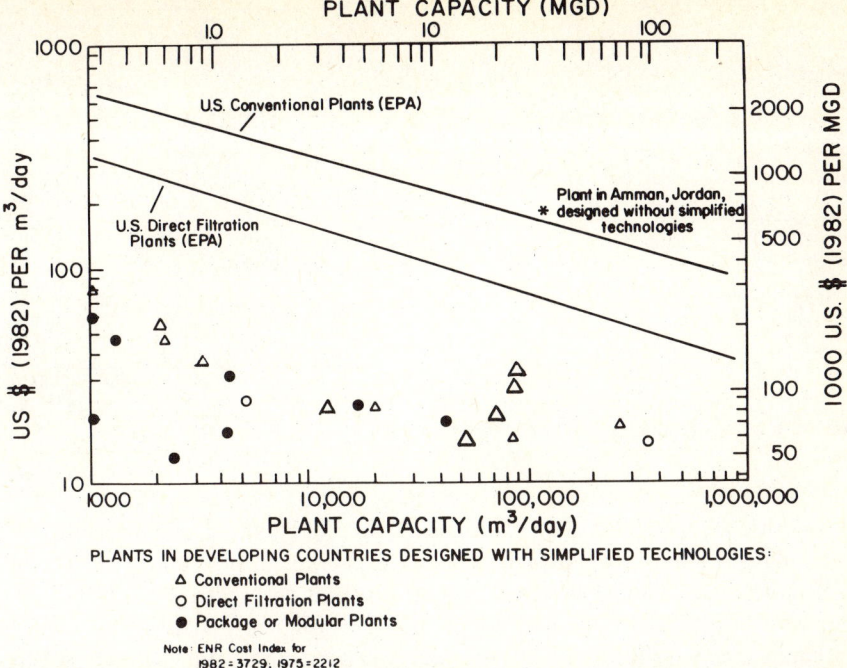

Figure 10.1. Comparative construction costs (1982 US$) of water treatment plants in the United States and in developing countries.

plants are about one order of magnitude lower than those for plants designed in the United States. For example, the conventional rapid filtration plant for the city of Cochabamba, Bolivia, has a capacity of 20,000 m³/day (5 MGD) and was built at a unit cost of U.S. $22 per m³/day (U.S. $83,300 per MGD); whereas the unit cost of an identically sized plant built in the United States is about U.S. $260 per m³/day (U.S.$980,000 per MGD)—12-fold larger.

Plants designed without using simplified technologies in developing countries, however, may have capital costs even higher than those in the United States. For example, a plant under construction in Amman, Jordan, designed with conventional technologies, is considerably more costly than in most developing countries, in part because of high costs in the Middle East (Wagner, 1982b). In most instances, conventional plants built in developing countries will cost less than similar plants in the United States. Kardile (1981) has shown 50% cost reductions between conventional and simplified plants in India.

Part of the cost difference between plants designed in the United States and the simplified plants in the developing countries can be explained by the lower labor costs and lower subsidized interest rates in developing countries. The primary reason, however, lies in the approach to design of the simplified plants; emphasizing low-cost, nonmechanized solutions that are compatible with socioeconomic and technical conditions in developing countries.

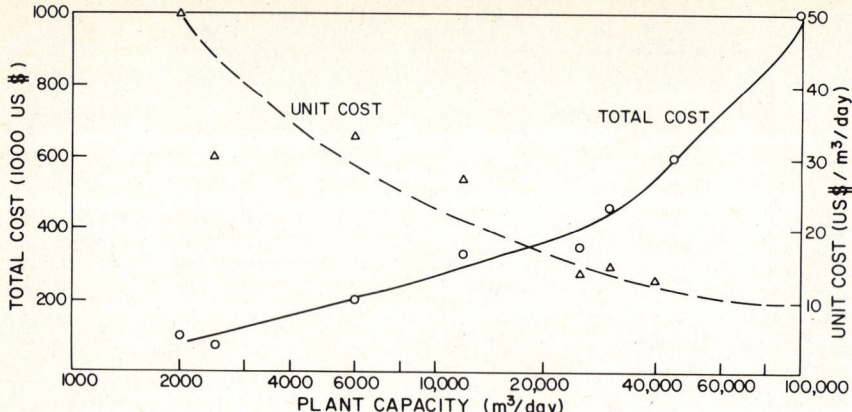

Figure 10.2.   Construction costs (1978 US$) of modular plants in Brazilian state of Parana. *Source:* Richter.

A cost study was conducted for the Brazilian state of Parana, based on construction costs of eight water treatment plants having capacities from 2,000 to 45,000 m³/ day (SANEPAR, 1979). The resulting cost curve is shown in Figure 10.2. The plants were similar to the modular plant shown in Figures 9.3 and 9.4, consisting of hydraulic mixing and flocculation, inclined-plate settling, and high-rate filtration with interfilter backwashing capabilities.

A similar Brazilian study based on package plants implemented in rural communities in the state of São Paulo (Azevedo-Netto) resulted in the cost curves shown in Figure 10.3. The plants were designed with a siphon-actuated backwash system for the filters.

In India, Kardile (1981) compared actual construction costs of 10 simplified

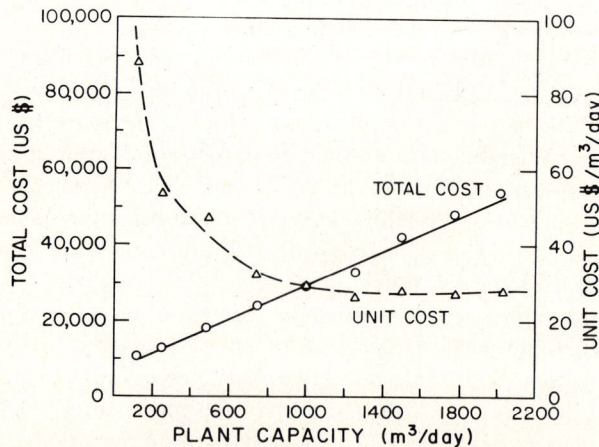

Figure 10.3.   Construction costs (1978 US$) of package plants in the Brazilian state of São Paulo. *Source:* Azevedo-Netto.

Table 10.3. Comparative construction costs of the Indian upflow-downflow plants and conventional plants

Location of Treatment Plant (Province)	Type of Plant[a]	Year of Construction	Capacity (m³/day)	Capacity (MGD)	Total Construction Costs[b] (U.S.$)	1982 Construction Costs per Unit Capacity[c] (U.S. $/m³/day)	1982 Construction Costs per Unit Capacity[c] (U.S. $/MGD)	Percent Reduction in Construction Costs of the Upflow-Downflow Plants
1) Ramtek* (Nagpur)	C	1973	2400	0.64	56,000	46	170,000	
	U-D				16,000	13	49,000	71
2) Surya Colony (Thana)	C	1976	660	0.17	25,000	59	230,000	
	U-D				13,000	31	120,000	48
3) Varangaon* (Jalgaon)	C	1977	4220	1.1	100,000	34	130,000	
	U-D				50,000	17	66,000	49
4) Kandla Port Trust (Gujarat)	C	1977	2000	0.53	44,000	32	120,000	
	U-D				25,000	18	68,000	43
5) Bhagur (Nasik)	C	1978	2000	0.53	44,000	30	110,000	
	U-D				24,000	16	61,000	45
6) Murbad (Thana)	C	1978	1000	0.26	31,000	42	110,000	
	U-D				19,000	26	98,000	39
7) Jejuri (Pane)	C	1978	2400	0.63	56,000	31	120,000	
	U-D				25,000	14	53,000	56
8) Akola (Nagpur)	C	1978	2400	0.53	25,000	17	63,000	
	U-D				13,000	8.7	33,000	48
9) Dhulia dairy (Dhulia)	C	1979	1500	0.40	38,000	31	120,000	
	U-D				19,000	16	59,000	51
10) Chandori* (Nasik)	C	1980	1000	0.26	31,000	36	140,000	
	U-D				19,000	22	84,000	40

Source: Adapted from Kardile, 1981.

* Plants described in this text.

[a] C = conventional; U-D = upflow-downflow.

[b] Construction costs for conventional plants were estimated.

[c] ENR Cost index for 1973 = 1895; 1976 = 2401; 1977 = 2577; 1978 = 2776; 1979 = 3003; 1982 = 3729.

upflow-downflow plants built in rural communities (see Chapter Eight, under "Up-flow-Downflow Filtration") with cost estimates for conventionally designed plants of the same capacities. The results are tabulated in Table 10.3. An average cost reduction of 50% results from the adoption of the simplified plants.

A detailed cost study on slow sand filtration in India by Paramasivam, Mhaisalkar, and Berthouex (1981) gave the results presented in Table 10.4. From the data shown in Table 10.4, the following cost equation was developed for slow sand filter beds:

$$C = 1220 A^{0.86} \tag{10.2}$$

where C = total construction cost (1980 Indian rupees); A = total area of the filter beds (m^2). Hence, the cost per square meter of slow sand filters in India in 1980 was 1220 Indian rupees; and the exponent (0.86) indicates little economy of scale.

A comparative study of capital, operation, and maintenance costs for rapid and slow sand filtration plants in India (Sundaresan and Paramasivam, 1981) gave the results shown in Table 10.5 and 10.6; the former showing costs for energy, chemicals, staff, and repairs; and the latter showing total capitalized costs for both types of

Table 10.4. **Construction Costs (1982 U.S.$) for a Given Area and Number of Slow Sand Filter Units in India**[a]

Area (m²)	Two Units (U.S.$)	Three Units (U.S.$)	Four Units (U.S.$)	Five Units (U.S.$)
50	4,400	4,800	5,200	5,400
100	7,600	8,000	8,500	8,900
150	10,000	11,000	12,000	12,000
200	13,000	14,000	14,000	16,000
300	18,000	19,000	20,000	20,000
400	24,000	24,000	25,000	26,000
500	29,000	30,000	31,000	31,000
600	34,000	35,000	36,000	37,000
700	38,000	40,000	41,000	42,000
800	43,000	44,000	46,000	47,000
900	48,000	49,000	50,000	52,000
1000	53,000	54,000	55,000	56,000
1200	62,000	64,000	65,000	67,000
1500	76,000	78,000	79,000	82,000
2000	98,000	100,000	100,000	110,000

Source: Adapted from Paramasivam, Mhaisalkar, and Berthouex, 1981.
[a]Rs. 8.8 = U.S.$1.00
ENR Cost index for 1981 = 3533; 1982 = 3729.

Table 10.5. Relative Costs of Rapid Filtration and Slow Sand Filtration in India[a] (1982 U.S.$ × 1,000)

Plant Capacity (m³/day)	Rapid Filtration[b]						Slow Sand Filtration			
	Construction	Energy	Chemical	Staff Salary	Repairs and Replacement	Annual Operation and Maintenance Cost[c]	Construction	Staff Salary	Repairs and Replacement	Annual Operation and Maintenance Cost[d]
1,000	54	0.11	2.1	4.9	1.1	8.1	38	1.5	0.42	1.9
1,900	110	0.11	4.1	4.9	2.1	11	66	1.5	0.63	2.1
2,300	120	0.11	5.0	5.7	2.3	13	79	2.2	0.84	3.1
6,700	150	0.32	15	6.5	6.5	24	260	2.2	2.6	4.9
15,000	340	0.64	33	12	8.7	54	590	7.3	5.9	13
30,000	570	1.3	65	18	14	98	1,200	10	12	22
45,000	910	1.9	98	18	20	140	1,800	140	18	29

Source: Adapted from Sundarasan and Paramasivam, 1981.
[a]ENR Cost Index for 1981 = 3533; 1982 = 3729.
[b]Includes rapid mixing, flocculation, sedimentation, and filtration units.
[c]Includes energy, chemical, staff, and repair charges.
[d]Includes staff and repair charges.

Table 10.6. Capitalized Cost Estimates for Different Capacities[a] (1982 U.S.$ × 1,000)

Plant Capacity (m³/day)	Rapid Filter				Slow Sand Filter			
	Construction	Annual Operation and Maintenance	Capitalized Operation and Maintenance[b]	Total Capitalized	Construction	Annual Operation and Maintenance[b]	Capitalized Operation and Maintenance[b]	Total Capitalized
1,000	54	8.1	62	120	38	1.9	14	52
1,900	110	11	85	200	66	2.1	16	82
2,300	120	13	99	220	79	3.1	23	100
6,700	150	24	190	340	260	4.9	37	300
15,000	340	54	410	750	590	13	100	690
30,000	570	98	740	1,300	1,200	22	170	1,370
45,000	910	140	1,000	1,900	1,800	29	220	2,000

Source: Adapted from Sundarasan and Paramasivam, 1981.
[a]ENR Cost Index for 1981 = 3533; 1982 = 3729.
[b]Capitalized value of operation and maintenance cost based upon 12% interest over 20 years.

smaller plants are less than those of equivalent capacity rapid filtration plants, but tend to become larger for larger plants.

OPERATION AND MAINTENANCE COSTS OF WATER TREATMENT PLANTS

Annual operation and maintenance (O&M) costs are highly variable among water treatment plants and much more difficult to estimate than construction costs. O&M costs depend upon labor costs, raw water quality, the extent of use of imported equipment and materials, and sophistication of the facilities. Furthermore, the operating costs of a treatment plant depend to a great extent on chemical and energy costs, which are extremely sensitive to changing market prices.

Because of the highly variable nature of O&M costs and the lack of reliable data on such costs in developing countries, cost curves for O&M are not included in the text.

O&M costs for water treatment plants are normally comprised of costs for the following elements: (1) chemicals; (2) energy; (3) personnel; and (4) maintenance materials requirements. Table 10.7 shows the variability of alum costs in developing countries. When alum has to be imported (which often is the case), its cost is much higher.

A Brazilian cost study (Macedo and Noguti, 1978) compared the cost-effectiveness of two types of chlorine used for disinfection in water treatment: liquid chlorine and sodium hypochlorite. Costs were compared on the basis of equipment, transportation, installation, and operation and maintenance for dosages and plant capacities ranging from 1 to 5 mg/l, and 170 to 83,400 m^3/day, respectively. The component costs and total costs for a chlorine dosage of 1 mg/l are presented in Table 10.8. Sodium hypochlorite was shown to be lower in cost at plant capacities below 500 m^3/day.

Table 10.7. Unit Costs of Alum for Several Plants in Developing Countries

City/Country	Alum (U.S. $/ton)
Cochabamba, Boliva	140
Linhares, Brazil	94
Prudentopolis, Brazil	110
Guatemala city, Guatemala	270
San Pedro Sula, Honduras	320
Ramtek, India	120
Amman, Jordan	350
Kano, Nigeria	400
Bamako, Mali	700

Table 10.8. Comparative Costs (1982 U.S.$) of Liquid Chlorine (40 kg and 900 kg containers) and Sodium Hypochlorite for a Chlorine Dosage of 1 mg/l—Brazil[a]

	Cost Component	Flow (m³/day)						
		170	430	810	4300	8600	43,200	86,400
Sodium Hypochlorite	Hypochlorite	7.8	19	39	190	390	1900	3900
	Equipment	34	34	34	34	34	34	34
	Transportation	21	52	100	521	1000	5200	10,000
	Container	2.1	4.2	7.3	32	65	330	650
	Installation	17	17	21	84	210	340	690
	Total	82	130	200	860	1700	7800	15,000
Liquid Chlorine (40 kg)	Chlorine	2.5	6.5	13	64	130	645	1300
	Equipment	100	100	100	100	100	100	100
	Transportation	.37	.98	2.0	9.7	20	198	200
	Container	11	11	11	39	72	360	720
	Installation	25	25	25	25	25	33	59
	Total	140	140	150	240	350	1200	2400
Liquid Chlorine (900 kg)	Chlorine	1.4	3.5	7.1	35	71	350	710
	Equipment	400	400	400	400	400	400	400
	Transportation	.27	.69	1.4	6.8	14	68	140
	Container	72	72	72	72	72	140	220
	Installation	84	84	84	84	84	84	84
	System for moving Cylinders	110	110	110	110	110	110	110
	Total	670	670	670	710	750	1200	1700

Source: Adapted from Macedo and Noguti, 1978.

[a]Brazil Cr. 170 = U.S.$1.00

ENR Cost Index for 1978 = 2776; for 1982 = 3729.

An interest rate of 12% has been assumed. Amortization periods of 20, 15, and 10 years have been taken for the equipment for liquid chlorine (900 kg), sodium hypochlorite, and liquid chlorine (40 kg), respectively; and one of 20 years for installation.

CONCLUSION

From the data presented in this chapter, the relative costs among types of treatment plants (rapid, slow sand, upflow-downflow, modular, package, direct filtration), and between simplified and sophisticated designs are useful. Two important conclusions can be drawn: (1) the design of treatment plants using simplified technologies can result in capital costs about one order of magnitude lower than those for plants designed with conventional technologies; and (2) capital and O&M costs can be lowered considerably by investigating alternative types of plants, such as slow sand filtration, direct filtration, or upflow-downflow plants, in situations where they are technically feasible.

Reliable cost data on water treatment projects in developing countries are difficult to obtain, primarily because national or state agencies responsible for the planning and development of such projects generally lack the resources to evaluate and make available realistic cost data. In fact, most of the cost data presented in this chapter originated in Brazil and India, which have fairly strong water supply infrastructures. In order to provide a greater diversity of cost data from other countries in future editions of this text, individuals or organizations who have access to such data are asked to submit this information to the authors at the Department of Environmental Sciences and Engineering of the University of North Carolina at Chapel Hill.

chapter eleven

HUMAN RESOURCES DEVELOPMENT

"Neither general programs nor even generous supplies of capital will accomplish much until the right technology, competent management, and manpower with the proper blend of skills are brought together and focused effectively on well-conceived projects." This statement, by George D. Woods, former president of the World Bank, made in an address to the UN Economic and Social Council on March 26, 1965, summarizes the dilemma that is faced in the provision of water in the developing world.

Instances are legion where even the most appropriately designed, properly constructed, water treatment facilities are found to be operating extremely poorly if at all after a few years service. The instances where instruments are not functioning, where stocks of chemicals have been exhausted, where pump motors have burned out, where sludge cake has been allowed to accumulate and solidify in sedimentation tanks, where sand has been flushed out of filters, and where laboratory equipment and materials have not been replaced are all too common. Accordingly, a water treatment facility cannot be considered to be complete, even if the construction is finished and the equipment has been installed and operations have begun, if the personnel necessary to assure continuous operation maintenance and management of facilities are not *already* qualified and in place.

Personnel requirements include not only the operators of the facility but the managers who are responsible for employing and deploying the personnel required, the technicians and craftsmen who are necessary for maintenance, and the laboratory personnel required for monitoring the operations, including the provision for their training. An ongoing training system must be instituted to provide for upgrading existing personnel and to train new personnel. Furthermore, the conditions of employment and career opportunities should be such as to retain qualified personnel.

One of the major problems is that most plants are quite small and cannot afford the quality of personnel to assure their proper operation. This problem has plagued the industrialized countries as well as the developing countries, and has led Britain to a regrouping of water supplies beginning in 1945 and culminating in 1974, which was based upon insuring that every water supply system would be large enough to employ the services of a qualified manager, an engineer, and a chemist. The population to be served for a system to afford such personnel was estimated to be about 150,000 (Okun, 1976). The population that would be required to afford adequate personnel in the developing world remains to be determined, because it varies substantially from place to place. However, there can be no question but that institutional development, through such devices as regional organizations, might be a device for providing the necessary qualified personnel to supervise and conduct operations at the smaller facilities.

Although the personnel problem constitutes one of the major constraints to proper water supply service to people in small communities in the United States, the situation in the developing world is considerably aggravated because of the inadequate institutional infrastructure for human resources development and very poor legacy in the field of general education left from the colonial period. The lead time required to develop the necessary personnel for the operation of water treatment plants in the absence of a sound basic educational resource is bound to be greater than the lead time required for designing and constructing the facilities to be operated. Nevertheless, little attention has been given to meeting this need for qualified personnel in a timely manner, either by external financing agencies or by the countries themselves. Far more attention is given to the technical and financial adequacy of the project than to the personnel upon whose shoulders the ultimate success of the project must rest.

This chapter discusses the personnel requirements for the types of treatment facilities presented in this text and something of the training required. However, the issue of human resources development for sustaining water supply projects deserves far more attention than a volume such as this can give.

One word of caution is appropriate. Because the facilities described are less sophisticated and involve less mechanical and electrical equipment than is generally found in water treatment plants in the industrialized world, this should not imply that the training may therefore be more modest. Those who would be employed in water treatment plants in the industrialized world have already had good education in the public schools of the country, and they have grown up in a mechanized setting, where young people are exposed to mechanical and electrical devices as a matter of course. The specialized training required for water treatment plant operations in such instances calls for only a relatively small amount of additional information, such as the chemistry of water treatment and the hydraulics appropriate to the facilities. Furthermore, should an operator have difficulty, it is only necessary to reach for the phone to get assistance from the state agency, the purveyors of the equipment, or other experts who are easily accessible.

The operators of treatment plants in a developing country, on the other hand, are drawn from a population with far less general education, little knowledge of mathematics and science, and with little experience with mechanical and electrical equipment. In the event of trouble, the resources upon which to call for assistance are not available. Technical assistance is not likely to be provided by the central government, and the purveyors of the equipment will be a continent away. Accordingly, the personnel responsible for water treatment facilities in developing countries must be far more self-reliant and qualified than their cohorts in the industrialized world.

The results of a detailed examination of training needs and strategies for meeting them, although beyond the scope of this text, may be summarized from a report on manpower development and training in the water sector for the World Bank (Okun, 1977).

The lack of qualified personnel constitutes a significant constraint to the successful operation of projects in the developing countries.

If the goals of the International Decade are to be approached, hundreds of thousands of trained personnel are required to service these facilities.

External financing agencies and the developing countries themselves, with but a few notable exceptions, have not yet given attention to human resources development, nor to training, at all consistent with the level of their investments in facilities, with the result that operations are generally poor. Water that should be safe is unsafe, extensions of service to unserved populations are slow, and service breakdowns are frequent.

Until attention to human resources development matches the priority given to technical and financial feasibility, particularly in the early stages of a project, little improvement can be expected.

Even where a commitment to human resources development is made, little training will be undertaken unless the external financing agencies are perceived as being committed to human resources development themselves. This would require that these agencies and the host countries develop institutional resources, including personnel, whose chief obligations in the water sector are to assure adequate human resources development.

A wide spectrum of skills is required for water supply: managers, engineers, chemists and biologists, technicians, and craftsmen.

The populations being served by these facilities need to be educated to the benefit of water service and safe water so they can play a role in assuring the appropriateness of the facilities to be provided for them and the quality of the operation of these facilities.

Institutions responsible for water supply must provide continuing personnel development and career planning to avoid the attrition so common in the field, where the most qualified individuals are drawn off to other sectors or to other countries. The commitment of the water agency to human resources development, and the initiation of training programs, would not only improve the technical competence of the staff but would demonstrate to the staff that the agency has an interest in their careers.

This chapter presents the various kinds of personnel required in the various water treatment facilities, their numbers in accordance with the type and size of facilities, and the training required for such personnel.

AN OVERVIEW OF MANPOWER DEVELOPMENT IN THE DEVELOPING COUNTRIES

To meet the United Nations target for the International Water Supply and Sanitation Decade (1981 to 1990), an additional 600 million people are to be provided with water supply. Based upon an estimated four employees per 1000 connections, and six people per connection, and assuming sufficient personnel for existing levels of service, this decade would require more than 400,000 additional trained personnel. The shortage of trained personnel is readily apparent in the water treatment plants of developing countries, where at least 50% of the installations lie idle in disrepair after construction (McGarry and Schiller, 1981). In fact, a 1970 WHO survey of 86 developing countries, regarding which of eight constraints to developing water supplies was most significant, found that lack of trained personnel was rated as the second most serious constraint after insufficient internal financing (WHO, 1973).

Statistics on personnel needs in the water supply and sanitation sector in developing countries are not available. One very approximate rule of thumb is that one employee is required for each 1000 population served. Where the proportion served with sanitation facilities is much less than served with water supply, the number of employees would, of course, be less. More reliable personnel surveys and estimates have been made for individual projects and countries.

Master plan studies for the Kaduna State Water Board in Nigeria in 1977–78 yielded a ratio of one employee per 1100 people served, primarily in the water supply sector (IBRD/WHO, 1979). This ratio was used to derive personnel requirements in the water sector throughout Nigeria. A study made for Nepal, a poorly developed, predominantly rural country of about 15 million people, found that 870 were working in the water sector in 1982 and that a total of some 5330 would be needed by 1990 (UNDP/IBRD, 1983). Of these, about 50% were engineers and supervisors, and some 10% water supply technicians. Great care is required in translating such estimates to other countries, even in the same region.

A manpower and training resources inventory was completed in 1975 (Carefoot, 1977). A summary of that inventory for five sector employers in Peru is presented in Table 11.1. Table 11.2 summarizes population and utilities employees as related to water and sewage service in Peru, determined for 1975 and estimated for 1980. Based on the 1975 numbers a manpower comparison index (ratio of utility personnel to population served) was calculated. On this basis, 6000 additional employees were

Table 11.1.   Summary of Manpower Inventory for Water and
Wastewater Utilities in Peru—December 1975

| General Classifications | Number of Workers | | | | | Percentage of Utility Work Force |
	ESAL[a]	ESAR[b]	DGOS[c]	DIS[d]	Total	
General managers and deputies	5	2	12	2	21	0.38
Advisors	7	2	2	1	12	0.22
Managers	100	21	89	31	241	4.42
Class 1	10	6	24	5	45	—
Class 2	20	8	37	10	75	—
Class 3	40	7	28	14	89	—
Class 4	30	0	0	2	32	—
Supervisors	0	10	157	9	170	3.22
Engineers	36	3	41	17	97	1.78
Other professionals	7	4	10	0	21	0.38
Architects	0	0	2	0	2	—
Accountants	0	1	6	0	7	—
Economists	1	0	0	0	1	—
Others	6	3	2	0	11	—
Technician 1	31	2	50	5	88	1.62
Technician 2	4	4	79	7	94	1.72
Qualified employees	137	26	117	28	338	6.18
Semiqualified employees	395	91	584	24	1094	20.03
Foremen	43	6	32	0	81	1.48
Qualified workers	135	0	5	3	143	2.62
Semiqualified workers	629	62	572	28	1291	23.83
Nonqualified workers	830	76	850	10	176	32.32
Subtotal	2359	309	2630	165	5463	100.00
Personnel of small communities— estimated	—	—	—	—	300	—
Personnel of rural areas—estimated	—	—	—	—	660	—
Total	—	—	—	—	6423	—

Source: Carefoot, 1977.
[a]Empresa de Saneamiento de Lima.
[b]Empresa de Saneamiento de Arequipa.
[c]Direccion General de Obras Sanitarias, Ministerio de Vivienda.
[d]Direccion de Ingeneria Sanitaria, Ministerio de Salud.

Table 11.2.   Demographic Data for Peru Related to Water and Sewage Service

Population by Category	1975 Actual	1980 Estimated
Total population	15,326,000	18,527,000
Population with water service	7,289,800	12,532,000
Population with sewage service	4,777,000	11,482,460
Population with water or sewage service	12,066,800	24,014,460
Number of water and wastewater employees	6,423	12,500

Source: Carefoot, 1977.

needed for the water and wastewater utilities in 1980.* This number does not include the manpower requirements to provide for the high attrition generally experienced in developing countries.

The following conclusions were drawn from the Peru exercise:

An estimated 80% of the 1975 work force required training to meet the demand of their jobs, representing a training backlog of about 5000.

The top professionals in the industry represent 2.8% of the total labor force. Originally, particularly in Latin America, the training focus had been almost exclusively on this group.

The total number of workers occupying semiqualified and nonqualified positions represent 76 percent of the personnel. None of the training institutions contacted offered courses for these employees.

There is a marked shortage of appropriate manuals and teaching aids for this sector. Training manuals do not exist for some of the subprofessional job classifications.

Uganda, a country of 13 million people, 92% rural, commissioned a manpower and training study for the water decade in 1981–82 (Uganda, 1983). Based upon 62 urban-type water supply systems and 18 sewerage and sewage disposal systems, either existing, under construction, or proposed, plus 6 semi-urban water supply systems, over 5000 boreholes, and some 950 storage tanks, it was found that a total of about 7000 people would need to be employed by 1991. The presently available and additional personnel, and their occupational categories, are shown in Table 11.3. The table indicates clearly that most of the additional personnel are required today, a condition common to most developing countries. The study detailed institutional, recruitment, staffing, and training recommendations that need to be

---

*The population served with sewerage is counted twice, because it can be assumed that those with sewerage also have water service. If it is assumed that those with water service but without sewerage have some form of sanitation, then the ratio of employees per 1000 population served with water supply and sewerage in 1975 was about 0.9, close to the rough estimate of 1 per 1000.

Table 11.3.   Personnel in the Water Sector in Uganda

		Numbers of Personnel	
	Actual	Required	Estimated Requirement
	1981	1982	1991
Professional-Technical			
Engineers	21	82	95
Chemists	1	2	3
Hydrotechnologists	3	10	14
Total	25	94	112
Professional-Administrative	16	59	68
Sub-Professionals			
Technicians	152	262	332
Supervisors	104	327	364
Total	256	589	696
Clerical	404	494	704
Operating Staff			
Operators, skilled	260	600	796
Semi-skilled workers	751	1136	1519
Unskilled workers	2002	2506	3086
Total	3013	4242	5401
Total	3714	5478	6981

Source: Adapted from Uganda, 1983.

implemented if the facilities being rehabilitated and constructed are to serve properly. One of the critical priorities identified was the establishment of a training center.

The development of human resources in developing countries is often characterized by the general conditions outlined below.

1. Government agencies in the least developed countries tend to employ expatriates for supervisory or technical positions on a temporary basis, with the expectation that local personnel can be adequately trained to take over these positions in the future. In too many instances, however, expensive expatriate contracts are renewed, because the expatriate has little interest in the training of individuals who would make him or her redundant, so the expatriate becomes a permanent fixture.

2. Technically skilled and experienced personnel are difficult to attract and to retain in supervisory positions in water supply agencies because salary scales are low. Furthermore, many plants experience high rates of employee turnover because of the opportunities available for qualified personnel to work in more "prestigious" and higher-paying jobs in the private sector and in other, richer, countries.

3.   There is a tendency to overman water treatment plants with unskilled personnel because of high unemployment and political pressures within the country. This results in poor performance, the underutilization of personnel, employment inflexibility, and a reluctance to learn new and improved job methods (Barker, 1976).

4.   Inadequate programs of preventive maintenance and a lack of spare parts characterize almost all plant operations. These are symptomatic not only of a shortage of trained operators but, more important, of supervisory and administrative personnel.

5.   With a few notable exceptions, as for example, Brazil, Tunisia, and India, training facilities are available for only a small fraction of the professional staff now employed, with almost no provisions for those to be employed in the future.

6.   Greater attention is now being given by financing agencies to so-called operator or technician training. This is modeled after training in the industrialized world, where such training has been perceived as being of high priority. This is the case in the industrialized world, where professional training is well-institutionalized and the professionals promote the training of subprofessionals. On the other hand, the great need in the least developed countries is for professionals who can initiate the planning, design, financing, and management of projects in the water sector. It is this group that will need to be responsible for institutionalizing training for subprofessionals. Too often, when the financing of training instituted by external financing agencies comes to an end, the training itself ends.

CLASSIFICATIONS OF PLANT PERSONNEL

The various kinds of personnel required to operate and maintain a water treatment plant are largely a function of the type and size of plant. Personnel required at central agency headquarters are not included here. Table 11.4 lists the kinds of personnel and resources required for four categories of water treatment. The table classifies manpower requirements in three groups: professional, skilled, and unskilled.

Professional personnel require a substantial amount of formal training, generally from a university. The superintendent of a large rapid filtration plant, for example, would fall into this group. Skilled, or subprofessional personnel require some formal training, generally a secondary school education, plus two to three years of specialized vocational training. Training facilities for subprofessionals in a sector are often maintained in a developing country by the agency of the government responsible for that sector in order to meet their specific requirements for manpower. Unskilled personnel, or common laborers, require little formal training; these individuals can be provided the necessary skills on the job, but even this requires organization and a highly qualified operating staff.

Table 11.4. **Kinds of Personnel and Resources Required for Water Treatment Plants**

Treatment Methods	Manpower Required for Operation			Resources Required			
	Unskilled	Skilled	Professional	Operation Equipment	Process Materials	Maintenance Supplies	Chemical Supplies
(Slow sand filtration (conventional, dynamic)	X				X	X	
Conventional rapid filtration (conventional, dual media, upflow-downflow)		X	X	X	X	X	X
Advanced rapid filtration (multi-media, inclined-plate, or tube settling; polyelectrolytes)		X	X	X	X	X	X
Disinfection (chlorination)		X		X	X	X	X

Source: Adapted from Reid and Coffey, 1978.

NUMBERS OF PLANT PERSONNEL

The numbers of personnel required for the operation and maintenance of water treatment plants depend in general, on the design, layout, size, and complexity of the facility. For example, chlorinated groundwater supplies may be operated by only one person, who would be responsible for checking pumps and chlorinators. Rapid filtration plants with continuous operation require a minimum of four operators: one chief operator and three shift operators (assisted by maintenance mechanics and laborers). For very large plants separate groups must handle pumping stations, chemical building, filter operating floor and laboratory; a minimum of four operators is needed by each group, supervised by a plant superintendent and assisted by maintenance mechanics and laborers (Cox, 1964).

The following factors are important in determining manpower requirements: (1) type of supply, surface, or groundwater; (2) location of the source of supply as related to location of the treatment facility; (3) variability of raw water characteristics with regard to flow and quality; (4) type and complexity of equipment; (5) employment of any special treatment; and (6) requirements for pumping. The simplified designs described in this text require a significant number of unskilled laborers periodically to handle various labor-intensive jobs such as, for example, the removal of sludge from settling basins, or the removal and washing of the dirty sand from the bed of a slow sand filter.

The Institution of Water Engineers and Scientists of England in its water practice manual on *Water Supply and Sanitation in Developing Countries* (Dangerfield, 1983) summarizes a survey of several African countries. Only part-time unskilled staff is provided for very small works, but it is supported by visits or part-time work by technicians from a central office. Plants over 2000 m^3/day all had a full-time staff on shift work. Plants of about 35,000 m^3/day had a skilled superintendent and three semiskilled and four unskilled workers. At major works of 250,000 m^3/day, the total operating staff would be on the order of 20 to 25 persons.

Groundskeepers are in general in ample supply, and in addition to the numbers shown above. Because the work is familiar, the care of grounds is often better than the care of the facilities.

Manpower requirements for various types of plants and population levels are summarized in Table 11.5; manpower requirements for cleaning of slow sand filters are shown in Table 11.6 for both manual and mechanical methods; and the laboratory staff required for plants of different capacities is shown in Table 11.7.

The numbers required to operate a treatment plant cannot be stated with any precision because, in addition to the variables cited above, the availability of central support services will determine both the local plant needs and the attention given to these needs. Last, the history and social structure of the community and the country will affect the assignment of personnel. In some countries, a major problem is the large number of unskilled workers that a water plant is obliged to employ for economic and political reasons.

Table 11.5. Operation and Maintenance Manpower Requirements for Water Treatment Plants

Type of Treatment	Size of Community	Manpower Required		
		Unskilled	Skilled	Professional
Slow sand filtration (conventional, dynamic)	500 to 2500	1		
	2500 to 15,000	2		
	15,000 to 50,000	5		
	50,000 to 100,000	8		
Conventional rapid filtration (conventional, dual-media, upflow-downflow)	500 to 2500	1	1	
	2500 to 15,000	1	1	1
	15,000 to 50,000	8	2	1
	50,000 to 100,000	10	3	1
Advanced rapid filtration (multi-media, inclined-plate or tube settling, polyelectrolytes)	500 to 15,000	1	1	1
	15,000 to 50,000	6	2	2
	50,000 to 100,000	10	5	2
Disinfection (chlorination)	500 to 2500	1		
	2500 to 15,000	1	1	
	15,000 to 50,000	2	1	1
	50,000 to 100,000	4	1	1

Source: Adapted from Reid and Coffey, 1978.

Table 11.6. Comparison of Requirements for Cleaning Slow Sand Filters (slow sand filter with an area of 2000 m²)

	Manual Method	Tractor Scrapers		Gentry Scraper Amsterdam	Hydraulic Method
		London	Berlin		
Number of hours required for					
draining	2	2	2	2	0
cleaning	9	4	5	3	6
refilling	5	5	5	5	0
reripening	24	24	24	4	4
Total time of service (hours)	40	35	36	14	10
Total number of men employed	8	4	2	2	1
Total number of man-hours involved	75	20	15	10	10

Source: Huisman and Wood, 1974.

Table 11.7.   Staff Required for
Water Treatment Plant Laboratories

Size of Plant (m^3/day)	Staff (full-time)
<20,000	1/2
20,000 to 80,000	1 to 2
80,000 to 200,000	2 to 5
>200,000	>5

Source: Adapted from Hudson, 1981.

TRAINING

The organization of training programs and the design of appropriate curricula are beyond the scope of this text. Many useful documents on various elements of training have been developed, and are readily available from several agencies, including the International Reference Center for Community Water supply, the U.S. Agency for International Development, and the World Health Organization. Of particular value is the *Basic Strategy Document on Human Resources Development for the Decade,* together with an "HRD Check List Package," published by WHO (1982), which emanated from a task force of the Decade Steering Committee representing the international, bilateral and nongovernmental organizations in the sector.

A number of points with regard to the approach to training need to be emphasized.

The lead time required for preparing qualified personnel for plant operation is greater than for the design and construction of the plant. Hence, the initiation of a training program should enjoy a high priority in both timing and funds. A most satisfactory approach is to identify and appoint the operating staff prior to beginning of plant construction so the staff can be employed and trained during construction. They can be used on construction inspection while being trained with the help of the resident engineering staff in charge of construction. Those responsible for design and construction can conduct classes regularly during construction, explaining the purpose and operation of all the elements of the plant.

The designing engineers should be obligated to prepare an operating manual translated into the local language for each plant. This manual would describe the units and their operation, with instruction for operation of valves, pumps, chemical feeders and the like. This manual should be available well before the plant is to go into operation and be used in the training program for plant personnel.

The organization and conduct of training programs require specialized skills, and are the role of specialists with water agencies. The grafting of training responsibilities onto an engineer with many other responsibilities is bound to result in training being inadequately served. If a training specialist is not available, an individual may be selected for the post and given the time and opportunity to prepare for training responsibilities.

The pedagogical approach must recognize the background of the individuals to be trained. Most will not have had much rewarding classroom experience, so formal classroom lecturing should be minimized and greater emphasis given to "hands-on" experiences. Accordingly, the training facility should be integrated with a suitable operating facility where feasible, and skilled operators who are identified as being articulate should be used in the training.

Before new training institutions are developed, a survey of available resources may reveal existing facilities which, with some assistance, can be adapted to the task. Such facilities would include universities, teacher training institutions, vocational training schools, or other specialized establishments. Universities should be encouraged to undertake responsibilities for subprofessional as well as professional training, because they have much to offer in facilities, personnel, and status.

One important benefit of training which is often overlooked in planning is that providing training for employees enhances the status of their positions and endows them with an aura of importance that increases interest and improves job performance. Associated with this view of training is the value of follow-up of trained personnel by periodic visits from managers to assure them of the importance of their work.

Continued education, in the several modes that are feasible, needs to be institutionalized for personnel at all levels.

Certification of operating personnel might be considered where appropriate.

Those responsible for human resources development must be adamant that training be institutionalized, and that a training component be an important element of every project in the water and sanitation sector.

appendix a

COMMON CHEMICALS USED
IN WATER TREATMENT
IN THE UNITED STATES

The design engineer and the superintendent of operation should ascertain which chemicals, and in what form, are available locally. The wide variety of packaging shown here is not likely to be available in most countries in Asia, Africa, and Latin America.

1. Aluminum Sulfate

Chemical Formula	$Al_2(SO_4)_3 \cdot 14H_2O$
Common Name	alum, filter alum, sulfate of aluminum
Available Forms	ground, rice, powder, lump form in 50- and 100-kg bags; 150 and 180-kg barrels; 10-, 50-, and 125 kg drums; and carloads
Appearance and Properties	light tan to gray-green; dusty, astringent, only slightly hygroscopic; 1% solution-pH 3.4
Weight	960 to 1200 kg/m^3
Commercial Strength	at least 17% Al_2O_3

(Information compiled from table prepared by B.I.F. Industries, a Division of New York Air Brake Company, Providence, Rhode Island, USA)

Feeding	fed dry in ground and rice form; maximum concentration when dissolved in water 60 g/l
Handling Material	handled dry in iron, steel, and concrete; wet in lead, rubber, asphalt, cypress

2.   Alum, Liquid

Chemical Formula	$Al_2(SO_4)_3 \cdot \chi H_2O$
Common Name	liquid alum
Available Forms	solution, manufactured near site and transported in 6000- to 8000-gal steel tank cars, 2000- to 4000-gal rubber-lined steel tank trucks; high freight costs preclude distant shipment
Appearances and Properties	light green to light brown; 1% solution-pH 3.4
Weight	1270 g/l
Commercial Strength	5.8 to 8.5% Al_2O_3
Feeding	Dilute to 3% solution concentration before applying to water
Handling Materials	Lead or rubber-lined tanks, hard rubber or plastic pipe

3.   Calcium Hydroxide

Chemical Formula	$Ca(OH)_2$
Common Name	hydrated lime, slaked lime
Available Forms	powder in 20-kg bags, 50-kg barrels, and carloads
Appearances and Properties	white caustic, dusty and irritant; saturated solution-pH 12.4
Weight	560 to 800 kg/m^3
Commercial Strength	62 to 74% CaO
Feeding	fed dry, 60 g/l maximum, and as slurry 110 g/l maximum
Handling Materials	rubber hose, iron, steel, asphalt, and concrete; must be stored in dry place

4.   Calcium Hypochlorite

Chemical Formula	$Ca(OCl)_2 \cdot 4H_2O$
Common Name	HTH (High-Test Hypochlorite), Perchloron, Pittchlor
Available Forms	powder, granules, and pellets in 50-kg barrels, 2-, 7-, 50-, and 140-kg cans, and 360-kg drums
Appearances and Properties	white or yellowish-white; nonhygroscopic, corrosive and odorous
Weight	800 to 900 kg/m^3

Commercial Strength	65 to 70% available Cl_2
Feeding	fed as solution up to 2% strength (30 g/l)
Handling Materials	ceramics, glass, plastics, and rubber-lined tanks; store dry and cool, avoid contact with organic matter

5.   Calcium Oxide

Chemical Formula	CaO
Common Name	quicklime, burnt lime, chemical lime, unslaked lime
Available Forms	lumps, pebbles, crushed or ground in 50-kg moistureproof bags, wooden barrels, and carloads
Appearances and Properties	white (light gray, tan); unstable, caustic, and irritating; slakes to calcium hydroxide with evolution of heat when water is added; saturated solution-pH 12.4
Weight	880 to 1100 kg/m³
Commercial Strength	70 to 96% CaO
Feeding	best fed dry as 2-cm pebbles or crushed to pass 2.5-cm ring; solution concentration ranges from 165 to 285 g/l, dilute after slaking to 110 g/l (10% solution)
Handling Materials	handled dry in iron, steel, concrete; handled slaked in iron, steel, rubber hose, and concrete; should not be stored for more than 60 days, even in tight container

6.   Chlorinated Lime

Chemical Formula	$CaO \cdot 2CaOCl_2 \cdot 3H_2O$
Common Name	bleaching powder, chloride of lime
Available Forms	powder in 45-, 135-, 350-kg drums
Appearances and Properties	white, corrosive and odorous, unstable, deteriorates
Weight	720 to 800 kg/m³
Commercial Strength	25 to 37% available Cl_2
Feeding	must be dissolved and mixed thoroughly in a solution tank, followed by a settling period to remove insoluble solids; about 30 g/l makes 1% solution of available Cl_2
Handling Materials	glass, rubber, stoneware, wood; must be stored dry

7.   Chlorine

Chemical Formula	Cl_2
Common Name	chlorine gas, liquid chlorine

Available Forms	liquified gas under pressure in 50- and 70-kg steel cylinders, 1-ton containers, cars with 15-ton containers, and tank cars of 16-, 30-, and 55-ton capacity
Appearances and Properties	greenish-yellow gas; pungent, noxious, corrosive gas heavier than air; health hazard
Weight	specific gravity with respect to air-2.49
Commercial Strength	99.8% Cl_2
Feeding	fed as gas vaporized from liquid and as aqueous solution (2.4 g/l or more) through gas feeder or chlorinator
Handling Materials	dry liquids or gas handled in black iron, copper and steel; wet gas in glass, silver, hard rubber

8. Copper Sulfate

Chemical Formula	$CuSO_4 \cdot 5H_2O$
Available Forms	ground and as powder or lumps in 50-kg bags and 200-kg barrels or drums
Appearances and Properties	clear blue crystals or pale blue powder; poisonous
Weight	1200 to 1400 kg/m^3 ground; 1200 to 1300 kg/m^3 as powder; and 1000 kg/m^3 as lumps
Commercial Strength	99% pure
Feeding	best fed ground and as powder; maximum concentration 30 gr/l
Handling Materials	stainless steel, asphalt, rubber, plastics, and ceramics

9. Ferric Chloride

Chemical Formula	$FeCl_3$ (anhydrous and as solution); $FeCl_3 \cdot 6H_2O$ (crystal)
Common Name	chloride of iron, ferrichlor
Available Forms	solution in 20- and 50-liter carboys and in tank trucks; lumps and granules in 45-, 180-, 200-kg kegs, and 65-, 160-, 280-kg drums
Appearance and Properties	solution—dark brown syrup; crystals—yellow-brown lumps; anhydrous—green, black; hygroscopic, very corrosive; 1% solution—pH 2.0
Weight	solution weighs about 1440 kg/m^3; crystals 1000 kg/m^3; anhydrous chemical 1440 kg/m^3
Commercial Strength	solution should contain 35 to 40%; crystals 60%, and anhydrous chemical 96 to 97% $FeCl_3$

Feeding	fed as solution containing up to 45% $FeCl_3$ (800 g/l for anhydrous form)
Handling Properties	rubber, glass, ceramics, and plastics

10. Ferric Sulfate

Chemical Formula	$Fe_2(SO_4)_3 \cdot 3H_2O$; and $Fe_2(SO_4)_3 \cdot 2H_2O$
Common Name	Ferrifloc, Ferriclear, iron sulfate
Available Forms	granules in 50-kg bags, 180- and 190-kg drums, and carloads
Appearances and Properties	$2H_2O$—red brown, $3H_2O$—red gray; hygroscopic, very corrosive
Weight	about 1150 kg/m^3
Commercial Strength	$3H_2O$ should contain 18.5% Fe; $2H_2O$ should contain 21% Fe
Feeding	best fed dry, 170 to 290 g/l, dissolver detention time 20 min
Handling Materials	stainless steel, rubber, lead, and ceramics; must be stored in tight container

11. Ferrous Sulfate

Chemical Formula	$FeSO_4 \cdot 7H_2O$
Common Name	copperas, iron sulfate, sugar sulfate, green vitriol
Available Forms	granules, crystals, powder, and lumps in 50-kg bags, 180-kg barrels, and bulk
Appearances and Properties	green to brownish-yellow; hygroscopic, very corrosive
Weight	about 1000 kg/m^3
Commercial Strength	20% Fe
Feeding	best fed as dry granules, 60 g/l, dissolver detention time 5 min
Handling Materials	handled dry in iron, steel, and concrete; wet in lead, rubber, iron, asphalt, cypress, and stainless steel; stored dry in tight containers

12. Sodium Carbonate

Chemical Formula	Na_2CO_3
Common Name	soda ash
Available Forms	crystals and powder in 50-kg bags, 50-kg barrels, 10-kg drums, and carloads

Appearances and Properties	white, alkaline, hygroscopic; 1% solution—pH 11.2
Weight	480 to 1000 kg/m^3, extra light to dense
Commercial Strength	58% Na_2O
Feeding	best fed as dense crystals, 30 g/l, dissolver detention time 10 min; longer for higher concentration
Handling Materials	iron, steel, and rubber hose

13.   Sodium Hypochlorite

Chemical Formula	NaOCl
Common Name	bleach liquor, chlorine bleach
Available Forms	liquid in 20-, 50-liter carboys; 4900-, 7500-liter tank trucks
Appearances and Properties	light yellow; corrosive and odorous; relatively stable, generally free from suspended solids
Weight	—
Commercial Strength	12 to 15% available Cl_2
Feeding	fed as solution diluted with water to a strength of 0.5 to 1.0% available Cl_2
Handling Materials	ceramic, glass, plastic, and rubber-lined tanks; store in cool place, protect from light

appendix b

HYDRAULIC CALCULATIONS FOR SELECTED UNIT PROCESSES

This appendix contains several practical examples for the design of selected unit processes for water treatment; including:

B-1.   Horizontal-flow baffled channel flocculator.
B-2.   Gravel bed flocculator.
B-3.   Staircase type heliocoidal-flow flocculator.
B-4.   Tube-settler modules in horizontal-flow settling basins.
B-5.   Inclined-plate settlers in horizontal-flow settling basins.

B-1.   HORIZONTAL-FLOW BAFFLED-CHANNEL FLOCCULATOR

Problem

Design a horizontal-flow baffled channel flocculator for a treatment plant of 10,000 m³/day capacity. The flocculation basin is to be divided into three sections of equal volume, each section having constant velocity gradients of 50, 35, and 25 s^{-1}, respectively. The total flocculation time is to be 21 min and the water temperature is 15°C. The timber baffles have a roughness coefficient of 0.3. A common wall is shared between the flocculation and sedimentation basins; hence the length of the flocculator is fixed at 10.0 m. A depth of 1.0 m is considered reasonable for horizontal-flow flocculators.

Solution: (1) Design the first flocculator section with a velocity gradient of 50 s^{-1} and detention time of 7 min.

1. Total volume of flocculation

$$V = (21/1440) \times (10,000) = 146 \text{ m}^3$$

2. Total width of flocculator:

$$W = \frac{V}{L \times H} = \frac{146}{(10.0)(1.0)} = 14.6 \text{ (say 15 m)}$$

3. Width of each section:

$$W = 15/3 = 5.0 \text{ m}$$

4. For water at 15°C (values obtained from Table 6.1)

$$\mu = 1.14 \times 10^{-3} \text{ kg/m} \cdot \text{s}$$

$$\rho \cong 1000 \text{ kg/m}^3$$

5. Number of baffles in first flocculator section (from eq. 6.4):

$$n = \left[\frac{2\,(1.14 \times 10^{-3})\,(7)\,(60)}{1000\,(1.44 + 0.3)} \left(\frac{1.0\,(10.0)\,(50)}{10,000/86,400} \right)^2 \right]^{1/3}$$

$$= 22$$

6. Spacing between baffles:

$$10.0/22 = 0.45 \text{ m (satisfies minimum spacing requirements,}$$
$$\text{as stipulated in Table 6.3)}$$

7. Head loss in the flocculator section (formula adapted from eq. 6.1):

$$h = \frac{\mu t}{\rho g} G^2$$

$$= \frac{(1.14 \times 10^{-3})\,(7)\,(60)}{1000\,(9.8)} (50)^2$$

$$= 0.12 \text{ m}$$

8. The same series of calculations is repeated for the remaining two flocculator sections. The results are as follows:

2d section —$G = 35$ s^{-1}
        —$t = 7$ min
        —number of baffles $= 17$, use 16 for the design configuration presented in Figure B.1
        —spacing between baffles $= 0.62$ m
        —head loss $= 0.06$ m

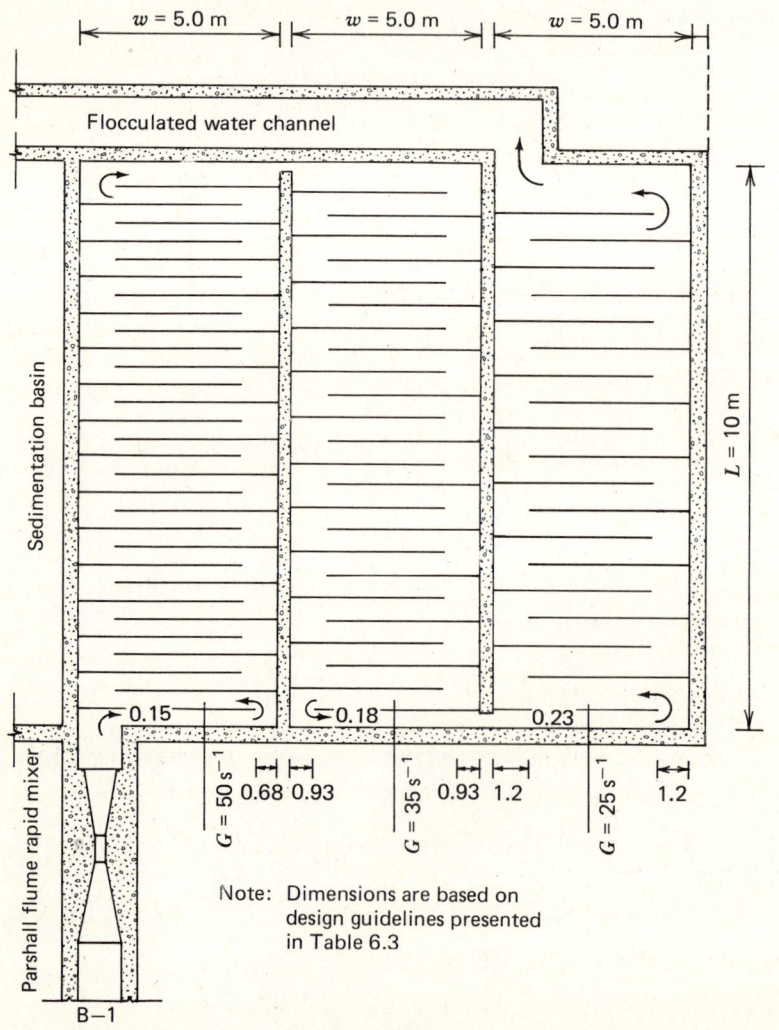

Figure B.1.   Horizontal-flow baffled channel flocculator for a plant of 10,000 m³/day capacity. (dimensions in meters.)

3d Section—$G = 25 \text{ s}^{-1}$
        $-T = 7 \text{ min}$
        —number of baffles = 14, use 13 for design configuration
        —spacing between baffles = 0.77 m
        —head loss = 0.03 m

(a) The total head loss in the flocculator:

$$h_T = 0.12 + 0.06 + 0.03 = 0.21 \text{ m}$$

The design of the horizontal-flow baffled channel flocculator is shown in Figure B.1.

B-2. STAIRCASE-TYPE HELIOCOIDAL-FLOW FLOCCULATOR

Problem

A plant having capacity of 12,960 m³/day contains a flocculator comprised of four square chambers with 25 min detention time. For one of the chambers design a staircase-type flocculator with a velocity gradient of 40 s⁻¹. The water temperature is 20°C and the chamber depth is 3.5 m.

Solution:

1. Volume of chamber:

$$V = (12{,}960)\,(25)/(4)\,(1440) = 56.3 \text{ m}^3$$

2. Cross-sectional area of chamber:

$$A = 56.3/3.5 = 16.1 \text{ m}^2$$

3. Length of one side of chamber:

$$L = (16.1)^{1/2} = 4.0 \text{ m}$$

4. For water at 20°C (values obtained from Table 6.1):

$$\rho = 998 \text{ kg/m}^3 \qquad \mu = 1.01 \times 10^{-3} \text{ kg/m} \cdot \text{s}$$

5. Head loss in the formula flocculator (formula adapted from eq. 6.6):

$$h = \left[\frac{2\rho KQ^3}{\mu L^4 G^2}\right]^{1/3}$$

$$= \left[\frac{2\,(998)\,(7.5)\,(12960/86400)^3}{(1.01\times10^{-3})\,(4.0)^4\,(40)^2}\right]^{1/3}$$

$$= 0.50\ \text{m}$$

6.   Number of helices in the flocculator:

$$3.5/0.50 = 7.0$$

7.   Value of pitch:

$$h = 3.5/7.0 = 0.5\ \text{m}$$

A staircase-type flocculator for a rectangular flocculation chamber is shown in Figure 6.11.

B-3.   Gravel-Bed Flocculator

Problem

A package water treatment plant having a capacity of 270 m³/day contains a gravel-bed flocculator that is comprised of five rectangular sections, each of which is successively larger in cross-sectional area, as shown in Figure 6.14. The dimensions of the flocculator sections and corresponding gravel sizes are given below. The gravel has a porosity of 0.4. Water temperature is 20°C ($\mu = 1.01 \times 10^{-3}$ kg/ms; $\rho = 998$ kg/m³).

Section	Length (m)	Width (m)	Height (m)	Gravel Size (cm)
1	1.0	0.05	0.20	0.5 to 1.0
2	1.0	0.13	0.20	0.5 to 1.0
3	1.0	0.23	0.20	0.5 to 1.0
4	1.0	0.35	0.20	1.0 to 2.0
5	1.0	0.51	0.20	1.0 to 2.0

Calculate the nominal flocculation time in the system and the velocity gradients and head loss for each section.

Solution: (1) *nominal flocculation time.*

1. Volume of flocculator:

$$V = 1.0 \times 0.2 \times (0.05 + 0.13 + 0.23 + 0.35 + 0.51) = 0.254 \text{ m}^3$$

2. Nominal flocculation time:

$$t = \frac{0.254}{270/86,400} = 81.3 \text{ sec } (1.35 \text{ min})$$

Solution: (2) *Head loss and velocity gradient for Section 1*

1. Face velocity:

$$v = \frac{270/86,400}{1.0 \times 0.05} = 0.06 \text{ m/sec}$$

2. Head loss in Section 1 (from eqs. 6.8 through 6.10):
   (a) Reynolds Number:

$$R_n = \frac{0.0075 \times 0.06 \times 998}{1.01 \times 10^{-3}} = 445$$

   (b) Friction factor:

$$f = 150 \left[\frac{1 - 0.4}{445} \right] + 1.75 = 1.95$$

   (c) Head loss:

$$h = \frac{1.95}{0.8} \times \frac{1 - 0.4}{(0.4)^3} \times \frac{0.2}{0.0075} \times \frac{(0.06)^2}{9.8} = 0.22 \text{ m}$$

3. Volume:

$$V = 1.0 \times 0.05 \times 0.20 = 0.01 \text{ m}^3$$

4. Velocity gradient (from eq. 6.7):

$$G = \left[\frac{0.22 \times 998 \times 9.8 \times 270}{(1.01 \times 10^{-3}) \times 0.4 \times 0.01 \times 86,400} \right]^{1/2} = 1290 \text{ sec}^{-1}$$

Solution: (3)   The same series of calculations is repeated for the remaining four sections of the flocculator. The results are as follows:

Section 1:	$h = 20$ cm	$G = 1290$ sec^{-1}
Section 2:	$h = 4.1$ cm	$G = 345$ sec^{-1}
Section 3:	$h = 1.7$ cm	$G = 165$ sec^{-1}
Section 4:	$h = 0.34$ cm	$G = 60$ sec^{-1}
Section 5:	$h = 0.16$ cm	$G = 35$ sec^{-1}

B-4.   TUBE-SETTLER MODULES IN HORIZONTAL-FLOW SETTLING BASINS

Problem

A water treatment plant having a capacity of 114,000 m^3/day includes two horizontal-flow settling basins, each of which is 24.4 m long, 18.3 m wide, and 3.7 m deep. Calculate (1) the actual surface loading rate (settling velocity) of the basins, and (2) the surface loading rate (settling velocity) that would be obtained if prefabricated modules comprised of square tubes inclined at 60° are installed the last 12.5 m of the basin. The modules are 61 cm high and the cross-sectional area of each tube is 5.1 × 5.1 cm.

Solution: (1)   *Surface loading rate for each basin without tube settlers*

1.   Surface loading rate (from eq. 7.3):

$$v_s = \frac{114,000}{(18.3)\,(24.4)\,(2)} = 128 \text{ m/day}$$

(2)   *Surface loading rate for each basin with tube settlers installed*

1.   Efficiency factor for square tube settling system:

$$K = 1.38$$

(*Note:* $K = 1.33$ for circular tubes;
$= 1.0$ for parallel plates)

2.   Relative settler length (from eq. 7.6)

$$L_R = 61/5.1 = 12.0 \text{ m}$$

3. Effective relative depth (from eq. 7.7)

$$L_u = 12.0 - (0.013)(280)$$

$$= 8.36 \text{ m}$$

4. Area of high-rate settling:

$$A = (12.5)(18.3) = 229 \text{ m}^2$$

5. Surface loading rate for area of high-rate settling:

$$v_o = 114,000/(229)(2) = 249 \text{ m/day}$$

6. Surface loading rate of tube settlers (from eq. 7.5):

$$v_{sc} = \frac{(1.38)(249)}{0.866 + (8.36)(0.5)}$$

$$= 68 \text{ m/day}$$

B-5. INCLINED-PLATE SETTLERS IN HORIZONTAL-FLOW SETTLING BASINS

Problem

The settling capacity of a water treatment plant is to be increased from 19,000 m³/day to 48,400 m³/day. There are three horizontal-flow settling basins, each of which is 23.5 m long, 12.0 m wide, and 4 m deep. Parallel plates are to be placed 5 cm apart at an angle of 60° from the horizontal. The plates are 2.4 m long, 1.0 m wide, and 1.0 cm thick. The water is being treated mainly for color removal, hence the surface loading rate should not exceed 30 m/day. Calculate the area required for high-rate settling.

Solution

1. Relative settler length (from eq. 7.6):

$$L_R = 100/5.0 = 20$$

2. Effective relative depth (from eq. 7.7):

$$L_u = 20 - (0.013)(280)$$

$$= 16.4$$

3. Total area required for high-rate settling (formula adapted from eq. 7.5 and 7-8):

$$A = \frac{QK}{v_{sc} (\sin \theta + L_u \cos \theta)} = \frac{(48,400)(1)}{30[0.866 + (16.4)(0.5)]} = 178 \text{ m}^2$$

4. Area required per basin:

$$178/3 = 59.3 \text{ say } 60 \text{ m}^2; \text{ or } 12 \text{ m} \times 5 \text{ m}$$

5. Number of plates needed:

$$5/0.05 = 100 \text{ plates per row of 2.4 m width}$$

6. Rows of plates needed per basin width:

$$12/2.4 = 5 \text{ rows}$$

7. Total length of basin that will be covered by the plates:

$$100 (0.01) + 5.0 = 6.0 \text{ m}$$

A diagram showing the installation of parallel plate settlers in the horizontal flow basin is presented in Figure B.2.

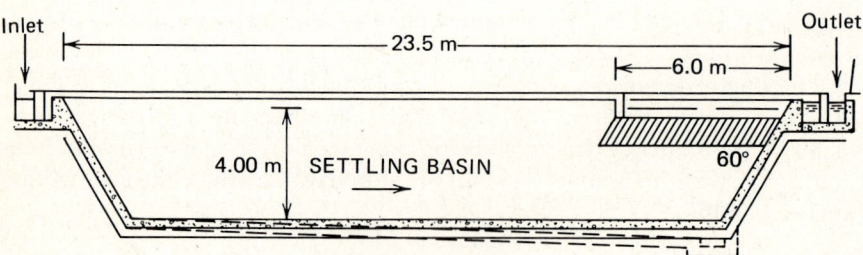

Figure B.2.   Installation of inclined plate settlers in a horizontal-flow settling basin.

appendix c

CHECKLIST FOR DESIGN OF WATER TREATMENT PROCESSES

CHECKLIST FOR DESIGN OF WATER-TREATMENT PLANTS

The purpose of this list is to assemble in an orderly manner the various items important in a water treatment plant designed to treat surface water. This permits utilizing the list as means of ensuring that essential points have not been overlooked, either in a preliminary or final design.

In general, the checklist is limited to the more commonly-used processes. An attempt has been made to separate the items into functional and operational considerations, as far as these apply. Items such as intakes and pumping stations, not included in this text, are covered for completeness.

PLANT AND BUILDING DATA

Plant Site: Distance from community . . . . ., access roads . . . . ., rail siding . . . . ., ground elevation . . . . ., protection against flooding . . . . ., size of property . . . . ., fencing . . . . ., landscaping . . . . ., outdoor lighting . . . . ., provision for future expansion . . . . . .

(adapted from Hardenburgh and Rodie, 1961)

Building: Type of structure . . . . ., size . . . . ., exterior finish . . . . . . **Chemical storage:** lime . . . . ., alum . . . . ., iron . . . . ., salt . . . . ., other . . . . ., unloading and handling methods . . . . . **Facilities:** drinking water . . . . ., toilet . . . . ., locker room . . . . ., washroom and shower . . . . ., lunchroom . . . . ., toolroom . . . . ., shop . . . . . **Laboratory: Bacteriological:** refrigerator . . . . ., incubator . . . . ., oven . . . . ., microscope . . . . ., balance . . . . ., still . . . . . **Chemical:** hood . . . . ., pH meter . . . . ., colorimeter . . . . ., residual chlorine . . . . ., reagents . . . . ., jar testing equipment . . . . . **General:** glassware . . . . ., sinks . . . . ., hot water . . . . ., vacuum . . . . ., electricity . . . . ., lighting . . . . ., gas . . . . ., air . . . . ., safety shower . . . . ., fire protection . . . . . **Material storage:** existing . . . . ., needed . . . . ., provided . . . . ., how . . . . . .

WATER SUPPLY

Source: Expected yield of source: average . . . . . m^3/day, safe yield in dry season . . . . . m^3/day

Population Served: Present . . . . ., design . . . . . . **Water Use:** present . . . . . m^3/day, design . . . . . m^3/day, average per day . . . . . m^3day, maximum month . . . . . m^3/day, maximum day . . . . . . m^3/day

Stream Flow: Average . . . . . m^3/day, maximum . . . . . m^3/day, minimum . . . . . m^3/day, high water level . . . . . m, low water level . . . . . m

Reservoir: Area . . . . ., depth . . . . ., high water level . . . . ., low water level . . . . . .

Range of Raw Water Quality: MPN . . . . ., BOD_5 . . . . ., pH . . . . ., total solids . . . . ., turbidity . . . . ., temperature . . . . ., color . . . . ., taste and odor . . . . ., alkalinity . . . . ., hardness . . . . ., algae . . . . . .

RAW WATER TRANSMISSION

Supply Line: Number . . . . ., size . . . . ., material . . . . ., length . . . . ., capacity . . . . . m^3/day, $C = $ . . . . ., gravity . . . . ., pumping . . . . ., pressure at plant . . . . ., head pumped against . . . . ., velocity in line for design flow . . . . ., corrosion protection . . . . ., interconnections . . . . ., air relief valves . . . . ., drains at low points . . . . ., isolation of sections for repairs . . . . ., access to right of way . . . . ., is line metered? . . . . .

Pumping Stations: Location . . . . ., number of pumps . . . . ., capacity of pumps . . . . ., size of suction lines . . . . ., size of discharge lines . . . . ., type of pumps . . . . ., efficiency . . . . ., motive power . . . . ., power requirements . . . . ., flood protection . . . . .

INTAKES AND SCREENS

Intakes: Number . . . . ., type . . . . ., size . . . . ., capacity . . . . ., head loss . . . . ., invert elevation of pipe out . . . . . **Elevation water surface:** high . . . . ., low . . . . ., average . . . . ., depth of water at intake . . . . ., distance intake from shore. . . . .

Screens: Where used . . . . ., what kind . . . . ., material . . . . ., mesh or opening size . . . . ., power requirements . . . . ., area of openings . . . . ., flow through screen . . . . ., are screens removable? . . . . ., are there duplicates? . . . . ., method of cleaning . . . . .

COAGULATION AND SEDIMENTATION

Chemicals: Kinds used . . . . ., design dosages . . . . . mg/l

Rapid Mix: Number of units . . . . ., type of mixer . . . . ., volume . . . . .m^3, retention . . . . . seconds, point of chemical feed . . . . .

Flocculator: Number of tanks . . . . ., tank length . . . . .m, width . . . . .m, depth . . . . .m, retention . . . . .min, type of mixer . . . .

Ports or Openings: Rapid mix to flocculator . . . . .m^2, velocity . . . . .m/s, flocculator to sedimentation . . . . .m^2, velocity . . . . .m/s, weir or baffle adjustment possible . . . . ., can tanks be drained? . . . . ., are walkways and guard rails provided? . . . . .

Feeders: Dry: Number . . . . ., capacity . . . . .kg/hr, **Liquid:** number . . . . ., capacity . . . . .m^3/hr

Settling Tanks: Number . . . . ., length . . . . .m, width . . . . .m, depth . . . . .m, diameter . . . . .m, retention time . . . . .hr, overflow rate . . . . .m/ day, flow line elevation . . . . .m, type of weirs (or launders) . . . . ., effluent pipe to . . . . ., can tanks be drained? . . . . ., where to? . . . . ., are walkways and guard rails provided? . . . . .

FILTERS

Units: Size . . . . ., number of units . . . . ., area . . . . .m^2, rate of filtration . . . . .m/hr, are walkways and guard rails provided? . . . . .

Filter Media: Fine medium: material . . . . ., effective size . . . . .mm, uniformity coefficient . . . . ., depth . . . . .m. **Coarse medium:** material . . . . ., sizes . . . . .

Underdrainage: type . . . . ., range of head losses . . . . .

Backwash: Rate . . . . .m/hr, water required . . . . .m³/day, where from . . . . ., lip elevation above filter surface . . . . ., surface wash? . . . . . **Pump:** size . . . . ., capacity . . . . .m³/day

Washwater Tank: Capacity . . . . ., elevation above filter . . . . .m, size of outlet pipe . . . . ., method of filling . . . . .

Filter Controls: Type . . . . ., location . . . . ., sewer to where . . . . ., cross-connection possible? . . . . .

CHLORINATION

Chlorinators: Number . . . . ., type . . . . ., capacity . . . .kg/day, where located . . . . ., point of chlorine application . . . . ., contact period provided . . . . .min, are chlorinators in separate building? . . . . ., is chlorine room isolated?. . . . ., scales . . . . .

Chlorine containers: Size . . . . ., storage for containers . . . . ., equipment for handling containers . . . . .

Safety precautions: Equipment provided . . . . ., adequate exhaust system . . . . ., louvers in door . . . . ., inside fixed window . . . . ., door opening outward . . . . ., light switch near door . . . . .

PLANT STORAGE AND PUMPING

Storage: Clear well: location . . . . ., capacity . . . . .m³, **Other low-level storage at plant:** type . . . . ., capacity . . . . .m³

Pumping: Where to?. . . . ., number of pumps . . . . ., sizes . . . . ., type . . . . ., drive . . . . ., controls . . . . ., standby power . . . . ., standby pump provided . . . . ., capacity . . . . ., power source . . . . ., disconnect switch for each pump?. . . . .

appendix d

SIMPLE PROCEDURES
FOR JAR TESTING

PURPOSE

Jar tests are used to determine the most effective doses of coagulant for a specific water in the control of coagulation and flocculation at a water treatment plant, especially when water quality fluctuates rapidly. They may also be used to determine settling velocities for the design of sedimentation basins, and to ascertain the potential of raw water for direct filtration.

EQUIPMENT REQUIRED

1. Jar tester (see Figure 4.1 in text)
2. 2-liter beakers—six
3. Pipettes—five 1-ml graduated (glass)
4. Erlenmeyer flasks—six 250-ml capacity
5. Funnels—six (ribbed, conical 6-cm diameter, short stemmed)
6. Graduated cylinders—two 100 ml (plastic or glass), one 1000 ml (plastic or glass)
7. Filter paper—11-cm Whatman #40 or other coarse filter paper
8. Scale—accurate to 1/20 of a gram
9. Bacteriological sampling bottles

(adapted from *Manual on Surface Water Treatment*, Ottawa Ministry of the Environment, 1976)

Date _____
Location _____
Raw Water Analysis _____
Run No. _____

Turbidity _____ NTU
Color _____
pH _____
Floc Temperature _____
Alkalinity _____ mg/l
Hardness _____ mg/l
Iron _____ mg/l
Chlorine _____ mg/l

	CHEMICALS					FLOC			SUPERNATANT
Jar	Alum mg/l	Lime mg/l	Polymer mg/l	Other mg/l	pH	Floc Form min	Size 20 Min mm	Settling Rate min	20 min Settled Turbidity NTU
1									
2									
3									
4									
5									
6									

FILTERED SUPERNATANT ANALYSIS

Jar	Turbidity NTU	Color	Al mg/l	Fe mg/l	Alk mg/l	pH
1						
2						
3						
4						
5						
6						

STIRRING:

Time	Speed	Gt/1000
___ Minutes @	___ RPM	___
___ Minutes @	___ RPM	___
___ Minutes @	___ RPM	___

Settling time: _____ Min

Figure D.1. Recording sheet for jar testing. *Source:* Ottawa Ministry of the Environment, 1976.

TREATABILITY TEST SHEETS

An example of a helpful guide for carrying out jar tests is illustrated in Figure D.1. The raw water analysis should be carried out and entered in the Raw Water Analysis section (upper right corner). With this analysis complete, the operator is ready to carry out the various jar test steps.

CHEMICAL SOLUTIONS

Stock solutions of coagulants, coagulant aids and other chemicals should be prepared at concentrations such that quantities suitable for use in coagulation tests can be measured accurately and conveniently.

Chemical	Conc. of Stock Sol.	Life	1 ml/l of Water Equivalent to
Alum	1%	1 month	10 mg/l
Ferric chloride	1%	2 months	10 mg/l
Lime[a]	1%	1 month[a]	10 mg/l
Polyelectrolyte[b]	0.05%	1 week	0.5 mg/l
Sulphuric Acid	0.1N	3 months	4.9 mg/l

[a]Lime slurries should be mixed by shaking every time they are used.
[b]Polyelectrolyte solutions should be used with the guidance of the manufacturers.

SOLUTION PREPARATION

Alum and Ferric Chloride

Both alum and ferric chloride are usually prepared as a 1% solution by weight, 1 g in 100 mls. The preparation for dry alum or ferric chloride is straightforward. However, with concentrated liquid solutions (e.g., liquid alum is 48.5% by weight) a dilution step is necessary. Any dilution step must consider the specific gravity (sg) of the solution being diluted (e.g., 48.5% liquid alum has an sg of 1.35). To make up a one percent solution from liquid alum the following steps should be carried out:

> 1 ml of 48.5% liquid alum weighs 1.35 g
> 1 ml contains 1.35 (0.485) = 0.65 g
> 1 g = 1/0.65 = 1.54 ml

Therefore, add 1.54 ml of liquid alum to 100 ml of water to make up a 1% solution.

Polymer

Care must be taken when polymer solutions for jar tests are derived from powdered polymers. Globs of polymer (fish eyes) can be formed if the polymer is added to the solution too quickly or if solutions greater than 1% are used. It is recommended that polymer concentrations of 0.05 or 0.1% be used for jar tests.

1 ml of 1% polymer added to 1000 ml water = 10 mg/l
1 ml of 0.1% polymer added to 1000 ml water = 1 mg/l

JAR TEST PROCEDURE FOR COAGULATION TREATMENT

Jar tests using coagulants require a 6-place laboratory stirrer or jar tester, as well as six 2-l beakers. The procedure for carrying out the test is:

1. Place a 2-l beaker under each stirring paddle.
2. Place into each beaker, from a graduated cylinder, exactly 2 l of a fresh sample of the raw water.
3. Note on the test sheet the amount of coagulant that is to be added to each beaker. This amount will vary from beaker to beaker.
4. With a measuring pipette, add the coagulant in increasing amounts to each successive beaker. For example, 10 mg/l to beaker #1, 20 mg/l to beaker #2, and so on.
5. Lower the stirring paddles into the beakers, start the stirrer, and operate it for 1 min at a speed of 60 to 80 rpm.
6. Reduce the speed to the selected degree of agitation (normally about 30 rpm) and allow the stirring to continue for about 15 min. The degree of agitation and the stirring period should attempt to match the plant operating conditions for flocculation.
7. Note how long it takes before a floc begins to form.
8. Note how well it withstands some stirring without breaking up.
9. After the stirring period is over, stop the stirrer and note how long it takes for the floc to settle to the bottom of the beaker.
10. After allowing the floc to settle for 20 min,, determine the color and the turbidity of the *supernatant* (the liquid above the floc).
11. Chemical dosages, mixing time and speed, pH, floc growth characteristics, and supernatant analysis should be noted on the operating sheets.
12. After allowing the floc 30 min to settle to the bottom, filter the supernatant through filter paper.

13. Filter another 100 to 150 ml of sample.
14. Perform turbidity, pH, color and, if necessary, residual aluminum tests on the filtrate.
15. The jar that gives the best results indicates the proper coagulant dosage for your plant.

JAR TESTS USING COAGULANTS PLUS COAGULATION AIDS

1. To determine if polyelectrolytes can help the coagulation-flocculation-sedimentation process, do the following: Repeat the jar test using the best coagulation dosage as determined from step 15 above (or slightly below this dosage), and *add* varying amounts of coagulant aid as described in step 4 above for the addition of the coagulant. Polyelectrolyte dosages rarely exceed 1 mg/l.
2. When determining the use of coagulation aids, keep one jar with alum only. Then compare the results (when using only alum) to the results obtained when a coagulation aid is added to the alum.

TEST FOR DETERMINING OPTIMAL pH

1. The second test involves the preparation of samples with the pH adjusted by using, for example, lime or sulfuric acid, so that the samples cover a range (e.g., pH - 5, 6, 7, and 8). The coagulant dose determined previously is added to each beaker. After this, the samples are examined and the optimum pH is determined. If necessary, a recheck of the minimum coagulant dose is done at the optimum pH.

When applying jar test results to the plant, it is sometimes found that the plant operates better at a chemical dosage slightly different from that indicated by the jar tests. The jar testing is very efficient both in stirring and settling. If the plant is not as efficient as the jar tests, a higher dosage of coagulant may be needed. The greatest value of the jar test is determining when changes in the raw water quality require changes in chemical dosages.

TEST FOR SETTLING

Tests for settling may be made in the test beaker by collecting samples by pipette at a preselected depth below the water surface, one sample just before mixing is stopped and the others at selected time intervals after mixing is stopped for a total of about 10 min. Analyses may be made on the samples for the concentrations of any substance selected as a tracer, such as Fe, Al, color, or turbidity. Cumulative frequency distribution curves of settling velocities may then be plotted against the concentration of the tracer, from which estimates may be made of removals to be expected in settling. (Cumulative frequency distribution curves for the design of settling tanks are described in Fair, Geyer and Okun, vol. II, 1978; Camp, 1946 and IRC, 1981b.)

appendix e

SIMPLIFIED PROCEDURES FOR WATER QUALITY ANALYSIS

These simplified methods are intended for use by operators of smaller treatment plants. Authorized standard methods should be followed at larger plants and by national laboratories. Suitable equipment for lab analyses and field testing of water quality in developing countries are described by Hutton (1983).

I.   pH

The pH of a solution is a measure of its free acidity or alkalinity. The pH scale ranges from 0.0 to 14.0, with 7.0 being neutral. The solution becomes increasingly acidic from 7.0 to 0.0 and more alkaline from 7.0 to 14.0. The method involves the addition of bromcresol and phenolphthalein indicators to produce a colored solution.

A. Preparation of Solutions

(1)   Bromcresol Purple Indicator Solution: Dissolve 0.1 gm of bromcresol purple indicator in a dropper bottle with 250 ml of distilled water.

(2) *Phenolphthalein Indicator Solution:*   Dissolve 0.30 gm of phenolphthalein indicator in a dropper bottle with 250 ml of 50 to 60% alcohol.
Label reagents as they are prepared.

B. Procedure

1. Place 15 ml of water sample into a square bottle. Add 1 drop of bromcresol purple solution and 1 drop of phenolphthalein indicator solution.
2. Observe the color of the solution and determine the pH range according to the following table:

Color of Solution	pH Range	Comment on pH
Yellow	<6.0	Too low
Blue	6.0–8.5	OK
Purplish-blue	8.5–9.5	Still OK
Red	>9.5	Too high

Note: If there is difficulty in identifying the red color, a standard red color solution can be prepared for comparison by repeating step 1, but in addition, adding a few drops of a caustic (NaOH) solution.

II.   TURBIDITY

The simplest method of measuring turbidity is to compare the sample with standards of known turbidity.

A. Preparation of Solutions

(1) *Stock Solution:*   Add 0.5 gm of fuller's earth to 50 ml of distilled water. This results in a stock solution with a turbidity of 1,000 NTU.
(2) *Standard Solutions:*   Shake the stock solution well. Dilute the stock solution with distilled water to get desired standards: 5 ml of stock solution to 100 ml of distilled water provides a 50 NTU standard, 5 ml of stock to 200 ml of water provides a 25 NTU standard, etc. Label the standards.
(3) *Preservation of Solutions:*   Add a few specks of mercuric chloride or a few drops of bleach to each standard solution. Standards should be prepared fresh each month.

B. Procedure

1. Fill a sample bottle completely with the water sample to be tested. The bottle should be the same type of bottle used to contain the standard solutions.
2. Shake the standards well.

3. Compare the samples with the standards, and determine if the sample is less than 2 NTU, between 25 and 50 NTU, or greater than 50 NTU. The comparison should be made by viewing the standards from the side, holding the bottles at eye level, while looking through both at some object and noting the distinctness with which it can be seen. Reading the samples may be easier if the bottles are placed in front of a background of black ruled lines on white paper.
4. Record the results.

III.   CHLORINE RESIDUAL

Chlorine is used more than any other disinfecting agent to kill disease-causing microorganisms in water. Chlorine also oxidizes organic matter. The efficacy of disinfection is a function of contact time and dose and is assured by the presence of a chlorine residual. From a public health standpoint, the proper application of chlorine to water is the most important water treatment process. Accordingly, the chlorine residual determination is the single most important measure of adequate treatment.

A. Preparation of Starch Solution

1. Measure 1 gm of clean starch into a container.
2. Add enough cold water while stirring the mixture to produce a thin paste.
3. Add about 100 ml of boiling water. Continue to stir.
4. Boil for 2 to 3 min.
5. Add several drops of chloroform (or formaldehyde) to preserve the solution.

Fresh solution should be prepared every one or two weeks.

B. Procedure

1. Fill a clean bottle to the neck with the sample of water to be tested. If the water is turbid, fill a second bottle for comparison later in the procedure.
2. Let the sample(s) stand undisturbed for 10 min.
3. Add two crystals of potassium iodide to the first bottle.
4. Add five drops of the starch solution to the bottles. (Add to both bottles if a control is used.)
5. Shake the samples vigorously and let stand for 5 min.

6.   Observe the solution for change of color.
     No color—absence of chlorine
     Faint blue color—correct amount of chlorine (0.15 mg/l)
     Dark blue color—excess chlorine (0.2 mg/l)
7.   Record the results.

IV.   COLIFORM ORGANISMS

The most serious health hazard is contamination of water with pathogens. Because determination of the presence of pathogenic microorganisms is difficult, the presence of indicator organisms is used as a measure of safety. These indicator organisms are members of a large group of bacteria called coliform bacteria. Most coliform bacteria do not cause disease themselves but, because they enter water supplies from the excreta of humans and animals, they indicate the possible presence of pathogens.

The procedure described here involves the application of a normal MPN (most probable number) coliform test. A series of culture bottles is prepared using media that can be made up with locally available ingredients. An indicator, bromcresol purple, is then added to the bottles. When coliforms are present, pH decreases, which induces a change of indicator color from purple to yellow. This change will occur after 48 hr of incubation. The number of coliform present can be estimated by the use of MPN tables.

A. Preparation of Media

Any of the following four methods may be used:

1.   Rice Broth: Place 25 gm of rice and 4 gm of powdered milk in 450 ml of water. Boil for about 5 min while stirring occasionally. Carefully decant the rice/milk broth into a glass bottle and discard the residue.

2.   Potato Broth: Place 50 gm of peeled or sliced potatoes (or sweet potatoes) with 4 gm of powdered milk in 450 ml of water. Boil for 15 min. Carefully decant the potato/milk broth into a glass bottle and discard the residue.

3.   Corn Meal Broth: Heat 400 ml of water to 70°C. Add 30 ml of corn meal and 4 gm of powdered milk. Stir frequently. Carefully decant the broth into a glass bottle and discard the residue.

4.   Lactose Broth: Dissolve approximately 1 gm of beef boullion and 4 gm of powdered milk in 250 ml of distilled water. Heat if necessary.

B. Preparation of Sterilized Culture Bottles

Introduce 15 ml of media into each of 10 clean screw-capped bottles. To each bottle add 5 drops of bromcresol purple indicator solution. Sterilize with the cap loosely placed on the mouth of the bottle. After the bottles have been allowed to cool slightly, tighten the caps.

C. Procedures

1. Introduce 10 ml of water sample into the first group of five sterilized culture bottles containing the media and indicator solution. A sterilized plastic or glass syringe or pipette should be used to add the sample to the bottles. It is important to label the bottles and to record the amount of water sample introduced into each bottle.

2. Introduce 1 ml of water sample to each bottle in the second group of five bottles.

3. Incubate the bottles at approximately 35°C for 48 hr.

4. After 48 hr, observe the bottles for color change. POSITIVE TEST—Bottles that have changed from purple to yellow.
   NEGATIVE TEST—No change in color

5. Record the number of bottles with positive results for each of the concentrations.

6. Use the MPN (most probable number) index below to estimate the number of coliforms per 100 ml of water.

MOST PROBABLE NUMBER (MPN) OF COLIFORM ORGANISMS

No. of Positive Tubes		MPN/100 ml of
10 ml	1 ml	water
0	1	2.0
1	0	2.2
1	1	4.4
2	0	5.0
2	1	7.6
3	0	8.9
3	1	12.0
4	0	15.0
4	1	21.0
5	0	39.0
5	1	Indeterminate

INDIVIDUALS AND INSTITUTIONS REFERRED TO IN THE TEXT

INDIVIDUALS

JORGE ARBOLEDA VALENCIA
Hidrosan Ltda.
Apartado Aereo 91247
Bogota, Colombia

JOHN AUSTIN
Senior Training Officer
Office of Health
US Agency for International
 Development
Washington, DC 20523 USA

SAMIA AL AZHARIA JAHN
Water Purification Project
PO Box 2681
Khartoum, Sudan

JOSE M. AZEVEDO NETTO
Consorcio Nacional de Engenheiros
 Consultore
Av. Alfredo Egydio de Souza Aranha,
 100
04726 Sao Paulo, SP, Brazil

A. G. BHOLE
Department of Civil Engineering
 (Public Health)
Visvesvaraya Regional College
Nagpur, India 440011

JOHN BRISCOE
Department of Environmental
 Sciences and Engineering
School of Public Health 201H
University of North Carolina
Chapel Hill, NC 27514 USA

ROLAND S. BURLINGAME
Senior Consultant
Camp Dresser and McKee, Inc.
710 South Broadway, Suite 201
Walnut Creek, CA 94596 USA

OCTAVIO CORDON
C. M. Cordon y Merida, Ings.
6A. Ave. 6-94. Zona 9
Guatemala City, Guatemala

FRANCIS A. DiGIANO
Department of Environmental
    Sciences and Engineering
School of Public Health 201H
University of North Carolina
Chapel Hill, NC 27514 USA

DAVID DONALDSON
Associate Director
WASH Project
1611 North Kent Street, ROOM 1002
Arlington, VA 22209 USA

NIGEL GRAHAM
Department of Civil Engineering
Imperial College
Imperial College Road
London SW7 2BU. England

HERBERT E. HUDSON, JR.
deceased; refer to
E. Glenn Wagner below

J. N. KARDILE
Research Officer
Environmental Engineering Research
    Division
Maharashtra Engineering Research
    Institute
Nasik-422 004 India

SUSUMU KAWAMURA
Senior Project Engineer
James M. Montgomery Consulting
    Engineers
555 East Walnut Street
Pasadena, CA 91101 USA

KAZUYOSHI KAWATA
The Johns Hopkins University
615 North Wolfe Street
Baltimore, MD 21205 USA

HERNAN SANDOVAL LORA
Civil Engineer
Apartado No. 4230
Cali, Colombia

JOSE M. PEREZ
Water Treatment Specialist
Centro Panamericano de Ingenieria
    Sanitaria y Ciencias del Ambiente
    (CEPIS)
Lima, Peru

RENHATO PINHEIRO
Design Engineer
pH Engenharia
Rio de Janeiro, Brazil

CARLOS RICHTER
Companhia de Saneamento do Parana
Curitiba, Brazil

E. GLENN WAGNER
Director
Water and Air Research, Inc.
6821 SW Archer Road
Gainesville, FL 32602 USA

JOHN WILLIS
Vice President
Camp Dresser and McKee, Inc.
1 Center Plaza
Boston, MA 02108 USA

INSTITUTIONS

AFDB*   African Development Bank
         Public Utilities Division
         Infrastructure and Industry Department
         P.O. Box 1387
         Abidjan, Ivory Coast

ADB*   Asian Development Bank
         Water Supply Division
         Infrastructure Department
         P.O. Box 789
         Manila, Philippines 2800

AID   U.S. Agency for International Development
         Community Water Supply and Sanitation Division
         Office of Health
         Washington, DC 20523 USA

AIT   Asian Institute of Technology
         PO Box 2754
         Bangkok, Thailand

AWWA   American Water Works Association
         6666 West Quincy Avenue
         Denver, Colorado 80235 USA

CAB   Companhia das Aguas da Beira
         (Water Authority for Beira)
         Caixa Postal 53
         Beira, Mozambique

CDB*   Carribean Development Bank
         P.O. Box 408
         Wilkey
         St. Michael
         Barbados, West Indies

CEPIS   Pan American Sanitary Engineering and Environmental Sciences
             Center (CEPIS)
         Casilla 4337
         Lima, 100, Peru

CLC   Claude Laval Corporation
         1911 N. Helm
         P.O. Box 6119
         Fresno, CA 93703 USA

*Institutions not referred to in the text.

ENSIC*   Environmental Sanitation Information Center
                Asian Institute of Technology
                PO Box 2754
                Bangkok, Thailand

GTZ   German Agency for Technical Cooperation
                Postfach 5180
                D-6236 Eschborn, West Germany

IBRD   International Bank for Reconstruction and Development
                Water Supply and Urban Development Department
                World Bank
                1818 H Street NW
                Washington, DC 20433 USA

ICCA   Instituto Costarico de Acueducto y Alcantarillado (Institute for
                    Water Supply and Sewerage)
                Apartado Postal 5120
                San Jose, Costa Rica

IDB*   Inter-American Development Bank
                Sanitary Engineering Section
                Project Analysis Department
                808 17th Street, N.W.
                Washington, DC 20577 USA

IDRC*   International Development Research Center
                PO Box 8500
                Ottawa, Canada K1G 3H9

IRC   International Reference Center for Community Water Supply
                PO Box 5500
                2280 HM Rijswijk
                The Netherlands

ITDG*   Intermediate Technology Development Group
                9 King Street
                London WC2E 8HN, England

IWSA*   International Water Supply Association
                1 Queen Anne's Gate
                London, SWIH 9BT, England

NEERI   National Environmental Engineering Research Institute
                Nehru Marg
                Nagpur-440 020, India

PAHO*   Pan American Health Organization
                525 23rd Street NW, Room 523
                Washington, DC 20037 USA

Peace Corps 806 Connecticut Avenue, NW
Washington, DC 20523 USA

UNC University of North Carolina
International Programs Office
Department of Environmental Sciences and Engineering
Chapel Hill, NC 27514 USA

UNDP United Nations Development Program
UNDP Division of Information
International Drinking Water Supply and Sanitation Decade
1 UN Plaza
New York, NY 10017 USA

UNEP* United Nations Environment Program
P.O. Box 30552
Nairobi, Kenya

UNESCO* United Nations Educational, Scientific and Cultural Organization
Division of Water Sciences
7, Place de Fontenoy
F-75700 Paris, France

UNICEF* United Nations Children's Fund
United Nations
New York, NY 10017 USA

UNV* United Nations Volunteers
Palais des Nations
CH-1211 Geneva, Switzerland

WASH Water and Sanitation for Health Project
1611 North Kent Street
Suite 1002
Arlington, VA 22209 USA

WHO World Health Organization
Environmental Health Technology and Support Division—GWS
1211 Geneva 27, Switzerland

UNITED STATES ORGANIZATIONS INVOLVED IN WATER SUPPLY IN DEVELOPING COUNTRIES

AFRICARE 1601 Connecticut Avenue, NW
Washington, DC 20009 USA

Agua del Pueblo Director for Research and Program
Development
320 45th Street
Oakland, CA 94609 USA

(Continued)

CARE	Coordination for American Relief Everywhere 660 First Avenue New York, NY 10016 USA
Catholic Relief Services	1011 1st Avenue New York, NY 10022 USA
Center for Population Activities	1717 Massachusetts Avenue NW, #202 Washington, DC 20036 USA
CODEL	Coordination in Development 79 Madison Avenue New York, NY 10157 USA
Lutheran World Relief	360 Park Avenue South New York, NY 10010 USA
MAP International, Inc.	P.O. Box 50 Wheaton, IL 60187 USA
Mission Health Foundation	Box 89 Independence, MO 64056 USA
National Council of Catholic Women	1312 Massachusetts Avenue NW Washington, DC 20005 USA
National Council for International Health	2121 Virginia Avenue, #303 Washington, DC 20037 USA
Save the Children Federation	48 Wilton Road Westport, CT 06880 USA
VITA	Volunteers in Technical Assistance 1815 N. Lynn Street, Suite 200 Arlington, VA 22209 USA

Note: A complete listing of institutions involved in the drinking water supply and sanitation sector, including donor agencies, banks and funds, international organizations, and voluntary service and nongovernmental organizations, is given in the WHO publication *Catalogue of External Support* (1983).

REFERENCES

Adeyemi, S.O. (1980). "Local Materials as Filter Media in Nigeria." 6th WEDC Conference. March 1980. Zaria, Nigeria. Loughborough University of Technology, Leicestershire, England.

Ahmad, S., M. T. Wais, and F. Y. R. Agha (1982). "Appropriate Technology to Improve Drinking Water Quality in Mosul, Iraq." *AQUA*, no. 4, pp. 439–444.

Ahmad, S. and M. T. Wais (1980). "Potential of Tube Settlers in Removing Raw Water Turbidity Prior to Coagulation." *AQUA*, no. 8, pp. 165–169.

AID UNC/IPSED (1966a). "Floating Platform Hypochlorite Solution Feeder." AID UNC/IPSED Series No. 7, University of North Carolina, Chapel Hill.

AID UNC/IPSED (1966b). "Float-Valve Hypochlorite Solution Feeder." AID UNC/IPSED Series No. 4, University of North Carolina, Chapel Hill.

AID UNC/IPSED (1966c). "A Proportional Chemical Feeder for Small Water Purification Plant." AID UNC/IPSED Series No. 5, University of North Carolina, Chapel Hill.

Ainsworth, L. D. (1984). "Innovative Direct Filtration Plant Inexpensive to Build and Operate." *Public Works*, vol. 115, no. 2, pp. 40–43.

Air Force Manual (1959). *Maintenance and Operation of Water Plants and Systems (AFM 85-13)*. Department of the Air Force, Washington, D.C.

Ametek-Schutte and Koerting, Div. (1976). "PVC and Penton Water Jet Eductors." Cornwell Heights, Pa.

Algarsamy, S. R. and M. Gandhirajan (1981). "Package Water Treatment Plants for Rural and Isolated Communities." *J. Indian Water Works Assoc.*, vol. 13, no. 1, pp. 73–80.

American Society of Civil Engineers (1940). *Water Treatment Plant Design*. Manual of Engineering Practice, No. 19, New York.

American Society of Civil Engineers, American Water Works Association (1969). *Water Treatment Plant Design*. New York.

American Water Works Association (1971). *Water Quality and Treatment*, 3d edition. McGraw-Hill, New York.

APS Technical Services, Ltd. (1982). "Containerised Water Treatment Plants." Information Sheet No. 1. Woking Surrey, England.

Arboleda, J. V. (1973). *Teoria Diseno y Control de los Procesos de Clarificacion del Agua.* Serie Tecnica 13. CEPIS, Lima, Peru.

Arboleda, J. V. (1974). "Hydraulic Control Systems of Constant and Declining Flow Rate in Filtration." *J. Am. Water Works Assoc.*, vol. 66, no. 2, pp. 87–93.

Arboleda, J. V. (1976). "Some Basic Ideas on Establishing a Water Treatment Technology." *International Training Seminar on Community Water Supply in Developing Countries*, Bulletin No. 10, IRC, Rijswijk, Netherlands.

Arboleda, J. V. (1983). "Latin American Experience in the Design and Construction of Water Treatment Plants." Paper presented at the Annual Conference of the American Water Works Association, Las Vegas, Nevada.

Azevedo-Netto, J. M. (1977). *Tecnica de Abastecimento e Tratamento de Agua. 2nd edition.* CETESB, Sao Paulo, Brazil, pp. 777–796.

Barker, H. W. (1976). "Assessment of Manpower Needs and Training Programs." *International Training Seminar on Community Water Supply in Developing Countries.* Bulletin No. 10. IRC, Rijswijk, Netherlands.

Barnes, D. and Mampitiyarachichi (1983). "Water Filtration Using Rice Hull Ash." *Waterlines*, vol. 2, pp. 21–23.

Baylis, J.R. (1935). "A Study of Filtering Material for Rapid Sand Filters." *Water Works Sewage*, vol. 82, no. 1, pp. 24–25.

Bernardo, L. and J. L. Cleasby (1980). "Declining Rate Versus Constant-Rate filtration." *J. Environ. Eng. Div. ASCE*, vol. 106, no. EE6, pp. 1023–1041.

Bhole, A. G. (1981). "Design and Fabrication of a Low Cost Package Water Treatment Plant for Rural Areas in India." *AQUA*, no. 5, pp. 315–320.

Bhole, A. G., J. M. Dhabadgaonkar, and R. W. Ingle (1977). "Package Water Treatment Plants for Rural Areas—Specific Problems of Treatment and Design Considerations." *J. Indian Water Works Assoc.*, vol. 9, no. 2.

Bhole, A. G. and S. D. Harne (1980). "Significance of Tapering of Time and Velocity Gradient on the Process of Flocculation." *J. Indian Water Works Assoc.*, vol. 12, no 1, pp. 57–63.

Bhole, A. G. and J. T. Nashikkar (1974). "Berry Seed Shell as Filter Media." *J. Inst. Eng.*, vol. 54, no. PH 2, p. 45 (India).

Bhole, A. G. and S. F. Rahate (1977). "Performance Study of a Dual-Cum-Mixed Media Filter." *J. Indian Water Works Assoc.*, vol. 9, no. 1., pp. 31–35.

Bhole, A. G., and V. A. Ughade (1981). "Study of a Surface Contact Flocculator." *J. Indian Water Works Assoc.*, vol. 13, no. 2, pp. 179–183.

Bhole, A. G., and S. Vaidyanathan (1980). "New Concepts in Low Cost Treatment Plants." *J. Indian Water Works Assoc.*, vol. 12, no. 1, pp. 49–56.

British Standards Institution, Council for Codes of Practice (1966). *The Design and Construction of Reinforced and Prestressed Concrete Structures for the Storage of Water and Other Aqueous Liquids* (CP 2007 with amendments). London.

Brown, J. C., and D. A. Okun (1968). "Criteria for the Incorporation of Mechanization and Automation in Treatment Plant Design." Paper presented to the Environmental Engineering Conference of the ASCE. ESE publication no. 183, University of North Carolina, Chapel Hill.

Brown, J. C. and D. A. Okun (1968). "Economic Design of Water Supply and Sewerage Systems." Paper presented at the 11th Congress of AIDIS. Quito, Ecuador. ESE publication no. 193. University of North Carolina, Chapel Hill.

Bulusu, K. R. and V. P. Sharma (1965). "Pilot Plant Studies on the Use of the Nirmali Seed as a Coagulant Aid." *Indian J. Environ. Health*, vol. 7, pp. 165–176.

Cady, P. D. and P. R. Groney (1976). "Hydraulic Cement Based on Lime Pozzolan (Silica) Mixtures for Cottage Level Cement Production." Unpublished paper. Pennsylvania State University, University Park.

Camp, T. R. and P. C. Stein (1943). "Velocity Gradients and Internal Work in Fluid Motors." *J. Boston Soc. Civil Eng.*, vol. 130, p. 219.

Camp, T. R. (1946). "Sedimentation and the Design of Settling Tanks," *Trans. Am. Soc. Civil Eng.*, vol. 3, p. 895.

Camp, T. R. (1970). "Towards a Rational Jar Test for Coagulation." *J. New England Water Works Assoc.*, vol. 84, no.3.

Carefoot, N. F. (1977). "Balance in Training for Latin American Water and Wastewater Utilities." *J. Am. Water Works Assoc.*, vol. 69, no. 12, pp. 641–643.

CEPIS (1982). *Modular Plants for Water Treatment*, vol. 1 and Vol. 2. Technical Document no. 8. Pan American Health Organization, Lima, Peru.

Central Public Health and Environmental Engineering Organization (1976). *Manual on Water Supply and Treatment.* Ministry of Works and Housing, New Delhi, India.

Chao, J. L. and B. G. Stone (1979). "Initial Mixing by Jet Injection Blending." *J. Am. Water Works Assoc.*, vol. 71, no. 10, pp. 570–573.

Ching, L. Y. (1979). "Appropriate Technological Applications to the Penang Water Works." *AQUA*, no. 1, pp. 2–9.

Chow, V. T. (1959). *Open Channel Hydraulics.* McGraw-Hill, New York.

Cleasby, J. L. and L. Bernardo (1980). "Hydraulic Considerations in Declining-Rate Filtration." *J. Environ. Eng. Div.*, ASCE, vol. 106, no. EE6, pp. 1043–1055.

Cohen, J. M., G. A. Rourke, and R. L. Woodward (1958). "Natural and Synthetic Polyelectrolytes as Coagulant Aids." *J. Am. Water Works Assoc.*, vol. 50, no. 3, pp. 463–478.

Cox, C. R. (1960). "Guide to the Design of Water Treatment Plants." *International Cooperation Administration*, Washington, D.C., 87 pp., mimeo.

Cox (1964). *Operation and Control of Water Treatment Processes.* WHO, Geneva.

Culp, G. L. and R. L. Culp (1974). *New Concepts in Water Purification.* Van Nostrand Reinhold, New York.

Culp, R. L. (1977). "Direct Filtration." *J. Am. Water Works Assoc.*, vol. 69, pp. 375–378.

Dallaire, G. (1981). "UN Launches International Water Decade, U.S. Role Uncertain." *Civ. Eng.*, ASCE, vol. 51, no. 3, pp. 59–62.

Dangerfield, B. J., ed. (1983). *Water Supply and Sanitation in Developing Countries.* The Institution of Water Engineers and Scientists, London, England.

Davis, C. V. (1952). *Handbook of Applied Hydraulics.* 2d edition. McGraw-Hill, New York.

Donaldson, D. (1976). "Planning Water and Sanitation Systems for Small Communities." *International Training Seminar on Community Water Supply in Developing Countries.* Bulletin No. 10. IRC. Rijswijk, Netherlands.

Engineering-Science Inc. (1977). "Oceanside Water Filtration Plant—Preliminary Design Report." Arcadia, Calif.

Environmental Protection Agency (1978). *Manual of Treatment Techniques for Meeting the Interim Primary Drinking Water Regulations.* (Publication no. EPA-600/8-77-005). Cincinnati, Ohio.

Fair, G. M., J. C. Geyer, and D. A. Okun (1968). *Water and Wastewater Engineering*, vol. 2. Wiley, New York.

Fair, G. M., J. C. Geyer, and D. A. Okun (1971). *Elements of Water Supply and Wastewater Disposal.* Wiley, New York.

Feachem, R., M. McGarry, and D. Mara (1977). *Water, Wastes, and Health in Hot Climates.* Wiley, London.

Flinn, Weston, and Bogart (1927). *Waterworks Handbook of Design, Construction, and Operation.* McGraw-Hill, New York.

Frankel, R. J. (1974). "Series Filtration Using Local Filter Media." *J. Am. Water Works Assoc.*, vol. 66, no. 2, pp. 124–127.

Frankel, R. J. (1977). *Manual for Design and Operation of the Coconut Fiber/Burnt Rice Husk Filter in Villages of Southeast Asia.* SEATEC International, Bangkok, Thailand.

Frankel, R. J. (1981). "Design, Construction and Operation of a New Filter Approach for Treatment of Surface Waters in Southeast Asia." *J. Hydrol.*, vol. 51, pp. 319–328 (Netherlands).

Gadkari, S. K., V. Raman, A. S. Gadkari, and N. W. Mirchandani (1980). "Studies of Direct Filtration of Raw Water." *Indian J. Environ. Health*, vol. 22, no. 1, pp. 57–65.

Grant, D. M. (1979). *Open Channel Flow Measurement Handbook.* Instrumentation Specialties Company, Lincoln, Nebraska.

Greenleaf, J. W., Jr. (1964). *Open Gravity Filters.* U.S. Patent No. 3,134, 735.

Hall, R. I. (1964). "A Works Evaluation of Sodium Alginate as a Coagulant Aid." *Proc. Soc. Water Treat.*, vol. 13, pp. 114–127 (London).

Hardenbergh, W. A. and E. R. Rodie (1961). *Water Supply and Waste Disposal.* International Textbook Co., Scranton, Pa.

Hofman, H. and R. van Kervoorden (1982). "A Low-Cost Ferrocement Rural Water Supply System—Treatment of Surface Water and Distribution," *J. Ferrocement*, vol. 12, no. 4, pp. 365–372.

Hsu, E. Y. (1950). "Hsu on Hydraulic Jump." *Trans. Am. Soc. Civ. Eng.*, vol. 115, pp. 988–991.

Hudson, H. E. (1974). "Physical Aspects of Flocculation." *AWWA Semin. Proc.*, pp. 157–175, June 15–16, Denver, Colorado.

Hudson, H. E. (1981). *Water Clarification Processes—Practical Design and Evaluation.* Van Nostrand Reinhold, New York.

Huisman, L. and W. E. Wood (1974). *Slow Sand Filtration.* WHO, Geneva.

Hutton, L. G. (1983). *Field Testing of Water in Developing Countries.* Water Research Centre, Medmenham, England.

Intermediate Technology Industrial Services (1982). Project Bulletin: *Water Chlorination: On-Site Generation of Sodium Hypochlorite.* Myson House Railway Terrace, Rugby, United Kingdom.

IBRD/WHO (1979). *Nigeria Water Supply and Waste Disposal Sector Study, Manpower and Training, Discussion Document*, June 20, 1979.

IDRC (1981). *Rural Water Supply in China.* International Development Research Centre, Ottawa, Canada.

IRC (1970). "Design Criteria for Rapid Gravity Filters." Meeting held at Dubrovnic, Yugoslavia, 7–14 October. Rijswijk, Netherlands.

IRC (1973). *Health Aspects Relating to the Use of Polyelectrolytes in Water Treatment for Community Water Supply.* Technical Paper No. 5, Rijswijk, Netherlands.

IRC (1977a). *Contributions to a Mail Survey on Practical Solutions in Drinking Water Supply and Wastes Disposal for Developing Countries.* Rijswijk, Netherlands (draft form).

IRC (1977b). *Global Workshop on Appropriate Water and Waste Water Treatment Technology for Developing Countries.* Bulletin Series No. 7. Rijswijk, Netherlands.

IRC (1977c). *Slow Sand Filtration for Community Water Supply in Developing Countries—A Selected and Annotated Bibliography.* Bulletin Series No. 9. The Hague, Netherlands.

IRC (1978). *Slow Sand Filtration for Community Water Supplies in Developing Countries—A Design and Construction Manual.* Technical Paper Series No. 11. J. C. Van Dijk and H. C. M. Oomen, au. Rijswijk, Netherlands.

IRC (1980). "A Study on Standard Water Purification Plants in Indonesia." Unpublished paper, Rijswijk, Netherlands.

IRC (1981a). *Report of a Regional Seminar on a Modular Approach in Small Water Supply Systems Design.* Jakarta, Indonesia, October 1980. Bulletin Series No. 17, Rijswijk, Netherlands.

IRC (1981b). *Small Community Water Supplies.* Technical Report No. 18. L. Huisman, J. M. Azevedo-Netto, B. B. Sundaresan, J. N. Lenoix, and E. H. Hofkes ed. Rijswijk, Netherlands.

Jahn, S. A. A. (1979). "African Plants Used for the Improvement of Drinking Water." *Curare*, vol. 2, pp. 183–199 (Wiesbaden, West Germany).

Jahn, S. A. A. (1981). *Traditional Water Purification in Tropical Developing Countries*. German Agency for Technical Cooperation. Eschborn, West Germany.

Jahn, S. A. A. and H. Dirar (1979). "Studies on Natural Water Coagulants in the Sudan, with Special Reference to Moringa Oleifera Seeds." *Water SA*, vol. 5, no. 2, pp. 90–97 (Pretoria, South Africa).

Kardile, J. N. (1972). "Crushed Coconut Shell as a New Filter Media for Dual and Multi-Layer Filters." *J. Indian Water Works Assoc.*, vol. 1, no. 1, p. 28.

Kardile, J. N. (1981). "Development of Simple and Economic Filtration Methods for Rural Water Supplies." *AQUA*, no. 1, pp. 226–229.

Kardile, J. N., and A. P. Chourasia (1981). "Augmentation of Water Treatment Plant at Nasik-Road (Maharashtra) by Application of New Techniques." *J. Indian Water Works Assoc.*, vol. 13, no. 2.

Kawamura, S., (1976). "Considerations on Improving Flocculation." *J. Am. Water Works Assoc.*, vol. 68, no. 6, pp. 328–336.

Kawamura, S. (1981). "Design of Water Treatment Plants in Developing Countries." *AQUA*, no. 1, pp. 223–225.

Kerkhoven, P. (1979). "Third World Tests for Sand Filters." *World Water*, vol. 2, no. 9, pp. 19–29.

Kirchmer, C. J., J. V. Arboleda, and M. L. Castro (1975). *Polimeros Naturales y su Aplicacion Como Ayndantes de Floculacion*. Serie Documentos Tecnicos 2. CEPIS, Lima, Peru.

Kuntschik, O. (1976). "Optimization of Surface Water Treatment by a Special Filtration Technique." *J. Am. Water Works Assoc.*, vol. 68, no. 10, pp. 546–551.

Levy, A. G. and J. W. Ellms (1927). "The Hydraulic Jump as a Mixing Device." *J. Am. Water Works Assoc.*, vol. 17, no. 1, pp. 1–23.

Lewis, W. M. (1980). *Developments in Water Treatment*, vol. 1 and vol. 2. Applied Science Publishers, London.

Logsdon, G. S., R. M. Clark, and C. H. Tate (1980). "Direct Filtration Treatment Plants, Costs and Capabilities." *J. Am. Water Works Assoc.*, March 1980, pp. 134–147.

MacDonald, D. V., and L. Streicher (1977). "Water Treatment Plant Design is Cost Effective." *Public Works*, vol. 108, no. 8, pp. 86–89.

Macedo, L. H. and M. Noguti (1978). "Custo Operacional da Desinfeccao com Cloro e Hipoclorito de Sodo." *Revista Dae*, vol. 42, no. 119 (Sao Paulo, Brazil).

McGarry, M. G. and E. J. Schiller (1981). "Manpower Development for Water and Sanitation Programs in Africa." *J. Am. Water Works Assoc.*, vol. 73, no. 6, pp. 282–287.

McJunkin, F. E. (1982). "Simple Water Treatment Methods." In *Water Supply and Sanitation in Developing Countries*, E. J., Schiller, and R. L. Droste, (eds.). Ann Arbor Science, Ann Arbor, Mich.

McJunkin, F. E. and P. A. Vesilind (1968). "Practical Hydraulics for the Public Works Engineer." *Public Works*, reprint.

Manual of British Water Engineering Practice, Volume 3: *Water Quality and Treatment*, 3d edition (1969). W. Heffer and Sons, Cambridge, England.

Mara, D. (1981). "Simple Bacteriological Analysis of Drinking Water Supplies." *Appropriate Technol.*, vol. 3, no. 3 England.

Marais, G. V. (1969). "Design of Small Grit Channels." *Water Supply and Sanitation in Developing Countries*. AID-UNC/IPSED Item No. 20. University of North Carolina, Chapel Hill.

Medina, L. E. and H. E. Hudson (1980). "Upgrading the Old While Building the New." *J. Am. Water Works Assoc.*, vol. 72, no. 12, pp. 666–671.

Nashikkar, J. T., A. G. Bhole, and R. Paramasivam. "Performance of a Dual-Media Filter." *J. Indian Water Works Assoc.*, vol. 8, no. 1, pp. 11–16.

NEERI (1971). "Survey of Water Treatment Plants." *Tech. Dig.*, no. 18 (Nagpur, India).

NEERI (1976). "Natural Coagulant Aids." *Tech. Dig.*, no. 52 (Nagpur, India).

Nelson, O. F. (1969). "Capping Sand Filters." *J. Am. Water Works Assoc.*, vol. 61, no. 10.

Okun, D. A. (1976). "Drinking Water for the Future." *Am. J. Public Health*, vol. 66, pp. 639–643.

Okun, D. A. (1977). "Manpower Development Training in the Water Sector—Responsibilities of the World Bank." World Bank Report No. PUN28, Washington, D.C.

Okun, D. A. (1982). "Water supply Around the World—Appropriate Technology." In *Water Supply and Sanitation in Developing Countries*, E. J. Schiller and R. L. Droste, eds. Ann Arbor Science, Ann Arbor, Mich.

Okun, D. A., and G. Ponghis (1975). *Community Wastewater Collection and Disposal*. WHO, Geneva.

Ontario Ministry of the Environment (1976). *Surface Water Treatment Workshop Manual*. Ministry of Government Services, Toronto, Canada.

Packham, R. F. (1967). "Polyelectrolytes in Water Clarification." *Proc. Soc. Water Treat. Exam.*, vol. 16, part 2, pp. 88–102 (London).

Paramasivam R. et al. (1973). "Bituminous Coal—A Substitute for Anthracite in Two Layer Filtration of Water." *Indian J. Environ. Health*, vol. 15, p. 178.

Paramasivam, R. and V. A. Mhaisalkar (1981). "Slow Sand Filtration—An Appropriate Technology for Medium and Small Water Supplies." *J. Indian Water Works Assoc.*, vol. 13, no. 2.

Paramasivam, R., V. A. Mhaisalkar, and P. M. Berthouex (1981). "Slow Sand Filter Design and Construction in Developing Countries." *J. Am. Water Works Assoc.*, vol. 743, no. 4, pp. 178–185.

Paul, B. K. and R. P. Pama (1978). *Ferrocement*. International Ferrocement Information Center, Asian Institute of Technology, Bangkok.

Peniston, Q. P. and Johnson, E. L. (1970). *Method for Treating an Aqueous Medium with Chitosan and Derivatives of Chitin to Remove an Impurity*. U.S. Patent No. 3,533,940.

Perez, J. (1980). *Filtros Dinamicus*. CEPIS, Lima, Peru.

Poblete, L. P. (1964). "Dosificacion de Alum-Cake en la Planta de Filtros de las Vizcachas de la Empresa de Aqua Potable de Santiago." *Engenharia Sanitaria*, vol. 15, no. 2, pp. 111–119 (Sao Paulo, Brazil).

Portland Cement Association (1963). "Concrete for Water Treatment Works." Information Sheet PA 069.01W, Skokie, Ill.

Portland Cement Association (1969). "Concrete for Hydraulic Structures." Information Sheet IS 012.03W, Skokie, Ill.

Ranade, S. V. and G. D. Agrawal (1974). "Use of Vegetable Wastes as a Filter Media." Presented at the Conference on Engineering Materials and Equipment. The Association of Engineers, Calcutta, India.

Ranade, S. V. and J. M. Gadgil (1981). "Design of Dual-Media Filters to Suit Existing Water Treatment Plants in India." *J. Indian Water Works Assoc.*, vol. 13, no. 1, pp. 81–85.

Rao, D.R.J. (1981). "Evolving High Rate Filter and Use of Crushed Stone as Filter Media." *J. Inst. Eng.*, vol. 61, pp. 92–96 (India).

Rao, C. V. and R. Paramasivam (1980). "High Rate Settlers—A Review." *IAWPC Tch*. Annual 6 and 7, pp. 91–106, (India).

Reid, G. and K. Coffey (1978). *Appropriate Methods of Treating Water and Wastewater in Developing Countries*. University of Oklahoma, Norman, Ok.

Richmond, A. C. (1981). "Chemical Treatment of Small Remote Water Supplies." *Water Serv.*, September.

Richter, C. A. (1980). "Fundamentos Teoricos de Floculacao en Meio Granular." *Engenharia Sanitaria*, vol. 429, pp. 20–24 (Brazil).

Richter, C. A. (1981). "Metodo Simplificado de Calculo de Floculadores Hidraulicos de Chicanas." Unpublished paper. Parana, Brazil.

Richter, C. A. and R. de Barro Moreira, (1981). "Floculadores de Pedras: Experiencias en Filtro Piloto." Unpublished paper (Brazil).

Rodriguez D. (1977). "Analisis de la Eficiencia De Filtros Dinamicos para Tratamiento de Agua Potable." Dept. de Ingenieria Sanitaria, Univ. de Chile, Santiago.

Salvato, J. (1982). *Environmental Engineering and Sanitation*, third edition. Wiley, New York.

SANEPAR (1979). "Water Treatment Plant for Small Community—Evaluation of the Araucaria Water Treatment Plant." Engineering report prepared by C. Richter, Parana, Brazil.

Sanks, R. (1978). *Water Treatment Plant Design: For The Practicing Engineer*. Ann Arbor Science Publishers, Ann Arbor, Mich.

Smethurst, G. (1979). *Basic Water Treatment for Application World-Wide*. Thomas Telford, Ltd., London, England.

Smethurst, B. (1983). "Water Quality and Treatment." In *Water Supply and Sanitation in Developing Countries*. B. J. Dangerfield, ed. The Institution of Water Engineers and Scientists, London.

Sperandio, O. A. and C. J. Perez (1976). *Evaluation of Lower Cost Methods of Water Treatment in Latin America*. Final report prepared by CEPIS.

Spink, John (1982). An Alternative to Traditional Disinfection Methods. Unpublished paper. Sienco, Inc., St. Louis, Mo.

Stone, R. (1950). "The Small-Scale Manufacture of Bleaching Powder in Backward Areas." *J. Am. Water Works Assoc.*, vol. 42, no. 3, pp. 283–285.

Sundaresan, B. B. and R. Paramasivam (1981). "Appropriate Technology in Water Supply Systems." *Proceedings First Southern Asia Regional Conference*, Bombay, October, pp. SI1–19.

Tam, D. M. (1981). "Water Quality Standards: An Updated Survey." *ENSIC* (Environmental Sanitation Information Center), Asian Inst. of Tech., vol. 3, no. 3, Bangkok, Thailand.

Tasgaonkar, S. K. (1981). "Report on a Typical Rural Water Supply Scheme." *Proc. First South. Asia Reg. Conf.*, Bombay, 19–23 October.

Thanh, N. C. (1978). *Functional Design of Water Supply for Rural Communities*. Asian Inst. of Tech., Bangkok, Thailand.

Thanh, N. C. and M. B. Pescod (1976). "Application of Slow Sand Filtration for Surface Water Treatment in Tropical Developing Countries." Final Report No. 65, Asian Inst. of Tech., Bangkok, Thailand.

Thanh, N. C. and J. P. A. Hettiaratchi (1982). *Surface Water Filtration for Rural Areas—Guidelines for Design, Construction, Operation and Maintenance*, ENSIC, Asian Inst. of Tech., Bangkok, Thailand.

Tripathi, P. N., M. Chaudhuri, and S. D. Bokil (1976). "Nirmali Seed—A Naturally Occuring Coagulant." *Indian J. Environ. Health*, vol. 18, no. 4, pp. 272–281.

Uganda, (1983). *Manpower and Training Study*, Ministry of Lands, Mineral and Water Resources, Final Report.

UNDP, IBRD (1983). *Manpower and Training Study for HM Government of Nepal*. NEP/79/032.

Vrale, L., and Jorden, R. M. (1971). "Rapid Mixing in Water Treatment." *J. Am. Water Works Assoc.*, vol. 63, p. 52.

Wagner, E. F. and J. N. Lanoix (1959). *Water Supply for Rural Areas and Small Communities*. WHO, Geneva.

Wagner, E. G. and H. E. Hudson (1982). "Low-Dosage High-Rate Direct Filtration." *J. Am. Water Works Assoc.*, vol. 74.

Wagner, G. (1982a). "Simplifying Design of Water Treatment Plants for Developing Countries." *J. Am. Water Works Assoc.*, vol. 75, no. 5, pp. 220–223.

Wagner, G. (1982b). "The Latin American Approach to Improving Water Supply." *J. Am. Water Works Assoc.*, vol. 74, no. 4, pp. 168–173.

Wegelin, M. (1982). *Slow Sand Filter Research Project*. Report 3, University of Dar es Salaam, Tanzania.

White, G. C. (1972). *A Handbook of Chlorination*. Van Nostrand Reinhold, New York.

WHO (1973). "Community Water Supply and Sewage Disposal in Developing Countries." *World Health Statistics*, vol. 26, no. 11.

WHO (1976a). *Surveillance of Drinking Water Quality*. World Health Organization. Geneva.

WHO (1976b). *Typical Designs for Engineering Components in Rural Water Supply*. South-East Asia Series, New Delhi, India.

WHO (1982). *Basic Strategy Document*. Interagency Task Force on Human Resources Development for the International Drinking Water Supply and Sanitation Decade, World Health Organization, Geneva.

WHO (1983) *Catalogue of External Support*. World Health Organization, Geneva.

WHO (1984a) *Guidelines for Drinking Water Quality, Volume I: Recommendations*. World Health Organization, Geneva.

WHO (1984b). *Guidelines for Drinking Water Quality, Volume II: Health Criteria and other Supporting Information*. World Health Organization, Geneva.

WHO (1984c) *Guidelines for Drinking Water Quality, Volume III: Drinking Water Quality Control in Small Communities*. World Health Organization, Geneva.

Wilder, C. R. (1971). *Concrete Sanitary Engineering Structures*. Report of ACI Committee 3650, Title No. 68–50, American Concrete Institute, Detroit, Mich.

Yao, K. M. (1973). "Design of High Rate Settlers." *J. Environ. Eng. Div.*, ASCE, vol. 99, no. EE5, p. 621.

SELECTED BIBLIOGRAPHY

American Society of Civil Engineers, American Water Works Association (1969). *Water Treatment Plant Design*. New York.

American Water Works Association (1971). *Water Quality and Treatment*, 3d edition. McGraw-Hill, New York.

Arboleda, J. V. (1973). *Teoria Diseno y Control de los Procesos de Clarificacion del Agua*. Serie Tecnica 13. CEPIS, Lima, Peru.

CEPIS (1982). *Modular Water Treatment Plants*, Volume 1 and Volume 2. Technical Document No. 8. Pan American Health Organization, Lima, Peru.

Cox, C. R. (1964). *Operation and Control of Water Treatment Processes*. WHO, Geneva.

Culp, G. L. and R. L. Culp, 1974. *New Concepts in Water Purification*. Van Nostrand Reinhold, New York.

ENSIC (1982). *Surface Water Filtration for Rural Areas: Guidelines for Design, Construction, and Maintenance*. Environmental Sanitation Information Center, Bangkok.

Fair, G. M., J. C. Geyer, and D. A. Okun (1968). *Water and Wastewater Engineering*, Volume 2. Wiley, New York.

Fair, G. M., J. C. Geyer, and D. A. Okun (1971). *Elements of Water Supply and Wastewater Disposal*. Wiley, New York.

Hudson, H. E. (1981). *Water Clarification Processes—Practical Design and Evaluation*. Van Nostrand Reinhold, New York.

Huisman, L. and W. E. Wood (1974). *Slow Sand Filtration*, WHO, Geneva.

IRC (1978). *Slow Sand Filtration for Community Water Supply—A Design and Construction Manual*. Technical Paper Series No. 11. J. C. Van Dijk and H. C. M. Oomen, aus. The Hague, Netherlands.

IRC (1981). *Small Community Water Supplies*. Technical Paper No. 18. L. Huisman, J. M. Azevedo-Netto, B. B. Sundaresan, J. N. Lanoix, and E. H. Hofkes., eds. The Hague, Netherlands.

Institution of Water Engineers and Sciences (1983). *Water Supply and Sanitation in Developing Countries*, B. J. Dangerfield, ed. London.

Jahn, S. A. A. (1981). *Traditional Water Purification in Tropical Developing Countries*. German Agency for Technical Cooperation, Eschborn, West Germany.

McJunkin, E. F. (1982). *Water and Human Health*. U.S. Agency for International Development, Washington, D.C.

Sanks, R. (1978). *Water Treatment Plant Design: For the Practicing Engineer*. Ann Arbor Science Publishers, Ann Arbor, Michigan.

Smethurst, G., (1979). *Basic Water Treatment for Application Worldwide*. Thomas Telford Ltd., London.

Swiss Association for Technical Assistance (1975). *Manual for Rural Water Supply*. Ministry of Agriculture, United Republic of Cameroon, Victoria/Buea.

Wagner, E. G. and J. N. Lanoix (1959). *Water Supply for Rural Areas and Small Communities*. WHO, Geneva.

AUTHOR INDEX

Agha, F.Y.R., 35
Agrawal, J.M., 154
Ahmad, S., 35
Ainsworth, L.D., 168
Ametek-Schutte and Koerting Div., 88, 89, 90
APS Technical Services, Ltd., 207
Arboleda, 10, 58, 60, 79, 93, 111, 113, 131, 132, 139, 144, 147, 156, 158, 160, 163, 168, 173, 174, 195, 202, 209
ASCE, 16, 17, 78, 106
AWWA, 32, 56, 57, 59, 69, 127, 147
Azevedo-Netto, J.M., 20, 106, 116, 130, 183, 184, 185, 186, 224

Barker, W., 239
Barnes, D., 155
Baylis, J.R., 165
Berthouex, P.M., 226
Bhole, 124, 139, 154, 155, 206
B.I.F. Industries, 246
Brown, J.C., 203
Burlingame, R., 84, 135, 145

Cady, P.D., 27
Camp, T.R., 21, 32, 54, 106, 129
Capitol Controls Co., 74, 75, 77
Carefoot, N., 235, 236, 237
Castro, M.L., 60
CEPIS, 36, 37, 41, 45, 102, 124, 186, 204, 209
Chao, J.L., 101
Ching, L.V., 171
Coffey, K., 240, 242
Cox, C.R., 50, 56, 59, 68, 69, 116, 127, 153, 241
Culp, G.L. & R.L., 71, 139, 141, 151

Dangerfield, B.J., 241
Dirar, H., 62

Donaldson, D., 204

Fair, G.M., 74, 92, 121, 127, 147, 149, 152, 157, 168, 197
Feachem, R., 191
Frankel, R.J., 40, 41

Gadgil, J.M., 152, 153
Geyer, J.C., 74, 92, 121, 127, 147, 149, 152, 157, 168, 197
Graham, N., 198
Grant, D.M., 99, 104
Greenleaf, J.W., 173
Groney, P.R., 27

Hardenbergh, W.A., 261
Hofman, H., 28
Houston, A.V., 37
Hudson, H.E., 21, 22, 27, 55, 126, 127, 128, 132, 133, 137, 139, 147, 156, 164, 165, 168, 173, 176, 179, 180, 181, 182, 244
Huisman, L., 31, 191, 193, 202, 243
Hutton, L.G., 272

IBRD, 14, 235
IDRC, 14, 15
IRC, 21, 28, 32, 59, 73, 76, 112, 127, 129, 135, 152, 161, 191, 193, 202, 203, 204, 213

Jahn, S.A.A., 60, 61, 62, 63, 65, 66, 67
Johnson, E.L., 66
Jorden, R.M., 92

Kardile, J.N., 106, 120, 124, 154, 186, 189, 192, 223, 224, 225
Kawamura, S., 66, 101
Kawata, K., 62, 63
Kirchmer, C.J., 60

Kuntschik, O., 41, 43

Laval, C., 48

MacDonald, D.V., 106, 111, 114, 115
Macedo, L.H., 229, 230
McGarry, M., 191, 235
Mampitiyarachichi, 155
Mara, D., 191
Marais, G.V., 48
Mhaisalkar, V.A., 194, 202, 226
Moore Fluid Equipment Co., 85, 86
Moreira, R., 106, 124

Nashikkar, J.T., 154
NEERI, 16, 51, 61
Nelson, O.F., 152
Noguti, M., 229, 230

Okun, D.A., 1, 2, 9, 22, 23, 25, 45, 74, 92, 94,
    121, 127, 147, 149, 152, 157, 168, 197, 203,
    233, 234
Ontario Ministry of the Environment, 52, 265

Packham, R.F., 61
Pama, R.P., 28
Paramasivam, R., 139, 153, 194, 202, 215, 226,
    227, 228
Paul, B.K., 28
Peniston, Q.P., 66
Perez, C.J., 173, 180, 183, 199, 209, 213
Piatt and Associates Consulting Engrs., 138
Pinheiro, R., 118, 119
Ponghis, 22, 23, 25, 45, 94

Rahate, S.F., 154, 155
Ranade, S.V., 152, 153, 154
Rao, C.V., 139
Rao, D.R.J., 155
Reid, G., 240, 242

Richmond, A.C., 82, 83
Richter, C.A., 106, 111, 123, 224
Rodie, E.R., 261
Rodriguez, D., 199, 202

Salvato, J., 50
Sanks, R., 126, 127, 132, 139, 147, 168, 216
SANEPAR, 224
Schiller, E.J., 235
Sienco, Inc., 72
Smethurst, G., 18, 32, 37, 38, 39, 108, 126,
    127, 128, 130, 139, 144
Sperandio, O.A., 173, 180, 183, 209, 213
Spink, J., 71
Stein, P.C., 106
Stone, B.G., 101
Stone, R., 72
Streicher, L., 106, 111, 114, 115
Sundaresan, B.B., 226, 227, 228

Tasgaonkar, S.K., 188, 190
Thanh, N.C., 41, 43, 44
Tripathi, P.N., 61

Ughade, V.A., 124

van Kervooden, R., 28
Vrale, L., 92

Wagner, E.G., 10, 55, 120, 123, 179, 180, 181,
    182, 223
Wais, M.T., 35
Wegelin, M., 39, 41
White, G.C., 68, 69, 70
WHO, 12, 13, 17, 235, 244, 282
Wilder, C.R., 22
Wood, W.E., 31, 191, 193, 202, 243
Woods, G.D., 232

Yao, K.M., 139

SUBJECT INDEX

Absorbents-weighting agents, 59
Alabama-type flocculators, 118-120
Algal problems, remedies, 34, 42, 50, 139
Alkalies, 56, 57
Alum, 56
    comparison with other coagulants, 56-67,
        62-67
    costs, 55, 180, 229
    dosages, 52, 55-56, 268
    feeding, 56, 79-86
    properties, handling, 55, 246-247
Amman, Jordan, 180, 223

Backwashing, 159. *See also* Filters, rapid
Baffled-channel flocculators, 109
Bahia, Brazil, 186
Beira, Mozambique, 71
Brasilia, Brazil, 180

Cali, Colombia, 176
Carmen-Kozeny equation, 121
Chandori, India, 186
Checklist, for design of treatment processes,
    261
Chemical feeders, 73
    chlorine, 74
        direct-gas feed, 76
        solution-feed, 74
    constant-rate, 79
    dry-chemical, 85
    eductors, 86
    proportional, 81
    saturated solution, 84
    solution-type, 79
Chemicals, 51
    alkalies, 57
    coagulant aids, 58
    coagulants, 53

    for control of microorganisms, 50
    disinfectants, 66
    types, properties, handling, storage, 51-52,
        246
China, 72
Chitosan, 63
Chittagong, Bangladesh, 145
Chlorinators, 74
    direct-gas, 76
    solution-feed, 74
Chlorine, 68
    diffusers, 78-79
    feeding of gaseous, 74
    hypochlorite compounds, 69
    measurement of residual, 274
    properties, handling, storage, 68, 248-249
    safety, 68-69
Clays, for coagulation, 59
Coagulant aids, 58
    natural, 58
        absorbents-weighting agents, 59
        British experience, 61
        Indian experience, 61
        polyelectrolytes, 59
        side effects, 66
    synthetic, 58, 59-60
Coagulants, 53
    alum, 55
    dosages, 52, 56
    feeders, 79-90
    ferric salts, 56
    jar testing, 52, 265
    natural, 62-66
Cochabamba, Bolivia, 175, 223
Coconut shells, husks, as filter media, 40-41,
    154, 188
Coliform organisms, measurement of, 275
Colon, Costa-Rica, 186

Concrete, 22
Constant-rate filtration, 168
Construction materials, for treatment plants, 22
Contact time, for disinfection, 68
Conversion factors, to U.S. customary units, 5
Copper sulphate, 50, 249
Costs, 214
    chemicals, 55, 59, 70-71, 72, 229
    construction, 48, 207, 213, 216
    currency conversion factors, 215
    ENR index, 215
    equations, 215, 226
    estimation of, 214
    factors affecting costs, 216
    operation and maintenance, 229
    rapid *vs.* slow sand filtration, 226-229
    treatment plant components, 37, 90, 152,
        216

Declining-rate filtration, 171
Delhi, India, 59
Density, variations with temperature, 107
Diffusers, 78, 102
    chlorine, 78-79
    coagulant, 98, 102-104
Direct filtration, 105, 179
    bench-scale, pilot plant studies, 179-180
    comparison with conventional filtration, 179,
        180-183
Disinfection, 66
    chlorination *vs.* hypochlorination, 68
    contact time, 68
    chlorine, gaseous, 68
    hypochlorite compounds, 69
    on-site hypochlorite generation, 69
    ozonation, 68
Dissipators, hydraulic, 101
Dual-media filters, 151, 180
Dynamic filtration, 199

Eductors, 86
Effective size, 152

Ferric coagulants, 56, 249-250
Ferrocement, 27-28, 34
Filters:
    rapid, 147
        auxiliary-scour wash systems, 165
        backwashing, 159
            with distribution-system pressure,
                161-162
            with elevated washwater tanks, 161
            interfilter-washing units, 162-164
            rates, 159-161
            washwater gullets, 159, 164-165

bottom and underdrains, 156
    main and lateral system, 157
    teepee-type bottom, 156-157
classifications of, 147
control systems, 165
    constant-rate, 168
    declining-rate, 171
    inlet rate controllers, 168-169
    outlet rate controllers, 169-171
design criteria, general, 148-149
dual-media, 151, 180
in industrialized countries and developing
        countries, 146-147
interfilter washing units, 162, 173
    backwashing arrangements, 162-164
    comparison with conventional filters,
        173-174
    design guidelines, 178
    with restricted declining flow rate,
        175-178
    with unrestricted declining flow rate, 175
media:
    coal, 153-154, 180
    coarse, 153
    coconut shells, husks, 40-41, 154, 188
    crushed stone, 155
    fine, 155
    gravel, 43, 156
    rice hull ash, 155
    sand, 151, 153
    sizing of, 152-153, 156
slow sand, 190
    advantages for developing countries,
        191-192
    cleaning, 190-191, 197-199
    comparison with rapid filters, 190
    control systems, 195-197
    costs, 226
    design criteria, 193
    dynamic filtration, 199
        operation and maintenance, 201
        shape of filter, 200
    inappropriate use of, 192
    information sources, 202
    mats, 198-199
    pretreatment, 31, 43, 192
    schmutzdecke, function of, 190-191, 198
upflow-downflow, 183
    comparison with conventional filters, 183
    costs, 224-226
    design criteria, 183-184
    evaluation of, 186, 188
    Indian experience, designs, 186-190
    Latin American experience, designs,
        184-186

Filtration, 146
   direct, 179
   rapid, 147
      constant-rate, 168
      declining-rate, 171
      vs. slow sand, 146, 148-149
   rates, 5, 39, 148, 173, 180, 184, 193
   roughing, 39
   slow sand, 190
   upflow-downflow, 183
Flocculation, 105
   design criteria, 106
   Gt values, 107-108, 114-115
   hydraulic *vs.* mechanical, 105-106
   process description, 105
   tapering, 108, 113-114, 122-123
Flocculators, 105
   baffled-channel, 109
      horizontal-flow, 109, 111-114, 252-255
      hydraulic calculations, 252
      tapering, 113-114
      vertical-flow, 109-111, 122
   gravel-bed, 120, 186-188, 206
      hydraulic calculations, 256
      tapering, 122-123
   hydraulic jet-action, 116
      Alabama-type, 118-120
      heliocoidal-flow, 116-118
      staircase-type, 118-255
   surface contact, 124
Flotation, of structures, 26
Flow measurement, in open channels, 104
Flumes, 45, 93, 96

Gravel-bed flocculators, 120, 186-188, 206,
      256
Grit chamber, 45
Gullets, filter, 159, 164-165

Heliocoidal-flow flocculators, 116-118
Human resources development, 232
   classifications of plant personnel, 239
   numbers of plant personnel, 241
   problems in implementation, 238-239
   in selected developing countries, 235-238
   training, 233-235, 244-245
   Water Decade requirements, 235
Hydraulic calculations, for selected treatment
      processes, 252
Hydraulic jump, 92, 96
Hypochlorite compounds, 69
   feeders, 79-90
   on-site generation of, 69
   properties, handling, storage, 69, 247-248,
      251

Institutions in water sector, 279
Intakes, 30
Interfilter-washing filtration systems, 173. *See
      also* Filters, rapid

Jar test, 52
   applications, 62-63, 66, 179-180
   equipment and procedures, 52, 265
Jedee-Thong, Thailand, 43

Laboratory, treatment plant, 272
   procedures, 265, 272
   staffing, 244
Launders, perforated, 135-137
Lime, 58, 84-85, 247-248
Lime-pozzolan cement, 27
Linhares, Brazil, 180

Maharashtra, India, 192
Manpower requirements, 235. *See also* Human
      resources development
Manuals, design, operator, 204, 244
Mats, for slow sand filters, 198-199
Media, *see* Filters, rapid
Modular treatment plants, 209
   costs, 213, 224
Moringa oleifera seeds, 62, 64
MPN test, 275

Nagpur, India, 61
Nairobi, Kenya, 55 98
Nirmali seeds, 60, 61, 62, 65

Oceanside, California, 114, 168
Orifices, hydraulics of, 172-173
Overloading, in settling basins, 126-127

Package treatment plants, 205
   costs, 207, 224
Palmer-bowlus flume, 96
Parshall flume, 45, 93
Penang, Malaysia, 102, 171
Perforated baffles, in settling basins, 133-135
pH, measurement of, 272
Pilot plants, studies, 21
   for coagulation, 265
   for filtration, 163, 179-180, 201-202
   for flocculation, 121, 124-125
   for pretreatment, 35, 41
   for sedimentation, 35, 129-130, 270-271
Pipe-flow rapid mixers, 101
Polyelectrolytes, as natural coagulant aids, 59
Plain sedimentation, 31
Plants, treatment:
   capacity, 9

Plants, treatment (*continued*)
   checklist for design, 261
   classifications of personnel, 239
   constraints in implementation, 10, 21
   construction materials, 22
     for small plants, 27
   design guidelines, 10
   economies of scale, 215
   flow diagram, 18
   modular, 209
   numbers of personnel, 241
   package, 205
   pretreatment methods, 31
   standardization, 203-205, 213
   structural design considerations, 26
Plate settling, inclined, 139
   hydraulic calculations, 259
   in presettling basins, 35
   sludge removal, 138
Plug-flow, 105
Pretreatment, 30
   control of microorganisms, 50
   definition, types of, 30-31
   grit removal, 45
   plain sedimentation, 31
   roughing filtration, 39
   storage, 37
Prudentopolis, Brazil, 209

Raleigh, North Carolina, 138
Ramtek, India, 186
Rapid mixers, 91
   baffled chambers, 100
   coagulant diffusers, 102-104
   eductors, 86
   flow measurement, 104
   hydraulic energy dissipators, 101
   hydraulic jump, 92
   Palmer-bowlus flumes, 96
   Parshall flumes, 93
   turbulent pipe flow, 101
   weirs, 97
Rapid mixing, 91
   design criteria, 91
   hydraulic *vs.* mechanical, 91-92
   open channel flow *vs.* pressure flow, 92
   process description, 91
Regionalization, in England, 233
Reservoirs, *see* Storage
Rio de Janeiro, Brazil, 103, 135
Roughing filters, 39
   horizontal-flow, 41
   vertical-flow, 40

Sanitary survey, 16, 17, 21
Schmutzdecke, 190, 191, 198
Sedimentation, 126
   basins, horizontal-flow, 127
   clarifiers, upflow, 144
   horizontal-flow *vs.* upflow, 126-127, 128
   inclined-plate and tube settling, 139
   plain, 31
   process description, 126
   settling rate of particles, 31-32
Separator, sand, 48
Settling basins:
   horizontal-flow, 127
     bench-scale tests, 35, 129-130, 270-271
     design criteria, 128
     inlet arrangement, 133
     manual sludge removal, 137
     mechanical *vs.* manual sludge removal, 127
     number of basins, 132-133
     outlet arrangement, 135
     process description, 127
   upflow, 144
Short-circuiting, 105-106, 116, 118, 129
Slow sand filtration, 190. *See also* Filters, slow
        sand
Sludge removal, settling basins, 137, 144-145
Soda ash, 58, 250
Sources, selection of, 16
Specific gravity, variations with temperature,
   107
Staircase-type flocculators, 118-155
Standards, drinking water, 12-16
Stilling wells, for flow measurement, 45, 104
Storage, 37
Superfiltration, 184-186
Surabaya, Indonesia, 84, 169
Surface-contact flocculators, 124
Surface loading rate, 5, 128-129, 130, 140

Tapering, 108
   in flocculators, 113-114, 122-123
   in settling basin outlets, 137
Technology, 1
   appropriate, summary of, 6
   capital-intensive *vs.* labor-intensive, 1-2, 10
   inappropriate, 1-4, 10
Teepee filter bottom, 156-157
Training, 233-235, 244-245. *See also* Human
     resources development
Treatment processes, selection of, 17
   checklist for design, 261
Tube settling, 139
   hydraulic calculations, 158

in presettling basins, 35
    sludge removal, 138
Turbidity, measurement of, 31, 273
Turnkey projects, 2

Uniformity coefficient, 152
Upflow clarifiers, 3, 126-127, 144
Upflow-downflow filters, 183
Upflow filters, 40, 183

Valves, 28
    butterfly, 28, 169, 176
    float, 29, 79, 196
    telescopic, 29, 196, 201
Varangaon, India, 186
Velocity gradients, 53-54, 91-92, 106-108

Viscosity, variations with temperature, 107

Water-proofing, of structures, 27
Water quality analysis, 272
Water quality criteria, 11
    in industrialized countries and developing
        countries, 11-12, 14-15
    for raw water sources, 17
    for small communities, 13-16
    WHO guidelines, 12-16
Weirs, 97
    grit chambers, 45
    rapid mixers, 97
    settling basins, 135-136

Zarzal, Colombia, 143